# THE RESPIRATORY MUSCLES: MECHANICS AND NEURAL CONTROL

# The Respiratory Muscles

## Mechanics and Neural Control


E. J. M. CAMPBELL
B.Sc., Ph.D., M.D., F.R.C.P.

*R. Samuel McLaughlin Professor and Chairman of the Department of Medicine, McMaster University, Hamilton, Ontario; formerly Senior Lecturer and Physician, Royal Postgraduate Medical School and Hammersmith Hospital, London*

E. AGOSTONI
M.D.

*Professor of Physiology and Chairman, Istituto di Fisiologia umana, Università di Ferrara, Ferrara*

and

J. NEWSOM DAVIS
M.A., M.D., M.R.C.P.

*Consultant Neurologist, National Hospital for Nervous Diseases and Royal Free Hospital, London; Honorary Lecturer, Institute of Neurology, National Hospital for Nervous Diseases, London*


*Second Edition*

LLOYD-LUKE (MEDICAL BOOKS) LTD

49 NEWMAN STREET

LONDON

1970

FIRST EDITION . . . 1958
SECOND EDITION. . . 1970

PRINTED AND BOUND IN ENGLAND BY
THE WHITEFRIARS PRESS LTD., LONDON & TONBRIDGE

ISBN 0 85324 086 8

# FOREWORD

THE first edition of this book, which appeared in 1958, was the first comprehensive review of the muscles of respiration in this century. It dealt briefly with respiratory movements of the thoracic cage and the mechanics of breathing, but was primarily concerned with the description of the behaviour of each individual respiratory muscle and a presentation of electromyographic studies. It was a superb one man effort by Dr. Campbell, based to a large extent on his own research.

The present edition is far more comprehensive, involving major contributions by Dr. Agostoni on the mechanical behaviour of the respiratory system and by Dr. Newsom Davis on muscle physiology and on neural control and organisation. An interesting chapter by Drs. Milic-Emili and Pengelly deals with the ventilatory effects of mechanical loading. Drs. Campbell and Newsom Davis finish with a section on the respiratory muscles in disease. Each of the authors is a recognised expert. The focusing of this expertise on the respiratory muscles illuminates the subject from unexpected angles. The book shows the power of the interdisciplinary approach.

Campbell and Newsom Davis jointly contribute a chapter on respiratory sensation in which they discuss dyspnoea. Two better men to discuss this important but highly subjective matter are not to be found anywhere since both have given it much careful thought in the light of highly specialised knowledge. The wonder is that any two men, approaching the subject from different backgrounds, could agree on any one chapter!

In the foreword to the first edition I stated: "Opinions and inferences as well as experimental findings and scientific facts find their place in this monograph." The present edition is considerably expanded, largely from established sub-disciplines within the field of physiology. The material is presented with mature confidence, backed by lengthy bibliographies and bears the stamp of authority. I recommend the present offering with enthusiasm, but with a touch of nostalgia. Times have changed and the format of this book has changed. Vast amounts of research have been performed in the twelve years between editions. The logarithmic growth phase is upon us. The day of the pioneer has been followed by the day of the organiser and developer. The authors have shown admirable capacity to adapt and have produced a valuable book with maximal scientific information and minimal loss of excitement.

RICHARD L. RILEY, M.D.

*Baltimore, Maryland*
*May*, 1970

# PREFACE TO THE SECOND EDITION

THIS book is successor to the small monograph published twelve years ago, but it is stretching convention to describe it simply as a second edition; hence the slight change of title. The field has developed very rapidly and is now too large to be encompassed by a single author. The mechanics of the chest wall and the neural control of the respiratory muscles have become topics of much interest and research; the change in title reflects this increased attention and also stresses their dual importance in understanding the respiratory muscles. The concepts and methods are also now becoming applicable to clinical problems.

An indication of the expansion of interest in what a decade or so ago was a no-man's land is that the bibliography of the first edition was only slightly larger than that cited for some individual chapters in the second.

The principal authors of each chapter are identified, but we must emphasise that all of us have read all chapters at each stage of their preparation and the final versions are the results of extensive consultation. Thus this book is not just a product of multiple authorship; it is truly collaborative. Finally, the developments since the first edition add force to the remark by Dr. W. O. Fenn quoted at the end of the Preface to the first edition.

MORAN CAMPBELL
EMILIO AGOSTONI
JOHN NEWSOM DAVIS

*April 1970*

# PREFACE TO THE FIRST EDITION

UNTIL the seventeenth century most of the great physicians and anatomists were interested in the respiratory muscles and the mechanics of breathing. Since then these muscles have been increasingly neglected, lying as they do in a no-man's land between anatomy and physiology. The description of their function in many textbooks of anatomy is apologetically in small print. In textbooks of physiology a few animal experiments are usually included to give a semblance of respectability to accounts which are really based only on anatomical observation. There has been no comprehensive review of the subject since those of Beau and Maissiat (1842, 1843) and Duchenne (1867).

In this age of the sodium pump and the ascending reticular formation such an obvious phenomenon as breathing is easily neglected in the belief that all that is worth knowing about it must already be known. Recent developments in both anatomy and physiology require that this neglect be remedied. Anatomists now demand a better understanding of the dynamics of muscle action than that supplied by the classical methods. Physiologists have devoted much attention to the mechanics of breathing and a better knowledge of the respiratory muscles is becoming increasingly desirable in this field. Understanding of the control of breathing would also be advanced if the results of neurophysiological experiments in animals were supplemented by a better knowledge of the behaviour of the human respiratory muscles.

Another important reason for the neglect of these muscles has been the lack of clinical problems requiring accurate knowledge of their physiology. The improved management of respiratory paralysis, the use of relaxants in anaesthesia, the practice of physiotherapy in chest disease and similar developments have changed this situation. There are also the unsolved problems of dyspnoea, part of whose explanation may be revealed by a better knowledge of the physiology of the respiratory muscles.

The interests prompting the studies which form the basis of this monograph were clinical, but the problems created by these interests proved to be mainly physiological and the monograph is offered primarily as a contribution to human physiology. Nevertheless it is hoped that it will also be of interest to anatomists; and particularly to those physicians and surgeons who are concerned with disorders of the lungs, heart or nervous system. Much of it may also interest anaesthetists, radiologists and physiotherapists. Although the title suggests that all muscles concerned with respiration might be considered, I have, in fact, discussed only those which directly supply the motive forces of breathing.

While preparing the text I have had three main aims: firstly, to gather together the scattered literature of the subject; secondly, to form a critical appraisal of previous work; and finally, to present an account which will be intelligible to non-specialist readers as well as to respiratory physiologists. The reconciliation of these aims has not always been easy.

The sections dealing with individual muscles are in greater detail than the more general sections later in the book. The reasons for this variation in detail are twofold. In the first place, I have studied the subject by investigating the behaviour of individual muscles, and in the second place, the more general sections are an attempt to fit studies of individual muscles into their broader context; inevitably this involves consideration of aspects about which our knowledge is at present inadequate. These general aspects are probably of more interest to physiologists, but they must be based on a sound knowledge of the behaviour of individual muscles.

It was originally intended to include a section on the applied aspects of the subject. However, there have as yet been few studies in this field and the draft contained so little data and so much speculation as to be quite out of harmony with the rest. It was, therefore, omitted and the data incorporated in the relevant sections of the text, the only exception being the inclusion of a short chapter on the clinical examination of the respiratory muscles.

I have not dealt with the use of the respiratory muscles in speech and singing. Studies of these problems demand a much better knowledge of the basic respiratory function of the muscles than at present exists. I hope that this book will partly meet this need.

Finally, I offer the following remark by Dr. W. O. Fenn: "The mechanics of breathing is a problem requiring on one hand the detailed knowledge of a classical anatomist and on the other hand the analytic understanding of an engineer."

*July 1957.* E. J. MORAN CAMPBELL.

# ACKNOWLEDGMENTS

We are grateful to the many people, institutions and other agencies who have contributed to this book both directly and indirectly. We particularly wish to thank Drs. S. Freedman, J. Milic-Emili, L. D. Pengelly and G. Sant'Ambrogio for their co-authorship of the chapters bearing their names. We would like to recognise our indebtedness to the following institutions: the Royal Postgraduate Medical School and Hammersmith Hospital, the Middlesex Hospital, the University of Ferrara, the University of Milan, the Institute of Neurology and the National Hospital for Nervous Diseases, London; to the following agencies who have supported our research: the Medical Research Council of Great Britain, the National Institute of Health (U.S.P.H.S.), the European Office of Aerospace Research (U.S.A.F.), the Italian National Research Council, and the Institute of Neurology; to the many scientific colleagues and friends who have read parts of this book, particularly to Drs. S. Freedman, J. Milic-Emili, L. D. Pengelly, G. Sant'Ambrogio, and T. A. Sears; to Mr. G. de Carli and Miss V. Fisher for preparing illustrations, and Mr. L. Morton for preparing the index; to our secretaries, Miss Sally Wood and Miss Ferdinanda Grappa; to Mrs. Rosemary Newsom Davis for editorial help; and, particularly, to Miss Ann Bourke who has masterminded the production. Finally we would like to thank Mr. Douglas Luke for his guidance and co-operation.

We thank the authors, editors and publishers for permission to reproduce illustrations from the following sources: Excerpta Medica Foundation, International Congress Series (Fig. 1); *British Medical Bulletin* (Fig. 2); *Journal of Physiology* (*London*) (Figs. 3, 5, 6, 60, 61, 62, 72, 74, 76, 77, 78, 80, 82, 83, 84, 96, 104); *Science Progress* (Fig. 4); *Gray's Anatomy*, published by Messrs Longmans Green and Co. (Fig. 8); *The Human Lung* by H. von Hayek, published by the Hafner Publishing Company (Fig. 9); *Handbook of Physiology*, American Physiological Society, Section 3, Volume I, Chapter 6: "Anatomy of the mammalian lung" (Dr. V. Krahl; Fig. 10), Chapter 13; "Statics of the respiratory system" (Dr. J. Mead; Figs. 23, 24, 28, 30); *Journal of Applied Physiology* (Figs. 13, 19–21, 29, 31, 32, 47, 56, 97, 98, 101, 102); *Respiration Physiology* (Figs. 18, 36); *Archives Internationales de Physiologie et Biochimie* (Fig. 58); *Annals of the New York Academy of Sciences* (Fig. 63); *Acta Physiologica Scandinavica* (Figs. 64, 75); *Archives of the Middlesex Hospital Journal* (Fig. 65); *Journal of Anatomy* (Figs, 66, 67, 68); *Nature* (Fig. 73); *Progress in Brain Research* (Fig. 81); *Journal of Neurophysiology* (Figs. 85, 87, 89, 90); *Experimental Neurology* (Figs. 86, 88); *Nobel Symposium I*, published by Almqvist and Wiksell (Fig. 92); *Journal of Neurology, Neurosurgery and Psychiatry* (Fig. 93); *American Journal of Physiology* (Fig. 100).

# CONTENTS

## INDIVIDUAL MUSCLES

## CONTROL AND ORGANISATION

## THE RESPIRATORY MUSCLES IN DISEASE

# Chapter I

# INTRODUCTION

J. NEWSOM DAVIS

THE effects of respiratory muscle contraction depend upon the mechanical properties of the rib cage and abdomen, the properties of the respiratory muscles themselves, and those of the neural control mechanism. These three aspects of respiratory muscle function will be independently described, but their interaction needs to be stressed. Altering the mechanical conditions during a breath may modify the pattern of excitation of the muscles, and in turn will lead to a change in the mechanical properties of the rib cage and abdomen.

This chapter gives an account of the basic features of striated muscle. In the first section, the morphological, electrical and contractile characteristics of individual fibres are described before the organisation of these fibres into functional units (the motor unit) and the mechanical properties of whole muscle are considered. In the second section, the proprioceptive innervation of skeletal muscle will be outlined.

## GENERAL PROPERTIES OF STRIATED MUSCLE

### Morphology

The respiratory muscles, like other skeletal muscles, are striated. Microscopically, two groups of fibres can be distinguished. The small red fibres have a high myoglobin content, and are rich in sarcoplasm with many mitochondria and fat droplets. The larger white fibres have a lower myoglobin content, less sarcoplasm, and fewer mitochondria and fat droplets. In addition, there are fibres with properties intermediate between those of the red and white fibres (Padykula and Gauthier, 1967). Muscles in which one fibre type predominates will show this in their colour to the naked eye. The soleus muscle, for example, is red in appearance and is composed largely of small red fibres.

With histochemical methods, differences in the oxidative and phosphorylase enzymatic activities of muscle fibres can be demonstrated (see Dubowitz and Pearse, 1961; Engel, 1962; Romanul, 1964). Fibres with high oxidative and low phophorylase activity (type 1 or C fibres) probably correspond to the red fibres and fibres in which these enzyme characteristics are reversed (type 2 or A fibres) to white fibres (Padykula and Gauthier, 1967). Fibres with intermediate enzymatic properties (type 3 or

B fibres) are also recognised. A simplified classification of the properties of striated muscle fibres is given in Table I.

### Electrical Properties

There is an electrical potential across the surface membrane of a skeletal muscle fibre in the resting (polarised state), the interior being charged about 75 mV negative with respect to its surroundings. In man, the reported mean values of the resting potential have varied among different groups of workers, ranging from 65 to 87·4 mV (see McComas *et al.*, 1968). In human intercostal muscle (Creese *et al.*, 1957; Elmqvist *et al.*, 1960; Hofmann *et al.*, 1966), the mean resting membrane potential is about −80 mV.

TABLE I

PROPERTIES OF MUSCLE FIBRES*

| Naked Eye | Microscopic | Histochemical | Contractile Properties |
|---|---|---|---|
| Red | High myoglobin content<br>Rich in sarcoplasm<br>Many mitochondria | High oxidative activity<br>Low phosphorylase activity<br>(Type 1 or C fibres) | Slow |
| | | Intermediate oxidative and phosphorylase activity<br>(Type 3 or B fibres) | Intermediate |
| White | Lower myoglobin content<br>Less sarcoplasm<br>Fewer mitochondria | Low oxidative activity<br>High phosphorylase activity<br>(Type 2 or A fibres) | Fast |

* This is a simplified classification. The precise correlation between microscopic, histochemical and contractile characteristics has not yet been established,

The arrival of a nerve impulse at the neuromuscular junction leads to the release of acetylcholine and the production of an end-plate potential, which gives rise to a spreading wave of depolarisation along the surface membrane of the muscle fibre (the action potential) whereby the inside of the membrane becomes temporarily positive and the contractile material in the interior of the fibre is activated (see Katz, 1966). It is these membrane changes which form the basis of the electrical activity recorded during electromyography.

### Contractile Properties

All muscles do not have the same speed of contraction. The speed of shortening is expressed by the twitch time (onset to peak tension of the contraction) to which it is inversely proportional (see Close, 1967). Fast contracting (phasic) muscles thus have a shorter contraction time than slow (tonic) muscles. A consequence of this is that fast muscles require a more rapid rate of stimulation than slow muscles to achieve a fused tetanus (Fig. 1) and a proportionally greater expenditure of energy in maintaining a contraction for a given time. The force achieved in a single twitch is much less than in a tetanus for both types of muscle because, owing to the series elastic component, there is insufficient time for the full tension to be developed before relaxation sets in (Hill, 1949).

The absolute values of contraction time which characterise a muscle as fast or slow are applicable only within a single species. Muscles from smaller animals are usually faster than the equivalent muscle from larger

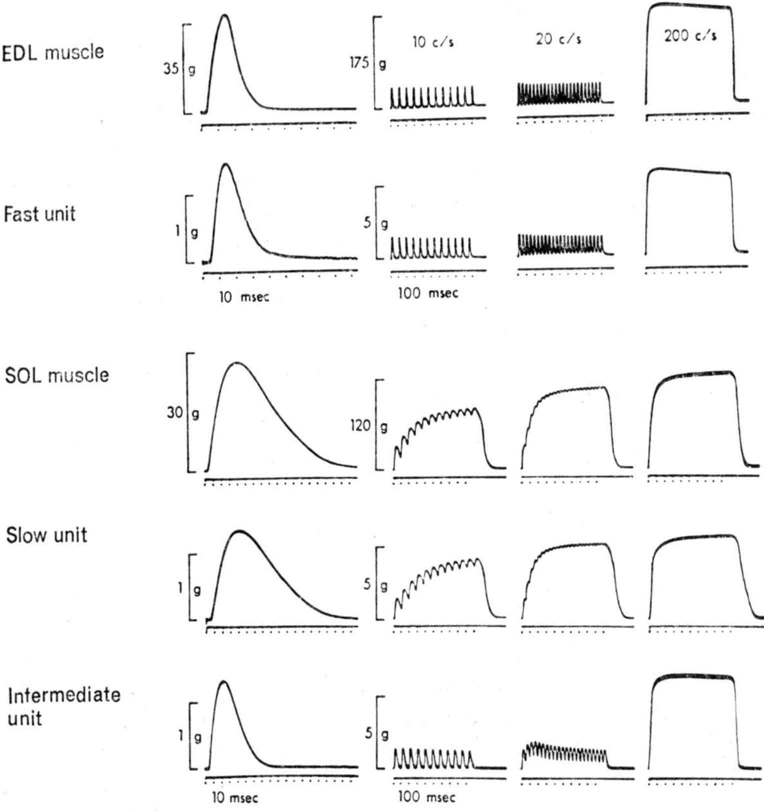


FIG. 1.—RECORDS OF ISOMETRIC CONTRACTIONS OF WHOLE MUSCLES AND MOTOR UNITS OF EXTENSOR DIGITORUM LONGUS (EDL) AND SOLEUS (SOL) MUSCLES OF 12-WEEK-OLD FEMALE RATS

The fast unit is from EDL and the slow and intermediate units are from SOL. The records are, from left to right, the isometric twitch and the responses to repetitive indirect stimulation at 10 c/s, 20 c/s and 200 c/s. Temperature = 35°C. (From Close, 1967.)

animals (Close, 1967). Differentiation between fast and slow muscles is not present in animals immature at birth, but develops during the postnatal period (see Close, 1964, 1967). The failure of slow muscles to differentiate when the lumbo-sacral cord is transected and deafferented indicates that there is an important central neural influence (Buller *et al.*, 1960*a*).

The contraction time of the individual muscle fibres in a fast or slow muscle is not uniform, however, because fast, slow and possibly intermediate fibres can be distinguished within a single muscle (Gordon and Phillips, 1953; Andersen and Sears, 1964; Wuerker *et al.*, 1965; Close, 1967). The contraction time of the whole muscle depends upon the relative proportions of fast and slow fibres. It is possible nevertheless to characterise a whole muscle as "fast" or "slow".

In man the mean contraction time of a fast muscle (frontalis) is 42·6 ± 4·3 SE msec (n = 10) and of a slow muscle (gastrocnemius) 117·6 ± 6·5 SE msec (n = 10) (McComas and Thomas, 1968). Fast and slow fibres in human muscle were first demonstrated in biopsies of intercostal muscle (Hofmann *et al.*, 1966). Recently, they have been shown in a number of other muscles (Eberstein and Goodgold, 1968). The pooled results of 29 biopsies in these muscles gave the mean contraction time for fast fibres as 64 ± 10 SD msec and slow fibres 120 ± 14 SD msec. It was noted that contraction times of human muscle were slower than those of the monkey, measured by the same technique, and than the values reported in cat limb muscles.

Fast muscle fibres broadly correspond to white muscle fibres and slow fibres to red fibres, although the precise correlation of function with morphological and histochemical features of muscle has yet to be established (see Table I).

The speed of contraction of muscle fibres is determined by the motoneurone which innervates them. When the nerve to a fast muscle (flexor digitorum longus) and to a slow muscle (soleus) is cut and cross-sutured so that the nerve to the fast muscle innervates the slow muscle and vice versa, the contraction time of the newly innervated muscle alters to a value close to that of the muscle originally supplied by the nerve, i.e. the soleus muscle is rendered fast and the flexor digitorum longus muscle rendered slow (Buller *et al.*, 1960*b*). This transformation fails, however, if the lumbo-sacral cord is isolated by transection and deafferentation. The means by which the motoneurone influences muscle contraction time is uncertain, but may be through the action of specific chemical substances travelling down the axon of the motoneurone into the muscle fibres (Buller *et al.*, 1960*b*; Eccles, 1967). Cross-innervation of fast and slow muscles has also been shown to alter the enzymatic characteristics of the muscle in parallel with the changes in its contractile properties (Dubowitz, 1967; Romanul and Van der Meulen, 1967).

## The Motor Unit

An individual motor nerve fibre and the muscle fibres it innervates constitute a motor unit (Sherrington, 1929); their overall ratio in the muscle (i.e. the average ratio of nerve to muscle fibres) is the "innervation ratio". The contractile properties of the muscle fibres in a given motor unit are homogeneous (Wuerker *et al.*, 1965), and a muscle can be considered functionally as consisting of motor units of different speeds of contraction. Fast units in general contain more muscle fibres than slow units. Their motoneurones have axons of larger diameter and faster conduction velocity, and the motoneurone itself may be presumed to be larger than that of slow units (Wuerker *et al.*, 1965). Fast motoneurones show a further difference in that their after-hyperpolarisation is shorter, and thus favours a faster rate of discharge (Eccles *et al.*, 1958). This characteristic of the motoneurone thus matches the twitch duration of the muscle fibres it innervates, for fast muscle fibres require a higher nerve discharge frequency than slow fibres to produce a fused tetanus.

## Characteristics of Tonic and Phasic Units

The tonic or phasic characteristic of a particular motoneurone remains unchanged by different forms of reflex activation, for example pinna twisting or the crossed extensor reflex; tonic units consistently respond with sustained firing, in contrast to phasic units (Granit *et al.*, 1957). Some features of their action potentials provide further evidence for a differentiation between tonic and phasic motoneurones (Eccles *et al.*, 1958).

Denny-Brown (1929) noted that slow (tonic) muscle had a lower threshold of activity in postural reflexes than fast (phasic) muscle. This threshold has since been investigated at the motor unit level in the cat. In response to stretch, small units are recruited before larger ones and, on release of stretch, cease firing in the reverse order (Henneman *et al.*, 1965). Units which are incapable of sustained firing with stretch, giving instead a phasic response, are usually large units. Henneman and his colleagues have argued that the recruitment order is dependent on the size of the motoneurone, and because small units are consequently the most "used" they must of necessity consist of slow muscle which is economic in its energy requirements and not subject to fatigue (Henneman *et al.*, 1965; Henneman and Olson, 1965). Large (fast) units develop more tension but fatigue quickly, being suited for situations where power and speed of contraction are important.

In man, voluntary activation of a muscle starts with a unit of small amplitude (recorded electromyographically) followed by units of progressively larger size (Kugelberg and Skoglund, 1946). The first unit to discharge is usually, but not always, the same in voluntary and reflex activation of the muscle, so that the inherent property of the motoneurone is not the only factor influencing the order of recruitment (Ashworth *et al.*, 1967). The recruitment order in a rapid voluntary movement may differ from that in a slowly initiated movement (Grimby and Hannerz, 1968).

**Mechanical Properties of Whole Muscle**

In mechanical terms, mammalian skeletal muscle consists of a contractile component with an elastic component in series and a further elastic component in parallel (Hill, 1949). The extensibility of the series elastic component is high at small loads becoming much less at large loads. The full isometric force produces in it an extension of about 10 per cent of the muscle length (Hill, 1949). The series elastic component has similar characteristics in different mammalian species (see Bahler, 1967). It has an important effect on the mechanical properties of whole muscle for it smoothes out rapid changes in force (Wilkie, 1956).

The contractile component, on excitation, passes into the "active state", the intensity of the active state at a given instant being defined as the isometric force which it can develop at that instant. Because of the series elastic component, the tension in the whole muscle (contractile component plus elastic component) follows a slower time course (Wilkie, 1956). The greater the compliance of the series elastic component, the longer the contraction time.

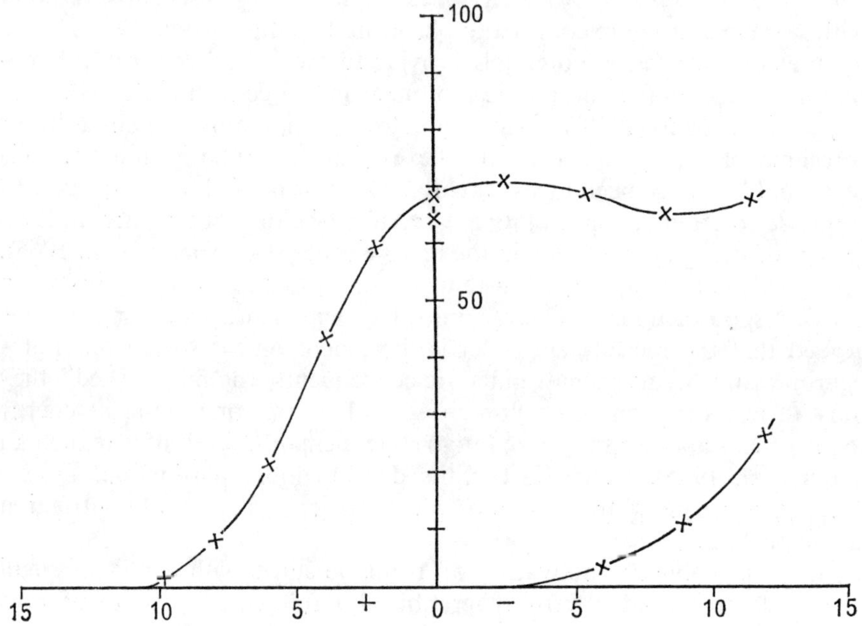


FIG. 2.—TETANIC FORCE-LENGTH CURVE IN STIMULATED MUSCLE (UPPER CURVE) AND STRESS-STRAIN CURVE FOR RESTING MUSCLE (LOWER CURVE)

Abscissa: muscle length (mm). Ordinate: tension (g.wt.). The zero of the abscissae is at the *in situ* length of the muscle. (From Wilkie, 1956.)

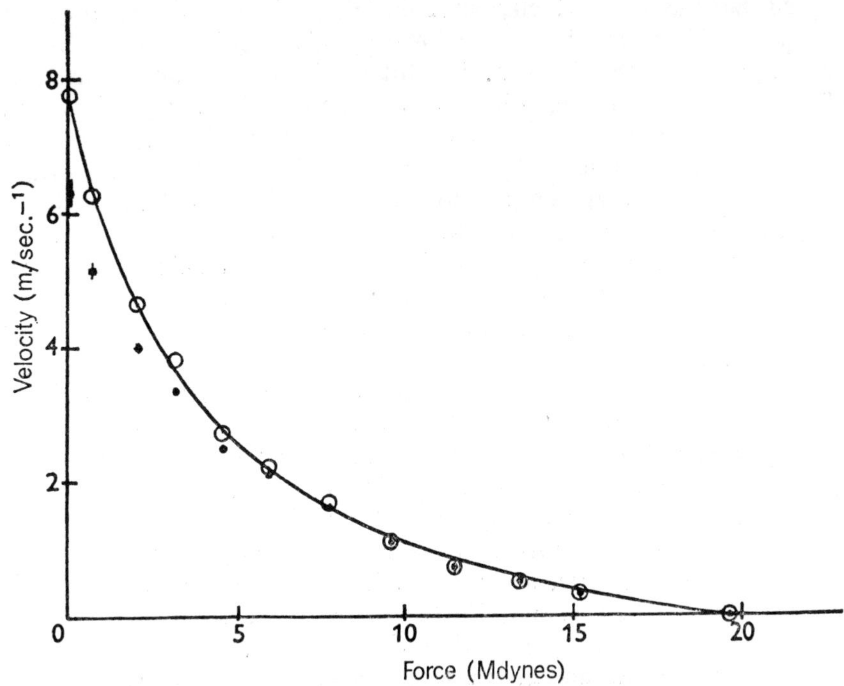


FIG. 3.—EXPERIMENTAL RELATION BETWEEN FORCE AND VELOCITY OF MUSCLE CONTRACTION (SUBJECT D.W.)

Dots: means of thirty determinations of velocity. Six times the standard error is plotted as a vertical bar through the mean; at many points it is too small to be visible on reproduction. Circles: experimental points after correction for inertia. Curve drawn from $(P + a)(V + b) = (P_0 + a)b$, with $a/P_0 = 0{\cdot}20$. (From Wilkie, 1950.)

In an isometric contraction, muscle length is held constant, while in an isotonic contraction, the load is kept constant and the muscle is allowed either to shorten (miometric contraction) or lengthen (pliometric contraction), depending on the size of the load. If the force is measured at different muscle lengths during a tetanus under isometric conditions, the force developed is seen to be a function of muscle length (Fig. 2) and the maximum force is developed when the muscle is approximately at its maximal unstretched length. An increase in external load, under isotonic conditions, leads to a longer latent period before length change occurs, a decrease in the maximum amount of muscle shortening, and a reduction in the velocity of shortening (Wilkie, 1956).

The relationship between the force of muscle contraction and the velocity of muscle shortening is shown in Fig. 3. This relationship is described by Hill's (1938) equation $(P + a)(V + b) = (P_0 + a)b$ where $P$ is the force at velocity $V$, $P_0$ is the force at zero velocity and $a$ and $b$ are constants with dimensions of force and velocity respectively. Wilkie (1950) showed that the force-velocity curve during maximal flexion of the elbow in human subjects could be represented by this equation, after correction for the inertia of the forearm. The force-velocity relationship in Fig. 3 refers to one particular length of muscle. To describe the relationship at other muscle lengths, a series of isolength force-velocity curves could be constructed. Fenn (1963) calculated from the data of Hill (1940) that the velocity of shortening required to reduce the force of human skeletal muscle to half its isometric value was 0·66 of the muscle length per second. In order to compare the force-velocity relationship of different muscles it is convenient to express the force as a fraction of the isometric force (i.e. when velocity is zero), and the velocity as a fraction of the maximal velocity (i.e. when the load is zero).

The force-velocity relationship is such that the same effort will raise small loads more rapidly than large loads. This may be accounted for by the "Fenn effect". Fenn (1923) showed that whenever a muscle shortens upon stimulation and does work in lifting a weight, an extra amount of energy is liberated which is not observed in an isometric contraction. The extra energy produced must depend upon some enzymic reaction which will have a limiting rate, so that less force can be maintained at high velocities than at low (Fenn, 1963). On the other hand, if a muscle is lengthened forcibly during the period of contraction, so that the work done is negative, the excess energy is also negative, i.e. less than that during an isometric contraction (Fenn, 1924; Hill, 1938). Further studies, reviewed by Hill (1960), have shown that the chemical reaction of contraction is reversed when a contracting muscle is stretched, in that the total heat produced in the muscle up to the time of full relaxation equalled in most experiments the total work done on the muscle. Therefore some of this work has been used for the chemical reaction of muscle contraction, and as a consequence the $O_2$ consumption for a given force exerted by a muscle decreases from pliometric to isometric and from isometric to miometric conditions.

The mechanical power developed by a muscle depends upon load and speed. Power at various speeds or at various loads is obtained from the force-velocity relationship by multiplying the force for a given speed by the corresponding velocity (Fig. 4). As pointed out by Hill (1950) the maximum power occurs at about 0·3 times the maximum speed, and about 0·3 times the maximum load. The mechanical efficiency, i.e. the ratio of work done to total energy used, also depends on the speed; its maximum value of about 25 per cent occurs at about 0·2 times the maximum speed and at about 0·5 times the maximum load. These maxima of power and efficiency are rather blunt so that it is possible to develop the maximum power with nearly maximum efficiency, but at high or low speed both power and efficiency fall off markedly (Fig. 4) (Hill, 1950).

The factors limiting the force that a muscle can develop during voluntary effort have been studied in muscles of the hand (Merton, 1954; Naess and Storm-Mathisen, 1955). The force developed during a maximal tetanus of the muscle (by electrical stimulation of its motor nerve) was equalled by the force which could be developed by voluntary effort. When a stimulus was applied to the motor nerve during maximal effort, no twitch of the muscle occurred although it caused a normal action potential indicating that the contractile material was fully activated. Fatigue caused a fall in muscle tension which could not be restored by an applied tetanus

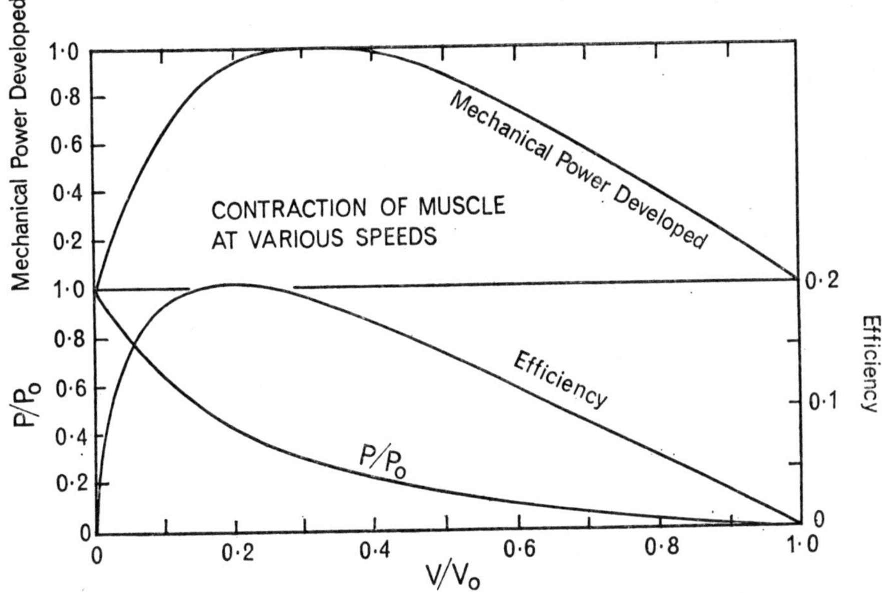


Fig. 4.—Relation between Various Mechanical Properties of Muscle and Speed of Shortening

Speed $V$ as a fraction of maximum speed $V_0$ under zero load. Force exerted $P$ as fraction of maximum force $P_0$ at zero speed. Efficiency = (mechanical work done)/(total energy used). Mechanical power = $PV$. From experiments on isolated muscles and on man. (From Hill, 1950.)

showing that the site of the fatigue is peripheral to the point of stimulation of the nerve. Since the amplitude of the action potentials remains unchanged during severe fatigue, it follows that failure of neuromuscular transmission is not contributing significantly to the decreased force which the muscle can develop (Merton, 1954). The fatigue may thus be presumed to be due to failure in the contractile component itself (see Kugelberg and Edstrom, 1968).

## Gradation of Contraction

When a muscle is relaxed, no motor unit activity can be recorded (Adrian and Bronk, 1929; Joseph *et al.*, 1955). At the onset of voluntary contraction, individual units appear discharging at a low frequency. As the contraction increases, the discharge frequency of active units increases and other units are recruited. From studies on human hand muscles, Bigland and Lippold (1954*b*) concluded that contraction was graded mainly by recruitment of units, with change in discharge frequency being important only at very low or very high contraction strengths. Some motor units are only recruited at strong or maximal contractions, and then discharge at a relatively low frequency. Motor unit discharge frequency does not usually exceed 50/sec (Lindsley, 1935; Bigland and Lippold, 1954*b*) but considerably higher frequencies have been recorded in the external ocular muscles (Björk and Kugelberg, 1953).

## EMG and Force of Contraction

There is a linear relationship between the integrated EMG* and muscle tension in man during isometric contraction (Lippold, 1952; Inman *et al.*, 1952); linearity is maintained during muscle shortening or

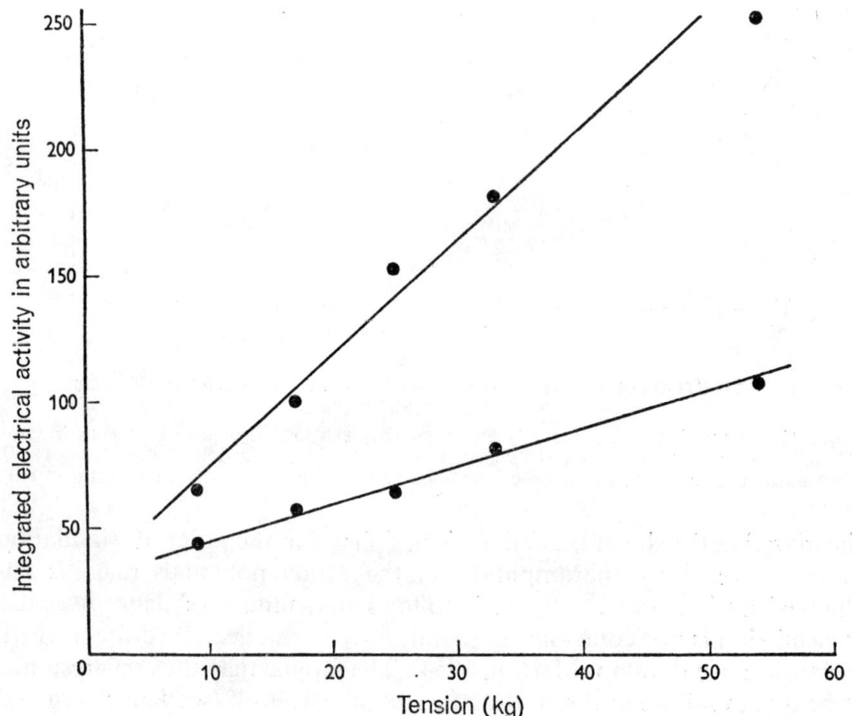


FIG. 5.—THE RELATION BETWEEN INTEGRATED ELECTRICAL ACTIVITY AND TENSION IN THE HUMAN CALF MUSCLES

Shortening at constant velocity (above) and lengthening at the same constant velocity (below). Each point is the mean of the first ten observations on one subject. Tension represents weight lifted, and is approximately 1/10 of the tension calculated in the tendon. (From Bigland and Lippold, 1954*a*.)

* Techniques for integrating the EMG are discussed in Appendix B, p. 201.

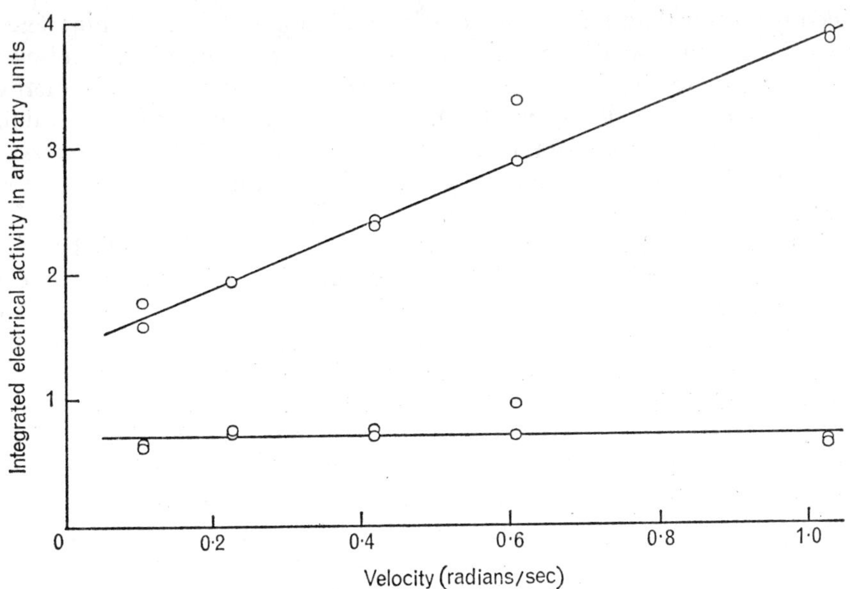


FIG. 6.—THE RELATION BETWEEN INTEGRATED ELECTRICAL ACTIVITY AND VELOCITY OF SHORTENING (ABOVE) AND LENGTHENING (BELOW) AT THE SAME TENSION (3·75 KG)

Each point is the mean of the first ten observations on one subject. (From Bigland and Lippold, 1954*a*.)

lengthening provided that the velocity of length change is constant (Bigland and Lippold, 1954*a*). During lengthening, the electrical activity is less than during shortening at the same tension (Fig. 5); in other words, the electrical activity on lowering a given weight is less than that on raising it. Over a range of velocities at constant tension, the electrical activity increases linearly with velocity of shortening, but remains independent of velocity when the muscle is lengthening (Fig. 6; Bigland and Lippold, 1954*a*). Thus force of contraction, velocity of change in length and integrated electrical activity are interdependent. In a fatigued muscle, the linear relationship between integrated EMG and tension is maintained, but the slope becomes steeper (Edwards and Lippold, 1956; Lenman, 1959).

The following points may thus be made about the interpretation of the EMG. Electrical activity recorded from a muscle does not by itself indicate either the force developed or whether an associated change of length has occurred in the muscle. If the length change is known or can be estimated, some conclusions can be drawn about the force that the muscle is exerting. Thus under isometric conditions an increase in the integrated EMG at any given recording site implies an increased force of contraction. For example: the abdominal muscles produce a certain level of electrical activity during the maintenance of an isometric expiratory pressure of 30 cm $H_2O$, which is greater than that recorded at a ventilation rate of 50 l/min; since in the latter manoeuvre the abdominal muscles are shortening, it follows that they are exerting less force here than when maintaining the isometric expiratory pressure of 30 cm $H_2O$. Such comparisons are only valid, however, if the data are obtained in the same individual from a constant recording site over a reasonably short period of time so that electrode conditions do not alter appreciably.

Participation of a muscle in a particular manoeuvre is implied by its associated electrical activity. Activity recorded in an intercostal muscle during inspiration, for instance, implies its participation in that phase of breathing, but it may be argued that the muscle should properly be described as "inspiratory" only if its mechanical action is directly to cause the intake of air into the lungs (Campbell, 1954).

*In summary*, grading of muscular contraction is brought about mainly by recruitment of motor units but also in part by change in discharge frequency. The integrated electrical activity of a muscle when it is exerting a force and shortening is greater than that when it exerts the same force isometrically, which in turn is greater than that when it is exerting the same force and lengthening. The importance of the rate and direction of change in muscle length in determining the electrical activity is a consequence of the force-velocity relationship of muscle in that the force developed increases as the velocity of shortening decreases.

## PROPRIOCEPTIVE INNERVATION OF STRIATED MUSCLE

Striated muscle contains two groups of specialised receptors which are concerned with proprioception: muscle spindles and tendon organs. There is general agreement about the structure of the *muscle spindle* (Fig. 7) (Barker, 1962; Matthews, 1964). It is an encapsulated organ lying in parallel with the main muscle fibres (extrafusal fibres) and containing its own muscle fibres (intrafusal fibres) which are of two types: nuclear bag fibres and nuclear chain fibres. Primary sensory endings (*synonym:* annulo-spiral endings) lie mainly on the nuclear bag fibres and are innervated by Group Ia fast conducting fibres; the secondary endings (*syn:* flower-spray endings) lie mainly on the nuclear chain fibres and are innervated by Group II slower conducting fibres. Mechanically, the arrangement may be considered as a sensory portion lying between two contractile poles, so that shortening of these muscular poles leads to stretching of the sensory portion. Primary endings are much more sensitive than secondary endings to the dynamic component of the mechanical stimulus.

The intrafusal muscle fibres are innervated by *fusimotor fibres* (*syn:* gamma motoneurones). These fusimotor fibres may be classified functionally as either "dynamic" or "static" with respect to their effects upon the response of the primary endings to stretch (Matthews, 1962). With the muscle at a constant length, stimulation of either static or dynamic fusi-

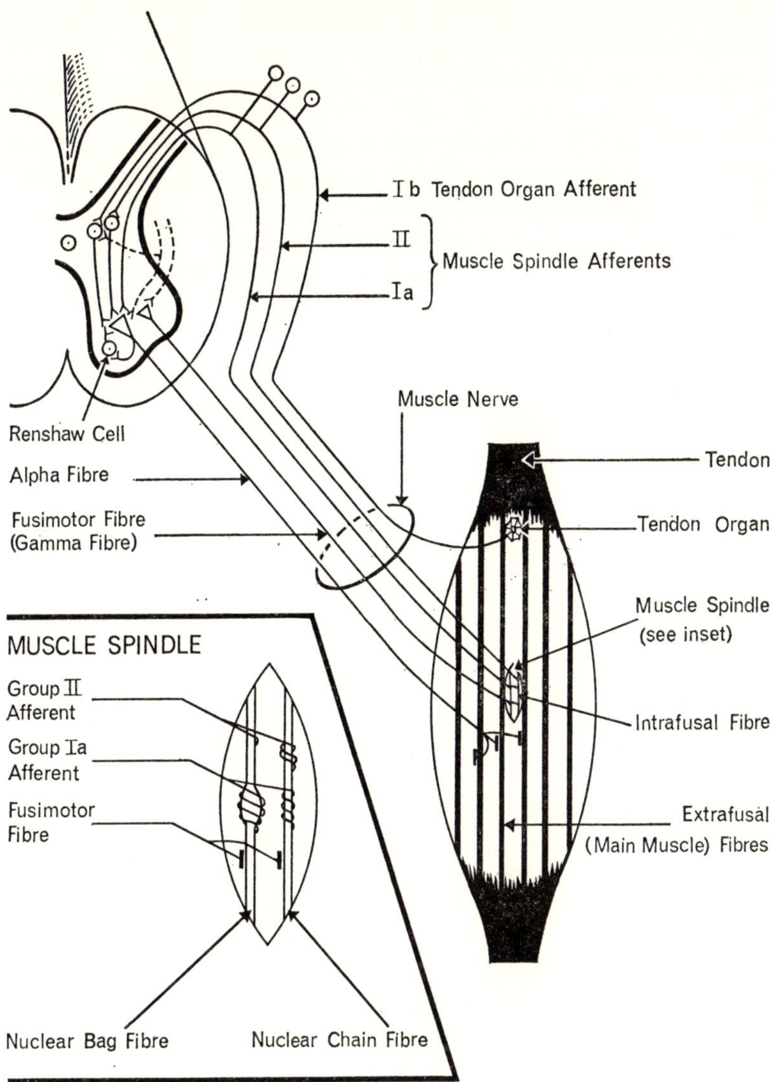


FIG. 7.—SCHEMATIC REPRESENTATION OF THE PROPRIOCEPTIVE INNERVATION OF SKELETAL MUSCLE

The descending projection to the segmental mechanism has been indicated by the broken lines. The two types of fusimotor neurones and the reflex connections to antagonist muscles have not been shown.

motor fibres increases the discharge of spindle primary endings. During stretching of the muscle, stimulation of static fusimotor fibres abolishes the dynamic response of the primary ending so that its response is more like that of a secondary ending, while stimulation of dynamic fusimotor fibres enhances its dynamic response. Secondary endings, on the other hand, respond to static fusimotor stimulation but are generally unaffected by dynamic fusimotor stimulation (Appelberg *et al.*, 1966).

In *tendon organs* the axon branches into small terminal tufts which are enclosed in a fine capsule (Cooper, 1960). The nerve fibres are fast-conducting (Group Ib). Tendon organs are situated at musculo-tendinous junctions and in relation to the aponeuroses which extend into the muscle. Studies in cat soleus and gastrocnemius muscles show that the distribution of tendon organs in the aponeuroses corresponds to the areas where muscle spindles have their attachment (Swett and Eldred, 1960).

**Reflex Connections**

The reflex connections of muscle receptors are shown diagrammatically in Fig. 7 (the two types of fusimotor neurones are not indicated). The alpha motoneurones and fusimotor neurones, which innervate respectively the main muscle and intrafusal muscle fibres, lie in the anterior horn of grey matter. Group Ia afferent fibres from the spindle primary endings connect monosynaptically with the alpha motoneurones of their own and of synergistic muscles, and polysynaptically with alpha motoneurones of antagonist muscles (not shown in Fig. 7). The Group II fibres from spindle secondary endings and Group Ib fibres from tendon organs make polysynaptic connections via interneurones which are themselves subject to central control (see Lundberg, 1964, for review). The term *gamma loop* indicates the circuit: fusimotor neurone—muscle spindle primary ending—Group Ia afferent fibre—alpha motoneurone.

**Functional Considerations**

*Muscle spindles.*—The main muscle fibres and the muscle spindles are in parallel so that stretching the muscle also stretches the spindles and evokes a discharge in the primary and secondary endings, primary endings showing greater dynamic sensitivity in their response than secondary endings. Conversely, if the main muscle shortens, afferent discharge from the spindles decreases or is abolished.

Primary spindle afferents are excitatory to their own alpha motoneurones and to those of synergistic muscles; they are usually inhibitory to antagonistic motoneurones (see Eccles, 1962) but not in the case of intercostal motoneurones (see p. 219). Thus stretching a muscle causes spindle primary discharge, reflex excitation of the alpha motoneurone and consequent muscle contraction to oppose the original stretch. This, of course, is the basis of the "stretch" reflex. The intrafusal (muscle spindle) fibres, however, are themselves under motor control by the fusimotor fibres. Fusimotor excitation causes shortening of the contractile poles of the muscle spindle which stretches the sensory portion and leads to discharge in the primary spindle afferent fibres with reflex excitation of the alpha motoneurone, as if stretching of the main muscle had occurred.

Thus there are two possible routes of excitation to the alpha motoneurone which will lead to muscle contraction: firstly, "direct" central excitation and secondly, "indirect" reflex excitation round the gamma loop. It is to the latter route that attention has been directed in recent years. If the central drive were predominantly to the gamma motoneurone, the system could be considered as a follow-up length servo in that main muscle length would tend to follow spindle length (see Merton, 1953; Eldred *et al.*, 1953; Hammond *et al.*, 1956). The spindle afferent discharge would then represent the error signal of the servo, that is the degree of misalignment between extrafusal and intrafusal fibre length. An alternative role for the fusimotor system would be the maintenance of the relationship between spindle length and main muscle length in the presence of shortening of the muscle so that reflex functions of the spindle would not be lost (see Hunt and Perl, 1960).

The discharge from spindle secondary endings in limb muscles has been thought to exert an excitatory effect on flexor motoneurones and an inhibition of extensor motoneurones whether the spindles lie in flexor or extensor muscles (see Eccles, 1962). Studies by Matthews (1969) now indicate that these receptors may contribute to the tonic stretch reflex of the decerebrate animal. The functional significance of these reflex effects in the intact animal is not yet clear.

*Tendon organs.*—Tendon organ afferents are inhibitory to the muscles in which they lie ("autogenetic inhibition"). Recently it has been shown that the tension threshold for these receptors is much lower during active muscular contraction than during passive stretch (Jansen and Rudjord, 1964; Houk and Hennemann, 1967), and is comparable with that of muscle spindles.

## REFERENCES


ADRIAN, E. D., and BRONK, D. W. (1929). The discharge of impulses in motor nerve fibres. Part II. The frequency of discharge in reflex and voluntary contractions. *J. Physiol.* (*Lond.*), **67**, 119–151.

ANDERSEN, P., and SEARS, T. A. (1964). The mechanical properties and innervation of fast and slow motor units in the intercostal muscles of the cat. *J. Physiol.* (*Lond.*), **173**, 114–129.

APPELBERG, B., BESSOU, P., and LAPORTE, Y. (1966). Action of static and dynamic fusimotor fibres on secondary endings of cat's spindles. *J. Physiol.* (*Lond.*), **185**, 160–171.

ASHWORTH, B., GRIMBY, L., and KUGELBERG, E. (1967). Comparison of voluntary and reflex activation of motor units. *J. Neurol. Neurosurg. Psychiat.*, **30**, 91–98.

BAHLER, A. S. (1967). Series elastic component of mammalian skeletal muscle. *Amer. J. Physiol.*, **213**, 1560–1564.

BARKER, D. (1962). The structure and distribution of muscle receptors. In: *Symposium on Muscle Receptors*, pp. 227–240. Ed. Barker, D. Hong Kong Univ. Press.

BIGLAND, BRENDA, and LIPPOLD, O. C. J. (1954*a*). The relation between force, velocity and integrated electrical activity in human muscles. *J. Physiol.* (*Lond.*), **123**, 214–224.

BIGLAND, BRENDA, and LIPPOLD, O. C. J. (1954*b*). Motor unit activity in the voluntary contraction of human muscle. *J. Physiol.* (*Lond.*), **125**, 322–335.

BJÖRK, A., and KUGELBERG, E. (1953). Motor unit activity in the human extra-ocular muscles. *Electroenceph. clin. Neurophysiol.*, **5**, 271–278.

BULLER, A. J., ECCLES, J. C., and ECCLES, R. M. (1960*a*). Differentiation of fast and slow muscles in the cat hind limb. *J. Physiol.* (*Lond.*), **150**, 399–416.

BULLER, A. J., ECCLES, J. C., and ECCLES, R. M. (1960*b*). Interactions between motoneurones and muscles in respect of the characteristic speeds of their responses. *J. Physiol.* (*Lond.*), **150**, 417–439.

CAMPBELL, E. J. M. (1954). The muscular control of breathing in man. Ph.D. Thesis, Univ. of London.

CLOSE, R. (1964). Dynamic properties of fast and slow skeletal muscles of the rat during development. *J. Physiol.* (*Lond.*), **173**, 74–95.

CLOSE, R. (1967). Dynamic properties of fast and slow skeletal muscles. In: *Exploratory Concepts in Muscular Dystrophy and Related Disorders*, pp. 142–149. Ed. Milhorat, A. T. Excerpta Medica Foundation Inter. Cong. Series no. 147.

COOPER, S. (1960). Muscle spindles and other muscle receptors. In: *The Structure and Function of Muscle*, **1**, 381–420. Ed. Bourne, G. H. New York: Academic Press.

CREESE, R., DILLON, J. B., MARSHALL, J., SABAWALA, P. B., SCHNEIDER, D. J., TAYLOR, D. B., and ZINN, D. E. (1957). The effect of neuromuscular blocking agents on isolated human intercostal muscles. *J. Pharmacol. exp. Ther.*, **119**, 485–494.

DENNY-BROWN, D. E. (1929). The histological features of striped muscle in relation to its functional activity. *Proc. roy. Soc. B.*, **104**, 371–411.

DUBOWITZ, V. (1967). Pathology of experimentally re-innervated skeletal muscle. *J. Neurol. Neurosurg. Psychiat*, **30**, 99–110.

DUBOWITZ, V., and PEARSE, A. G. E. (1961). Enzymic activity of normal and dystrophic human muscle: a histochemical study. *J. Path. Bact.*, **81**, 365–378.

EBERSTEIN, A., and GOODGOLD, J. (1968). Slow and fast twitch fibres in human skeletal muscle. *Amer. J. Physiol.*, **215**, 535–541.

ECCLES, J. C. (1962). Central connections of muscle afferent fibres. In: *Symposium on Muscle Receptors*, pp. 81–101. Ed. Barker, D. Hong Kong Univ. Press.

ECCLES, J. C. (1967). The effects of nerve cross-union on muscle contraction. In: *Exploratory Concepts in Muscular Dystrophy and Related Disorders*, pp. 151–160. Ed. Milhorat, A. T. Excerpta Medica Foundation Inter. Congr. Series No. 147.

ECCLES, J. C., ECCLES, R. M., and LUNDBERG, A. (1958). The action potentials of the alpha motoneurones supplying fast and slow muscles. *J. Physiol.* (*Lond.*), **142**, 275–291.

EDWARDS, R. G., and LIPPOLD, O. C. J. (1956). The relation between force and integrated electrical activity in fatigued muscle. *J. Physiol.* (*Lond.*), **132**, 677–681.

ELDRED, E., GRANIT, R., and MERTON, P. A. (1953). Supraspinal control of the muscle spindles and its significance. *J. Physiol.* (*Lond.*), **122**, 498–523.

ELMQVIST, D., JOHNS, T. R., and THESLEFF, S. (1960). A study of some electrophysiological properties of human intercostal muscle. *J. Physiol.* (*Lond.*), **154**, 602–607.

ENGEL, W. K. (1962). The essentiality of histo- and cytochemical studies of skeletal muscle in the investigation of neuromuscular disease. *Neurology* (*Minneap.*), **12**, 778–794.

FENN, W. O. (1923). A quantitative comparison between the energy liberated and the work performed by the isolated sartorius muscle of the frog. *J. Physiol.* (*Lond.*), **58**, 175–203.

FENN, W. O. (1924). The relation between the work performed and the energy liberated in muscular contraction. *J. Physiol.* (*Lond.*), **58,** 373–395.

FENN, W. O. (1963). A comparison of respiratory and skeletal muscles. In: *Perspectives in Biology*, pp. 293–300. Eds. Cori, C. F., Foglia, V.G., Leloir, L. F. and Ochoa, S. Amsterdam: Elsevier.

GORDON, G., and PHILLIPS, C. G. (1953). Slow and rapid components in a flexor muscle. *Quart. J. exp. Physiol.*, **38**, 35–45.

GRANIT, R., PHILLIPS, C. G., SKOGLUND, S., and STEG, G. (1957). Differentiation of tonic from phasic alpha ventral horn cells by stretch, pinna and crossed extensor reflexes. *J. Neurophysiol.*, **20**, 470–481.

GRIMBY, L., and HANNERZ, J. (1968). Recruitment order of motor units on voluntary contraction: changes induced by proprioceptive afferent activity. *J. Neurol. Neurosurg. Psychiat.*, **31**, 565–573.

HAMMOND, P. H., MERTON, P. A., and SUTTON, G. G. (1956). Nervous gradation of muscular contraction. *Brit. med. Bull.*, **12,** 214–218.

HENNEMAN, E., and OLSON, C. B. (1965). Relations between structure and function in the design of skeletal muscles. *J. Neurophysiol.*, **28**, 581–598.

HENNEMAN, E., SOMJEN, G., and CARPENTER, D. O. (1965). Functional significance of cell size in spinal motoneurones. *J. Neurophysiol.*, **28**, 560–580.

HILL, A. V. (1938). The heat of shortening and the dynamic constants of muscle. *Proc. roy. Soc. B*, **126**, 136–195.

HILL, A. V. (1940). The dynamic constants of human muscle. *Proc. roy. Soc. B*, **128**, 263–274.

HILL, A. V. (1949). The abrupt transition from rest to activity in muscle. *Proc. roy. Soc. B*, **136**, 399–420.

HILL, A. V. (1950). The dimensions of animals and their muscular dynamics. *Sci. Progr.*, **38**, 209–230.

HILL, A. V. (1960). Production and absorption of work by muscle. *Science*, **131**, 897–903.

HOFMANN, W. W., ALSTON, W., and ROWE, G. (1966). A study of individual neuromuscular junctions in myotonia. *Electroenceph. clin. Neurophysiol.*, **21**, 521–537.

HOUK, J., and HENNEMAN, E. (1967). Responses of Golgi tendon organs to active contractions of the soleus muscle of the cat. *J. Neurophysiol.*, **30**, 466–481.

HUNT, C. C., and PERL, E. R. (1960). Spinal reflex mechanisms concerned with skeletal muscle. *Physiol. Rev.*, **40**, 538–579.

Inman, V. T., Ralston, H. J., Saunders, J. B. de C. M., Feinstein, B., and Wright, E. W. (1952). Relation of human electromyogram to muscular tension. *Electroenceph. clin. Neurophysiol.*, **4**, 187–194.

Jansen, J. K. S., and Rudjord, T. (1964). On the silent period and Golgi tendon organs of the soleus muscle of the cat. *Acta physiol. scand.*, **62**, 364–379.

Joseph, J., Nightingale, A., and Williams, P. L. (1955). A detailed study of the electrical potentials recorded over some postural muscles while relaxed and standing. *J. Physiol.* (*Lond.*), **127**, 617–625.

Katz, B. (1966). *Nerve, Muscle, and Synapse.* New York: McGraw-Hill.

Kugelberg, E., and Edstrom, L. (1968). Differential histochemical effects of muscle contractions on phosphorylase and glycogen in various types of fibres: relation to fatigue. *J. Neurol. Neurosurg. Psychiat.*, **31**, 415–423.

Kugelberg, E., and Skoglund, C. R. (1946). Natural and artificial activation of motor units—a comparison. *J. Neurophysiol.*, **9**, 399–412.

Lenman, J. A. R. (1959). Quantitative electromyographic changes associated with muscular weakness. *J. Neurol. Neurosurg. Psychiat.*, **22**, 306–310.

Lindsley, D. B. (1935). Electrical activity of human motor units during voluntary contraction. *Amer. J. Physiol.*, **114**, 90–99.

Lippold, O. C. J. (1952). The relation between integrated action potentials in a human muscle and its isometric tension. *J. Physiol.* (*Lond.*), **117**, 492–499.

Lundberg, A. (1964). Supraspinal control of transmission in reflex paths to motoneurones and primary afferents. *Progr. Brain Res.*, **12**, 197–219.

McComas, A. J., Mrozek, K., Gardner-Medwin, D., and Stanton, W. H. (1968). Electrical properties of muscle fibre membranes in man. *J. Neurol. Neurosurg. Psychiat.*, **31**, 434–440.

McComas, A. J., and Thomas, H. C. (1968). Fast and slow twitch muscles in man. *J. neurol. Sci.*, **7**, 301–307.

Matthews, P. B. C. (1962). The differentiation of two types of fusimotor fibre by their effects on the dynamic response of muscle spindle primary endings. *Quart. J. exp. Physiol.*, **47**, 324–333.

Matthews, P. B. C. (1964). Muscle spindles and their motor control. *Physiol. Rev.*, **44**, 219–288.

Matthews, P. B. C. (1969). Evidence that the secondary as well as the primary endings of the muscle spindles may be responsible for the tonic stretch reflex of the decerebrate cat. *J. Physiol.* (*Lond.*), **204**, 365–393.

Merton, P. A. (1953). Speculations on the servo-control of movement. In: Ciba Fdn. Symp. on *The Spinal Cord*, pp. 247–255. Eds. Malcolm, J. L., and Gray, J. A. B. London: J. & A. Churchill.

Merton, P. A. (1954). Voluntary strength and fatigue. *J. Physiol.* (*Lond.*), **123**, 553–564.

Naess, K., and Storm-Mathisen, A. (1955). Fatigue of sustained tetanic contractions. *Acta physiol. scand.*, **34**, 351–366.

Padykula, H. A., and Gauthier, G. F. (1967). Morphological and cytochemical characteristics of fibre types in normal mammalian skeletal muscle. In: *Exploratory Concepts in Muscular Dystrophy and Related Disorders*, pp. 117–128. Ed. Milhorat, A. T. Excerpta Medica Foundation. Inter. Congr. Series no. 147.

ROMANUL, F. C. A. (1964). Enzymes in muscle. I. Histochemical studies of enzymes in individual muscle fibres. *Arch. Neurol.* (*Chic.*), **11**, 355–368.

ROMANUL, F. C. A., and MEULEN, J. P. van der (1967). Slow and fast muscles after cross innervation. *Arch. Neurol.* (*Chic.*), **17**, 387–402.

SHERRINGTON, C. S. (1929). Some functional problems attaching to convergence. *Proc. roy. Soc. B*, **105**, 332–362.

SWETT, J. E., and ELDRED, E. (1960). Distribution and numbers of stretch receptors in medial gastrocnemius and soleus muscles of the cat. *Anat. Rec.*, **137**, 453–460.

WILKIE, D. R. (1950). The relation between force and velocity in human muscle. *J. Physiol.* (*Lond.*), **110**, 249–280.

WILKIE, D. R. (1956). The mechanical properties of muscle. *Brit. med. Bull.*, **12**, 177–182.

WUERKER, R. B., MCPHEDRAN, A. M., and HENNEMAN, E. (1965). Properties of motor units in a heterogeneous pale muscle (M gastrocnemius) of the cat. *J. Neurophysiol.*, **28**, 85–99.

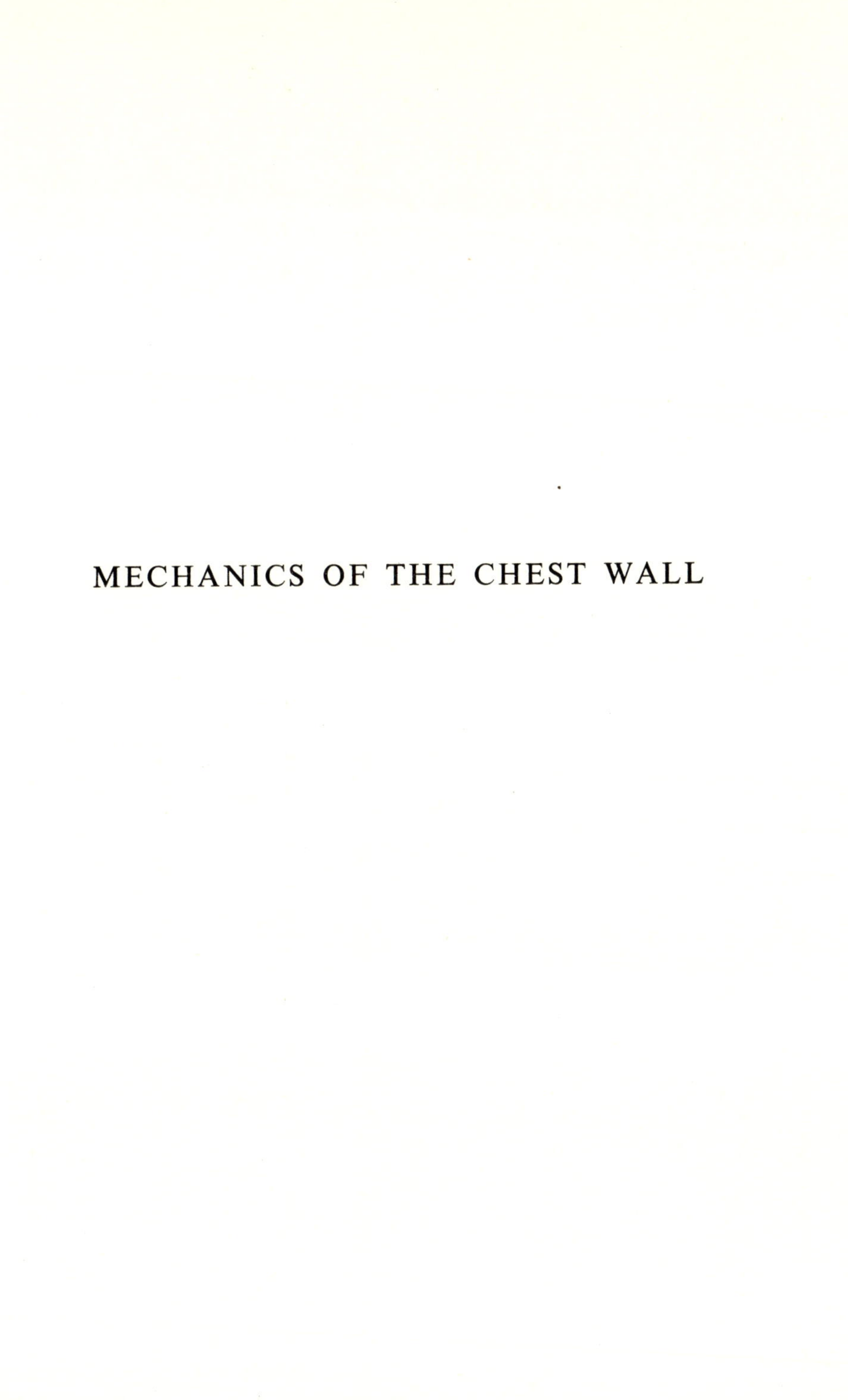

# MECHANICS OF THE CHEST WALL

# Chapter II

# KINEMATICS

E. Agostoni

## FUNCTIONAL ANATOMY OF THE RIB CAGE

Although there are wide differences between species, the shape of the rib cage may be approximated to that of a truncated cone. The main differences among mammals are due to posture and size. In quadrupeds the rib cage is laterally flattened particularly cephalad, whereas in mammals that maintain an upright posture the rib cage is dorso-ventrally flattened. The number of ribs range among species from 9 to 25 pairs (Krahl, 1964).

Owing to the articulation and morphology of the ribs, both the dorso-ventral and lateral diameters of the rib cage increase when they are raised. As a general consequence, therefore, the muscles tending to raise the ribs are inspiratory, those tending to lower the ribs are expiratory. In some subjects, however, the lateral diameter at large lung volume does not increase or may even decrease when the lung volume is increased (see p. 29).

The movement of the ribs takes place around the axis of their necks. The greatest movement should occur, therefore, at the point where the radial distance from the neck axis is greatest (Luciani, 1911) (Fig. 9). The axes of rotation of the two corresponding ribs decussate in front, forming angles that decrease from the first to the tenth rib from 125° to 88°; as a consequence, the part of the rib undergoing the largest movement is placed more ventrally in the upper ribs and more laterally in the lower ribs. Owing to the particular features of the first costo-vertebral joint, the only movement of the operculum of the thorax is that of raising or lowering the sternum; in a normal subject this does not occur during quiet breathing. This widely accepted description of the movements of the ribs has been criticised by Polgar (1949), who suggested that the excursion of the ribs takes place around two axes: one horizontal and one vertical. Jordanoglou (1969 and 1970), however, by means of a vector analysis of the rib movement, has shown that the ribs move essentially around their neck axis. This movement is oblique with three components: sagittal (forward), transverse (outward) and vertical (upward). The forward and upward components constitute the so called "pump-handle" movement, while the outward and upward components constitute the "bucket-handle" movement.

The rib cage undergoes a marked change of shape during growth. The

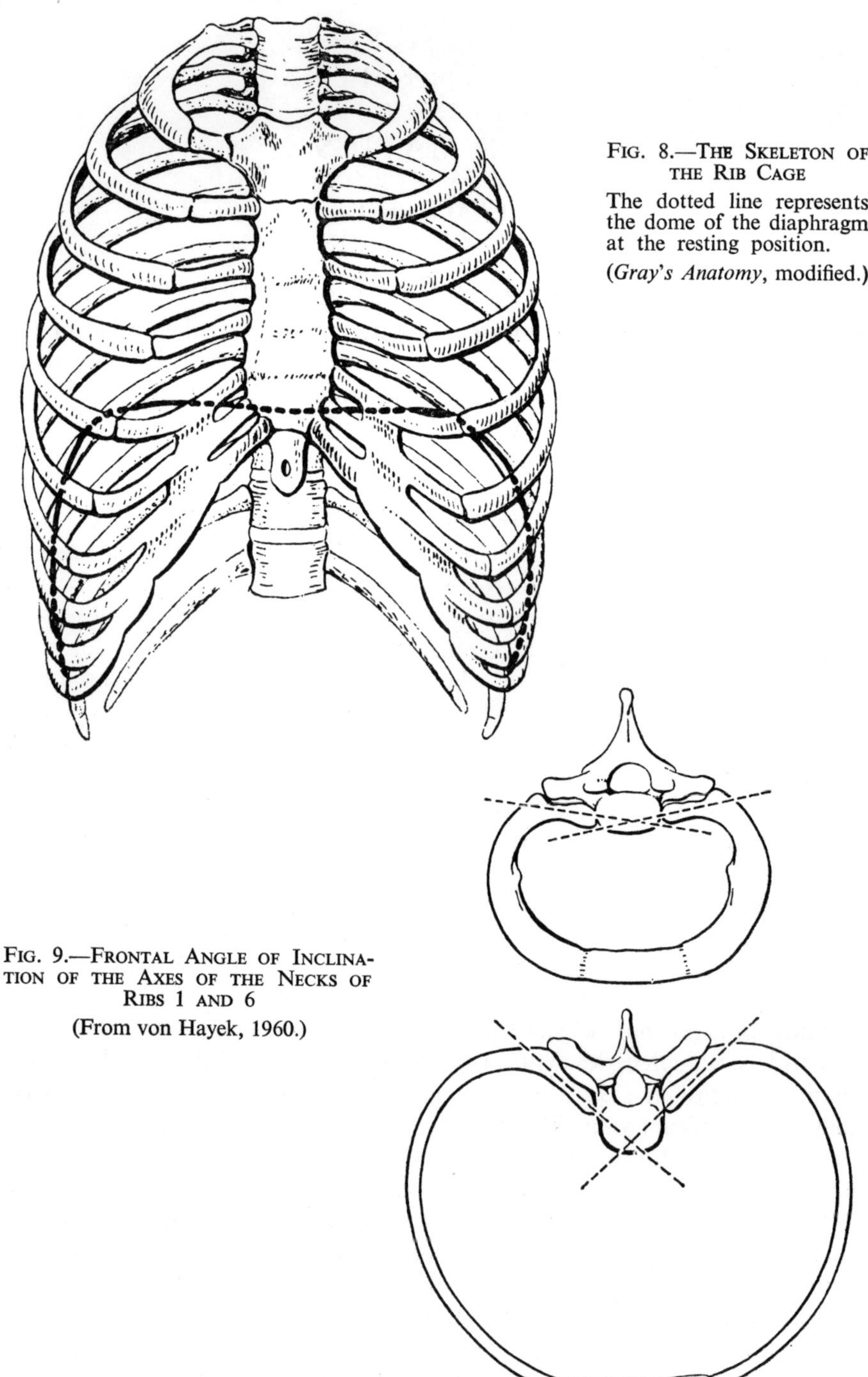

Fig. 8.—The Skeleton of the Rib Cage

The dotted line represents the dome of the diaphragm at the resting position.

(*Gray's Anatomy*, modified.)

Fig. 9.—Frontal Angle of Inclination of the Axes of the Necks of Ribs 1 and 6

(From von Hayek, 1960.)

difference in the horizontal section is illustrated in Fig. 10. Furthermore the ribs in the newborn human run horizontally. The changes in shape during growth are probably due to the effects of gravity and the action of respiratory muscles.

Changes of volume of the rib cage provide one mechanism to change the lung volume. The other mechanism is provided by the piston action of the diaphragm (see p. 35). Movements of the bony part of the chest wall also contribute to lung volume change indirectly by changing the distance between rib cage and pelvis and hence acting on the statics of the abdomen (see Chapter III). The distance between rib cage and pelvis is changed by movements of flexion and extension of the vertebral column, by movements of the shoulder girdle and by movements of the ribs.

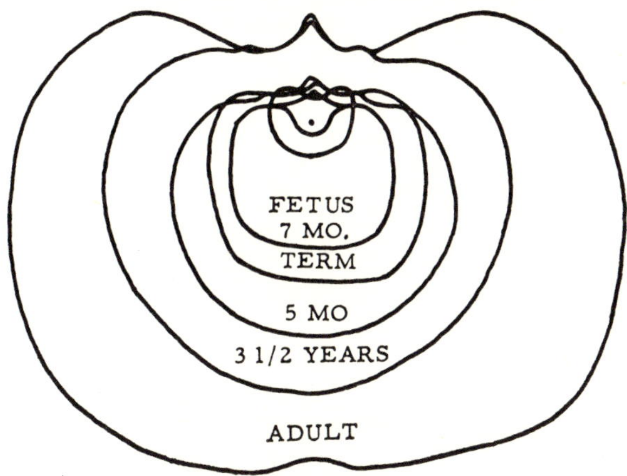


Fig. 10.—Superimposed Outlines of Thoracic Skeletal Contours at Various Ages

Outlines are aligned with respect to the centre of the vertebral body. (From Krahl, 1964.)

## ANALYSIS OF THE MOVEMENTS OF THE CHEST WALL

The movements of the rib cage and of the abdominal wall have been extensively studied in the past, but without systematic reference to lung volume changes. Those data are therefore of little value in a quantitative analysis of the mechanics of the chest wall. The first measurements of the movements of the rib cage related to changes of lung volume were made by Herxheimer (1949) using the mechanical thoracometer devised by Verzàr (1946) to measure rib cage circumference change. Today these measurements can be made more conveniently by means of a mercury-in-rubber transducer (Wade, 1954; Agostoni *et al.*, 1965). Changes in the diameters of the rib cage may be measured with pairs of differential transformers (Konno and Mead, 1967), provided care is taken to avoid artefacts due to movements of soft tissues. The difficulties encountered during muscular exercise may be overcome by using magnetometers (Mead *et al.*, 1967; Grimby, Bunn and Mead, 1968).

The movements of the dome of the diaphragm have been studied since the advent of radiology (Dally, 1908; Hitzenberger, 1927; Herxheimer, 1948; Hasselwander, 1949; Wade and Gilson, 1951; Wade, 1954); simultaneous measurements of rib cage circumference changes and of displacements of the diaphragmatic dome relative to the rib cage over the inspiratory capacity and the expiratory reserve volume were first undertaken by Wade (1954).

The studies in which the movements of a part of the chest wall have been related to the changes of lung volume have been done mainly on

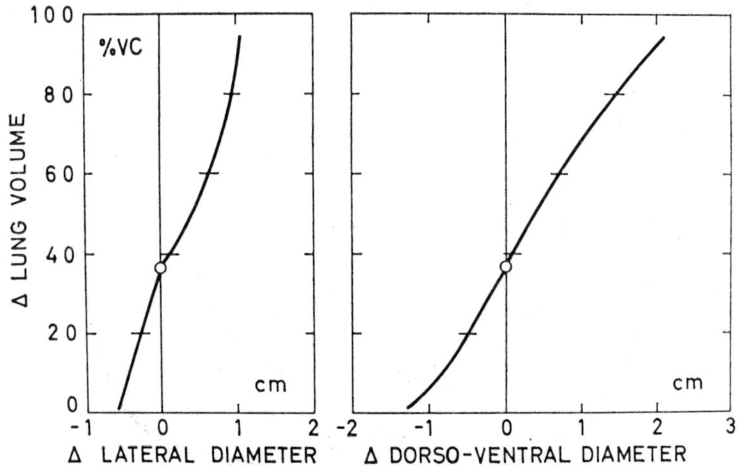


FIG. 11.—STATIC RELATIONS BETWEEN THE LUNG VOLUME AND THE CHANGES OF THE LATERAL AND DORSO-VENTRAL DIAMETER OF THE RIB CAGE AT THE XIPHOID LEVEL DURING RELAXATION

The circles indicate the resting volume (end of spontaneous expirations). Mean of four subjects in the sitting position, the horizontal bars indicate the standard error. (From data of Agostoni and Mognoni, 1966, and unpublished data.)

males. In those in which females also were studied (Herxheimer, 1949; Grimby *et al.*, 1968) no systematic differences between the sexes have been pointed out.

## Movements of the Rib Cage

*Diameters and circumference of the rib cage during relaxation.*—The static relations between lung volume and the lateral and dorso-ventral diameters of the rib cage at the xiphoid level during relaxation (with closed airways) in the sitting position are shown in Fig. 11. The change of the lateral diameter over the vital capacity range (1·6 cm) is about half that of the dorso-ventral one (3·4 cm). The change of the lateral diameter is even smaller if the percentage change of the diameters is considered, as the lateral diameter is larger than the dorso-ventral one: at FRC the ratio between the lateral and the dorso-ventral diameter is about 1·34 in the sitting position and 1·45 in the supine position (Agostoni *et al.*, 1965). In the standing position the average change of lateral diameter over the VC range during relaxation found by Konno and Mead (1967) is 1·2 cm, that of the dorso-ventral 2·8 cm.

When the subject moves from the sitting to the standing position at FRC the lung volume increases by about 2 per cent of the VC, whereas the rib cage circumference, at the xiphoid level, decreases by about 18 per cent of its maximum change in the sitting position (see below) (Agostoni *et al.*, 1965). The main respiratory muscles are relaxed in both positions and this difference is due to the different relation between rib cage and abdomen in the two positions. When the subject is sitting, the rib cage lies in part on the abdominal contents, which push the ribs out. When the subject stands, the rib cage, supported by antigravity muscles, no longer lies on the abdominal contents; its distance from the pelvis is increased (see p. 35) and the abdominal wall, because of the increased gravitational effect, pulls the ribs down and in.

When the subject moves from the sitting to the supine position at FRC the lung volume decreases by about 15 per cent of the VC, whereas the rib cage circumference increases by about 5 per cent of its maximum change in the sitting position (Agostoni *et al.*, 1965) as may be expected owing to the different gravitational effect of the abdomen (see Chapter III). These effects depend also upon the fact that when the subject changes from the erect to the horizontal position about 500 ml of blood move from the legs to the thoracic cavity decreasing the lung volume and increasing the volume of the chest wall. Mains *et al.*, (1968) were able to show this effect independently from changes of posture: they drew blood from the thorax of supine subjects by applying a negative pressure of 40 mm Hg to the legs and found an increase of lung volume and a decrease of the circumference of the rib cage and of the abdomen.

Unfortunately the changes of circumference of the rib cage are not a completely reliable index of the changes of cross section, because the rib cage may change its shape; if it becomes more circular an increase of cross section could be accompanied by a decrease of circumference, and vice versa. For most purposes, however, the changes of the cross sectional area of the rib cage may be approximated from measurements of the lateral and dorso-ventral diameters, assuming the shape of an ellipse for the cross section.

The approximate cross sectional area of the rib cage at the xiphoid level and the lung volume at FRC in the standing, sitting and supine position are illustrated by Fig. 12. These changes in the size of the rib cage show the distorting effect of gravity and emphasise the difficulty of defining the pure elastic recoil of the rib cage (see Chapter III).

These data point out that the length of the relaxed muscles of the rib

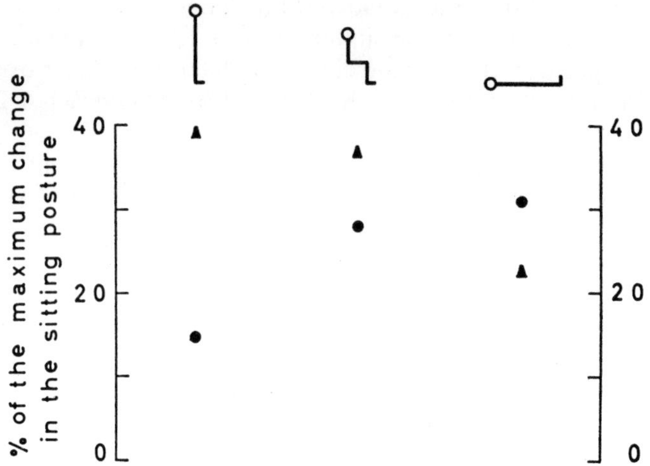


FIG. 12.—LUNG VOLUME (▲) AND CROSS SECTIONAL AREA OF THE RIB CAGE (●) AT FRC IN VARIOUS POSTURES

cage at FRC may be different at similar lung volumes (standing-sitting) or almost the same at different lung volumes (sitting-supine), because of different gravitational effects.

*Diameters and circumference of the rib cage during activity of the respiratory muscles.*—The average change of the circumference of the rib cage at the xiphoid level between the extremes of lung volume held actively with open airway corresponds roughly to 9 per cent of the rib cage circum-

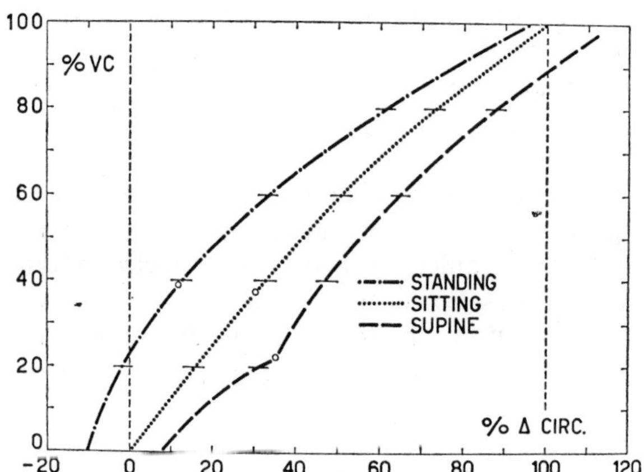


FIG. 13.—STATIC RELATION BETWEEN CHANGES OF LUNG VOLUME AND OF RIB CAGE CIRCUMFERENCE AT THE XIPHOID LEVEL DURING RELAXATION

Both are expressed as per cent of the maximum changes in the sitting position: mean of nine subjects in the standing, sitting, and supine position. Circles indicate the resting volume (end of a spontaneous expiration). The horizontal bars indicate the standard error. (From Agostoni *et al.*, 1965.)

ference at FRC. Wade (1954) in 10 male subjects, with an average VC of 4·8 litres BTPS, found an average change of 7·4 and 7·7 cm respectively in the standing and supine positions. Agostoni *et al.* (1965) found, in 9 male subjects with a VC of 5·3 litres, an average change ($\pm$ S.E.) of 8·2 $\pm$ 0·2; 7·7 $\pm$ 0·2; 7·9 $\pm$ 0·3 cm, respectively, in the standing, sitting and supine positions. The differences between positions are not significant.

The static relation between lung volume and rib cage circumference at the xiphoid level when the subject holds a given lung volume with open airways is shown in Fig. 13. These relations, as well as those between lung volume and rib cage diameters illustrated below, are not unique because the circumference and the diameters, at a given lung volume, are somewhat affected by the degree of extension of the vertebral column. The relations determined are those spontaneously occurring in a given position (Agostoni *et al.*, 1965).

The change of the rib cage circumference, expressed as per cent of the maximum, over half of the VC may be markedly different among individuals even in the supine position when the gravitational effect of the abdomen is about the same in all the subjects (Agostoni *et al.*, 1965). These differences suggest that the mechanical features of the rib cage vary widely in normal subjects.

For an increase of lung volume of 10 per cent of the VC from FRC, i.e. similar to that occurring during spontaneous inspiration at rest, the increase of rib cage circumference in the supine position is only about 65 per cent of that in the sitting and standing positions (Fig. 13). Hence, within this lung volume range, the volume displaced by the rib cage is smaller in the supine position (see p. 43). This phenomenon may be accounted for by the lower compliance of the rib cage and the higher compliance of the abdomen in the supine position within this volume range. The different mechanical features of the relaxed rib cage and of the abdomen in the two postures seem sufficient to account for the different slope of the relation between lung volume and rib cage circumference in the two postures; change of activity of the inspiratory muscles cannot be ruled out, but there is at present no information on this point (Agostoni *et al.*, 1965).

The average changes of the lateral and dorso-ventral diameters between the extremes of lung volume found by Agostoni and Mognoni (1966) in four male subjects in the sitting posture, were 3 and 3·3 cm respectively, and those found by Konno and Mead (1967) in six standing male subjects were 1·4 and 2·9 cm, respectively. The relation between lung volume and the dorso-ventral and lateral diameters of the rib cage when the subject holds a given lung volume with open airways, and that during relaxation against closed airways are shown in Fig. 14. In some subjects when the lung volume is increased above 60 per cent VC the lateral diameter does not increase and may even decrease. Above FRC the percentage change of the minor diameter (dorso-ventral) is greater than that of the major one, hence the change of area of the cross section per unit change of circumference is larger than if the cross section were circular (Agostoni *et al.*, 1965).

The differences between the diameters at a given lung volume when the subject holds it with open airways and when he relaxes against closed airways indicate that the action of the muscles produces a change of shape of the rib cage. This deformation shows that the forces resulting from the contraction of the respiratory muscles and acting at right angle to the surface of the chest wall are not evenly distributed (see Chapter III).

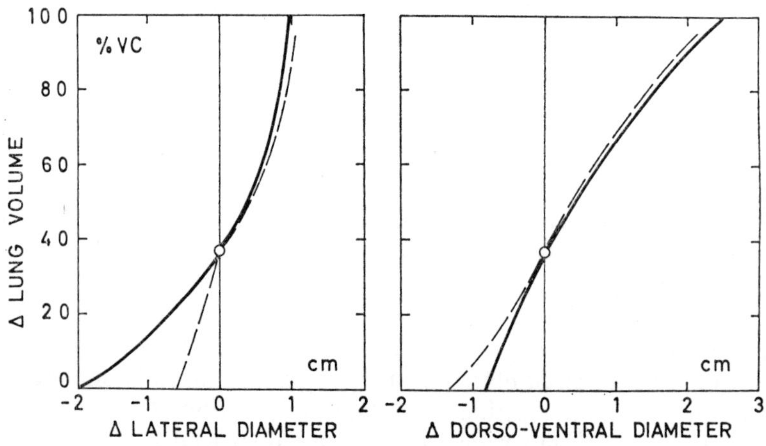


FIG. 14.—STATIC RELATIONS BETWEEN THE LUNG VOLUME AND THE CHANGES OF THE LATERAL AND DORSO-VENTRAL DIAMETERS OF THE RIB CAGE AT THE XIPHOID LEVEL WHEN THE SUBJECT HOLDS A GIVEN LUNG VOLUME WITH OPEN AIRWAYS (continuous line)

The circles indicate the resting volume (end of spontaneous expirations). The broken line indicates the relations during relaxation (see Fig. 11). (From data of Agostoni and Mognoni, 1966, and unpublished data.)

During static respiratory efforts the lateral diameter changes in the same direction as the lung volume whereas the dorso-ventral one changes, generally, in the opposite direction. During expiratory efforts the cross section of the rib cage becomes more circular than during relaxation at the same lung volume, whereas during inspiratory efforts it becomes, generally, more elliptical. These changes of shape show that the forces resulting from the activity of the respiratory muscles and acting at right angle to the surface of the chest wall operate predominantly on the lateral parts of the rib cage, the frontal part being mainly driven by the pressure acting across it and by the movement of the lateral parts (Agostoni and Mognoni, 1966; Agostoni *et al.*, 1966).

During maximum expiratory effort against closed airways performed at mid-lung volume the lateral diameter in some subjects becomes small as it is at RV, notwithstanding that the cross sectional area of the rib cage is obviously larger (Agostoni *et al.*, 1966).

These changes of shape of the cross section of the rib cage indicate that the relation between linear motion and volume of the rib cage is not unique if the respiratory muscles contract strongly. In other words the rib cage has more than one degree of freedom: the principal one related to the change of its volume, the secondary one related to the change of shape of its cross section.

The above findings imply that: (*a*) the horizontal cross section of the rib cage is more elliptical during inspiration than during expiration in most subjects, and (*b*) the motions of the diameters of the rib cage are out of phase during the breathing cycle when the mechanical load is increased because of increased ventilation and/or increased resistance (Agostoni and Mognoni, 1966).

The motion of the rib cage diameters is not a simple sinusoidal function of time, particularly when the respiratory system is loaded, but as a first approximation it may be considered such in order to study the phase shift caused by the action of the respiratory muscles. If two sinusoidals in phase are plotted one against the other in a diagram a straight line is obtained; this line has a slope of 1 provided the amplitude of each sinusoidal is made to span the same length on the coordinates. If the sinusoidals are out of phase, but still have the same frequency, a loop is obtained instead of a line, the loop tending towards a circle as the phase shift tends towards 90°. When the phase shift is 180° a line is formed again but with a slope of −1. By plotting on a diagram the motion of one diameter of the rib cage versus that of the other one, or by displaying the corresponding signals on a X − Y recorder, a fairly straight line with a slope of about 1 is obtained during quiet breathing, i.e. the phase shift is essentially nil in this condition. The loop at ventilatory rates of 20–30 litres/min (Fig. 15*a*) shows that the motion of the dorso-ventral diameter lags behind that of the lateral one during inspiration; during expiration there is no phase shift, as would be expected because the expiratory muscles do not contract at this load. When the load increases, the expiratory muscles contract and a phase shift appears also during expiration (Fig. 15*b* and *c*) and may become greater than during inspiration (Fig. 15*d*). The phase shift increases with the load as shown in Fig. 16 by the relation between the average phase shift throughout the breathing cycle and the average change of the esophageal pressure, taking this as an index of the load (Agostoni and Mognoni, 1966). During maximum exercise ventilation on the cycle-ergometer the phase shift between the motion of the diameters ranges from 15 to 30° according to the subject; during maximum voluntary ventilation in the sitting and standing position it ranges from 70 to 120° (Agostoni and Torri, 1967).

The change of the dorso-ventral diameter lags behind those of the lung volume and of the rib cage circumference, the phase shift being about 40

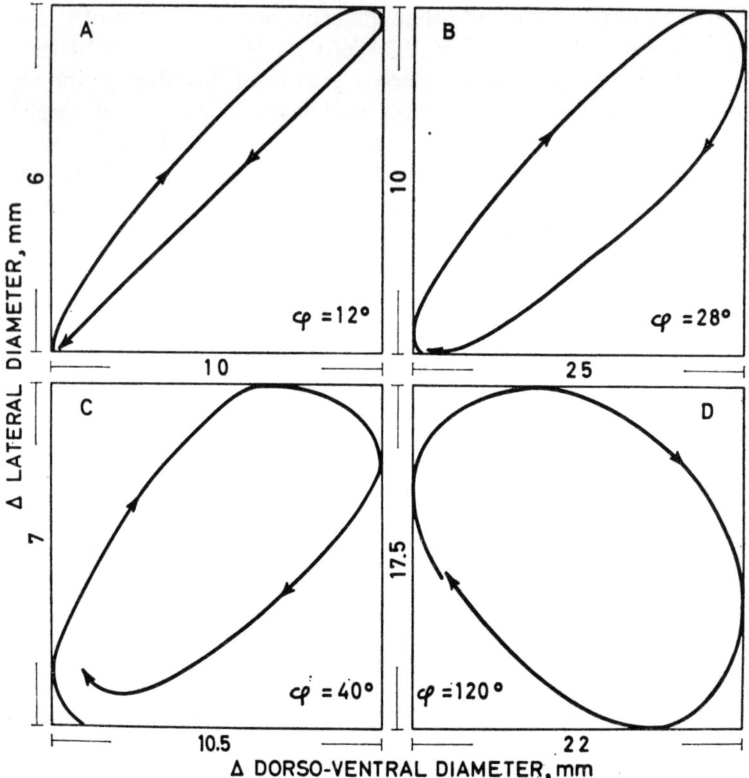


FIG. 15.—RELATIVE CHANGES OF LATERAL AND DORSO-VENTRAL DIAMETERS OF THE RIB CAGE AT THE XIPHOID LEVEL DURING DIFFERENT TYPES OF BREATHING CYCLES

The approximate angle of phase shift between the motion of the diameters is indicated by $\varphi$. Same subject in the sitting position. The beginning and end of the loops correspond to the end of expiration. A: re-breathing, $\dot{V} = 28{\cdot}2$ litres (BTPS)/min, $f = 20{\cdot}4$ cycles/min, $ERV = 35$ per cent VC. B: Heavy exercise on the cycle-ergometer, $\dot{V} = 97{\cdot}8$ litres/min, $f = 27$ cycles/min, $ERV = 20$ per cent VC. C: rebreathing through a resistance (17 cm $H_2O$/litres/sec), $\dot{V} = 24$ litres/min, $f = 15$ cycles/min, $ERV = 38$ per cent VC. D: maximum voluntary ventilation (rebreathing), $\dot{V} = 172$ litres/min, $f = 88$ cycles/min, $ERV = 38$ per cent VC. (From published and unpublished data of Agostoni and Mognoni, 1966, and Agostoni and Torri, 1967.)

per cent and, respectively, 60 per cent of that between the two diameters. The phase shift between the changes of the dorso-ventral diameter and those of the rib cage circumference at a level about 7 cm above the xiphoid is similar to that measured at the xiphoid. The phase shift between the change of the rib cage circumference and that of the dorso-ventral diameter in the midclavicular line is more than half that occurring with the dorso-ventral diameter in the midline; hence the part lagging behind is not limited to a narrow central band (Agostoni and Mognoni, 1966).

When the phase shift between the two main diameters is about 90° the tidal change of the cross sectional area of the rib cage is reduced to about 60 per cent of what would have occurred if the changes of the diameters were in phase. This suggests that the rib cage contribution to the tidal volume decreases when the breathing load is great. This decrease indeed occurs in some subjects, but in others the tidal changes of the rib cage diameters, for a given tidal volume, increase under load to such an extent that the tidal changes of the cross sectional area of the rib cage do not decrease, in spite of the great phase shift between the motion of the diameters (Agostoni and Torri, 1967).

The deformation of the rib cage occurring under load implies that: (*a*) some muscles lengthen instead of shortening and vice versa; (*b*) informa-

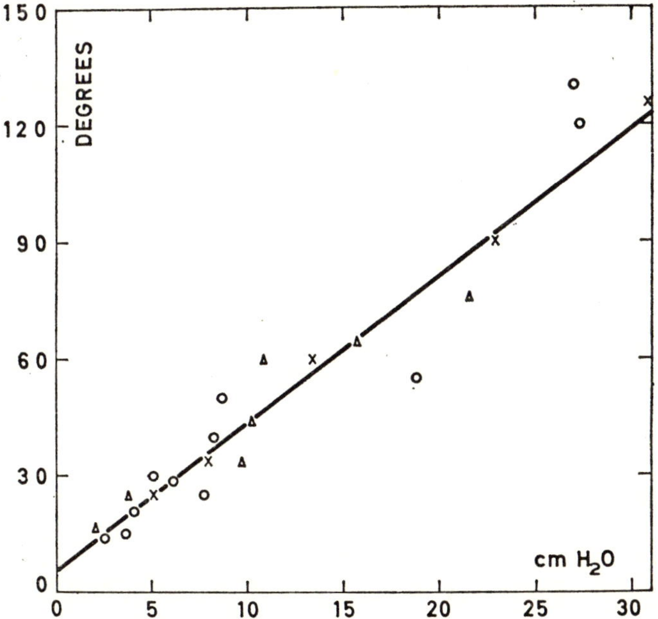


FIG. 16.—AVERAGE ANGLE OF PHASE SHIFT BETWEEN THE CHANGES OF THE LATERAL AND DORSOVENTRAL DIAMETERS OF THE RIB CAGE AT VARIOUS LOADS

The average change of the oesophageal pressure above and below the end-expiratory value throughout the breathing cycle has been taken as an index of the load. The symbols refer to three subjects during rebreathing, rebreathing through a resistance and voluntary hyperventilation while rebreathing. (Modified from Agostoni and Mognoni, 1966.)

tion from muscle, joint and lung receptors is out of phase (this may provide information about the degree of the breathing effort; see Chapter XIV); (*c*) the work of breathing under load is slightly larger than that calculated on the volume-pressure diagram (see Chapter V) (Agostoni and Mognoni, 1966).

During moderate increase of ventilation produced voluntarily by hypercapnia, hypoxia or exercise on a cycle-ergometer the circumference and the diameters of the rib cage at the end of expiration and inspiration maintain the same relationship with the lung volume as under static conditions.

During maximum exercise ventilation the lateral and the dorso-ventral diameters, for a given lung volume, are respectively slightly decreased and increased relatively to the static relationship. These shifts are probably due to the sustained contraction of the abdominal muscles produced by the exercise (Agostoni and Torri, 1967). Wade (1954) found that the end-expiratory rib cage circumference is increased and its tidal changes reduced during "forced breathing", but Agostoni and Torri only observed this in one of three subjects studied during maximum voluntary ventilation. The reduction of the tidal changes of the circumference is mainly related to the decrease of the tidal changes of the lateral diameter.

*Vertical displacement of the rib cage.*—The vertical displacement of the frontal and lateral parts of the rib cage depends in part upon the move-

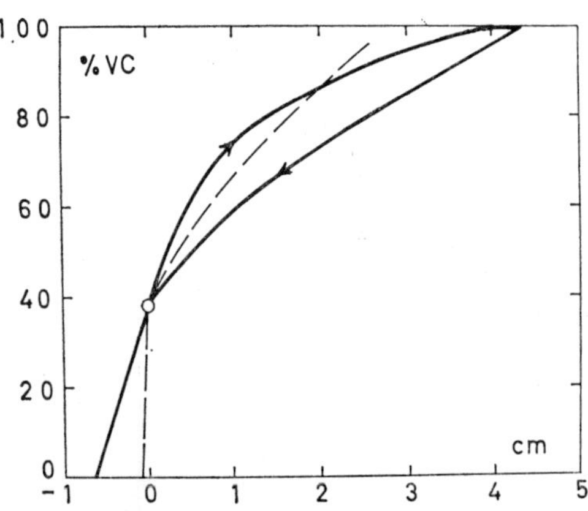


FIG. 17.—RELATIONS BETWEEN THE LUNG VOLUME AND THE CHANGE OF THE DISTANCE BETWEEN RIB CAGE AND PELVIS 1–2 CM VENTRALLY TO THE AXILLARY LINE

The continuous line refers to the quasi-static condition. The broken line refers to relaxation and the circle to the resting end-expiratory value. Average of three subjects in the sitting position.

ments of the ribs, and therefore upon the degree of expansion of the rib cage, but predominantly upon the degree of extension and flexion of the vertebral column. These movements are affected by the tone of the abdominal wall and by any force clothing may exert on the abdomen; when these factors make the abdominal wall more stiff, the rib cage rises more during inspiration.

Wade (1954), measuring the vertical movements of the sterno-xiphisternal joint relative to a point outside the body, found that over the inspiratory capacity the lifting occurred mainly near full inspiration, whereas the lowering was steady throughout the corresponding expiration. He found that extension and flexion of the vertebral column caused much variation in the rib movements in the standing position.

As an index of the vertical movement of the rib cage Agostoni *et al.* (1965) measured the change of the distance between the rib cage and the pelvis 1–2 cm frontally to the axillary line in the sitting and supine positions. Over the inspiratory capacity this distance changes about 4 cm both in the sitting and supine position; since in the supine position no unnecessary movements of the vertebral column are made, it seems likely also that the data collected in the sitting position are not affected by unnecessary movements of the vertebral column. In the sitting position (Fig. 17), as noted by Wade in standing, the lifting of the rib cage occurs mainly at large volumes, whereas the lowering is almost linearly related to the lung volume. This pattern may vary when the manoeuvres are performed rapidly. Over the ERV the distance between rib cage and pelvis changes less than 1 cm both in the sitting and supine positions. When the subject relaxes at full inspiration the distance in both positions decreases about 1 cm taking into account the concomitant change of lung volume due to the change of gas pressure; at full expiration the distance increases by about 0·5 cm (Fig. 17). When the subject at FRC moves from the sitting to the supine position the distance between rib cage and pelvis increases 6–8 cm. Lifting and lowering of the rib cage without changing lung volume in the sitting position produces, respectively, a decrease and an increase of the rib cage circumference.

## Movements of the Diaphragm

The movements of the diaphragmatic dome relative to the rib cage have been studied in the standing and supine positions by Wade (1954), who measured radiologically the displacements of the dome correcting for the cranio-caudal movements of the sterno-xiphisternal joint. The average displacement of the diaphragmatic dome over the VC is about 9·5 cm in both positions (Wade, 1954; see also Hasselwander, 1959). The data of Wade (1954) and of Agostoni *et al.* (1965) on the vertical motion of the rib cage indicate that about half diaphragm motion is not transmitted to the abdomen because of the opposite vertical movement of the rib cage (see above). The average displacements of the diaphragm over the IC are 5·5 cm in the standing position and 7·7 in the supine position (Wade, 1954). The tidal displacements during quiet breathing measured by Wade (1954) and referred to an equal volume change in both positions are about 1·5 cm in the standing position and 1·7 cm in the supine position.

## Movements of the Abdomen

The displacements of the abdominal wall are not uniquely related to those of the diaphragm because of the vertical motion of the rib cage. Depending on the action of other muscles a descent of the diaphragm relative to its insertion may be associated with an outward movement of the abdominal wall, no movement, or even an inward movement. If, for instance, the diaphragm descends to an extent equal to the rise of the rib cage the abdominal wall will not move.

Owing to the structure of its wall the shape of the relaxed abdomen changes markedly with posture. By studying the dorso-ventral motion of different points of the front of the abdomen relative to that of a reference

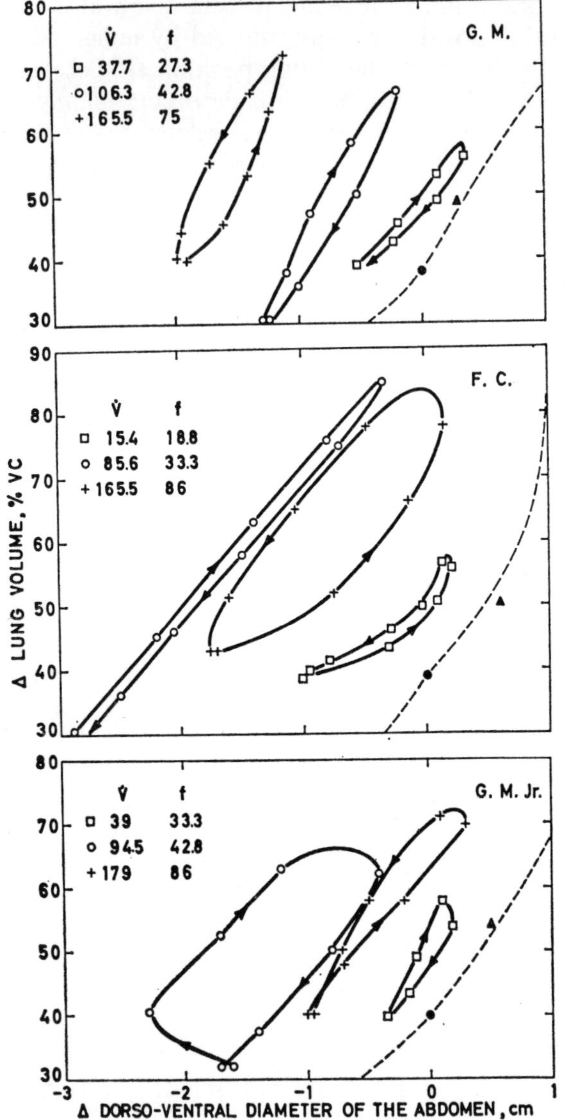


FIG. 18.—CHANGES OF THE LUNG VOLUME AGAINST THOSE OF THE DORSO-VENTRAL DIAMETER OF THE UPPER ABDOMEN AT DIFFERENT VALUES OF VENTILATION

Zero on the abscissa refers to the resting end-expiratory value, which is indicated by the closed circle. The broken line indicates the relationship during relaxation and the triangle indicates the end-inspiratory value during quiet breathing. Each loop refers to a typical cycle at increased ventilation. The values of ventilation in l/min and the breathing frequency in c/min are indicated in the diagrams. The periods between two consecutive squares, open circles, and crosses are respectively 0·4; 0·2; 0·1 sec. Standing posture. (From Agostoni and Torri, 1967.)

point, under conditions of light muscular activity, Konno and Mead (1967) showed that the front of the abdomen moves almost as a unit, but to a lesser degree than the front of the rib cage. During strong contraction of the abdominal muscles the horizontal cross section of the abdomen becomes more rounded even than that of the rib cage.

Changes of lung volume are plotted against those of the dorso-ventral diameter of the upper abdomen in Fig. 18. When the ventilation is increased, either voluntarily or by hypercapnia, the relation is shifted to the left suggesting a sustained rise of the rib cage and hence a sustained change of shape of the chest wall. At high ventilation the end-expiratory diameter of the abdomen in some subjects approaches values reached at residual volume and the end-inspiratory diameter is similar to the end-expiratory one at rest (Agostoni and Torri, 1967).

**Analysis of the Simultaneous Movements of the Rib Cage and of the Abdomen at Iso-lung Volume**

When the degree of muscular contraction is small, the deformation of the rib cage and of the abdominal wall is negligible. Taking advantage of this condition Konno and Mead (1967) considered the chest wall, in a given posture, as a system with two degrees of freedom, made up of two parts, separated by the costal margin, that may be moved independently. When the system is closed, i.e. the airways are closed, the system has only one degree of freedom left, hence the volume change of the rib cage must be equal and opposite to that of the abdomen. On this basis Konno and Mead made an elegant analysis of the combined movements of the rib cage and of the abdomen and measured the volume displaced by these two parts (see p. 39).

Their approach is as follows: the subject first closed the airways at a given lung volume and relaxed; then he displaced volume between rib cage and abdomen without flexing or extending the spine and without exceeding an alveolar pressure of $\pm 20$ cm $H_2O$. The changes of the dorso-ventral diameter of the rib cage were recorded against those of the dorso-ventral diameter of the abdomen at 20, 40, 60 and 80 per cent of the VC. Konno and Mead (1967), as illustrated in Fig. 19, showed that: (1) there is a unique relation between motion of the rib cage and of the abdomen at a given lung volume (provided this lung volume is associated with a constant degree of extension of the vertebral column), indicating that the chest wall has basically two moving parts when muscular contraction is small; (2) the iso-lung volume lines are nearly parallel, indicating that the relation is not influenced by the volume of a part; (3) the isopleths are approximately equi-distant for equal volume increments, so that the relation between motion and volume change for either part is approximately linear.

The antero-posterior motion of the abdomen is larger than that of the rib cage as volume is shifted between them. The total motion of the abdomen, within the experimental limits of pressure changes, is greater than that occurring over the VC range, whereas the total motion of the rib cage, within these limits, is about equal to that occurring over the VC range. At the extremes of lung volume the configurations are unique and the limits imposed are overridden because alveolar pressure exceeds $\pm 20$ cm $H_2O$ (Konno and Mead, 1967).

Points to the left of the relaxation line indicate a prevailing inspiratory activity on the rib cage and/or a prevailing expiratory activity on the

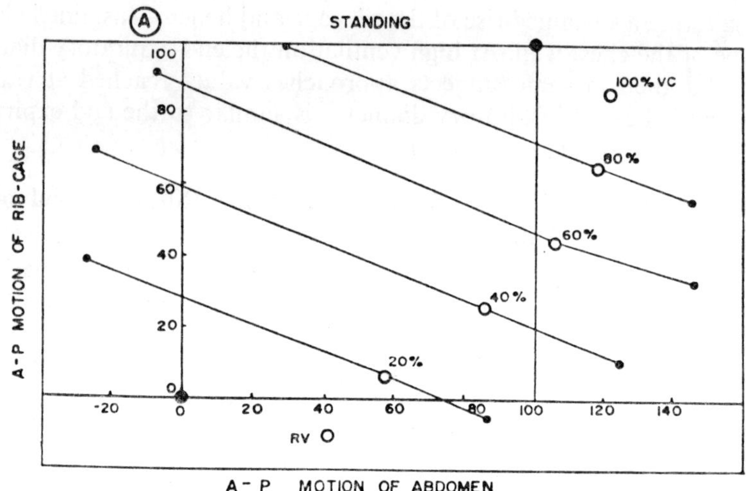


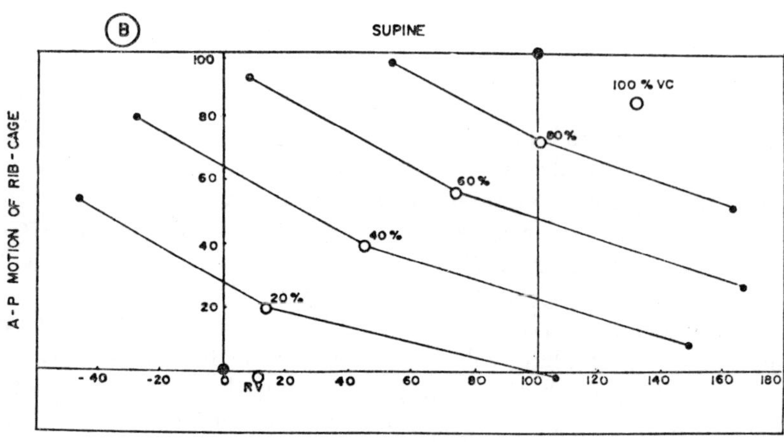


FIG. 19.—DORSO-VENTRAL MOTION OF THE RIB CAGE AGAINST THAT OF THE ABDOMEN

Both expressed as per cent of the total excursion over the VC. The motion of the rib cage is measured at the nipple level, that of the abdomen at the umbilicus level, both on a vertical line midway between the right nipple and the midline. The open circles refer to the relaxed state. The iso-lung volume lines are obtained by voluntary displacement of volume back and forth between the rib cage and abdomen without flexing and extending the vertebral column and maintaining alveolar pressure between $\pm 20$ cm $H_2O$. Mean of six subjects. (From Konno and Mead, 1967.)

abdomen (Fig. 19). Points to the right indicate a prevailing expiratory activity on the rib cage and/or a prevailing inspiratory activity on the abdomen. Quiet breathing in the supine position does not take place along the relaxation line, but the abdominal motion is relatively greater. During exercise the relationship is shifted to the left of the relaxation line (Grimby,

Bunn and Mead, 1968). During talking in the standing posture the pathways are deviated to the left of the relaxation curve: the rib cage is sustained at 50–60 per cent VC, while changes in volume are produced by movement of the abdomen; in the supine position the pathways are near the relaxation curve. During singing in the standing position the pathways are deviated to the left even more than during talking (Konno and Mead, 1967) (see Phonation, in Chapter IV).

**Volume Displaced by the Rib Cage, the Diaphragm and the Abdomen**

From the analysis of the simultaneous motion of the rib cage and of the abdomen at iso-lung volume Konno and Mead (1967) determined the volume displaced by these two parts. As is apparent from the previous analysis (see also "Movements of the abdomen"), this approach gives the volume displaced by the abdomen, not that displaced by the diaphragm, hence the fraction of this latter corresponding to the vertical movement of the rib cage is computed as volume displaced by the rib cage. This must be remembered when discussing the volume displacement of the rib cage.

The relationship between the linear motion of each part and its volume displacement in the diagrams of Fig. 19 is given by the intercepts with the iso-volume lines of a line parallel to a coordinate. This procedure is in fact the graphical equivalent of moving one part keeping the other fixed (Konno and Mead, 1967). The possible volume changes of the two parts during the manoeuvre described are represented in Fig. 20. During quiet breathing in the standing posture the volume displaced by the abdomen is about 40 per cent of the tidal, whereas in the supine position it is about 70 per cent of the tidal. The greater tidal displacement of the abdomen in the supine posture agrees with the finding of Agostoni *et al.* (1965) based on the simultaneous measurements of lung volume and rib cage circumference changes. The mechanical basis of this difference has been previously explained (see p. 29). The volume displacement of the abdomen over the VC range is about 30 per cent of the VC in both positions (Fig. 20). During exercise on the cycle-ergometer the relative contribution of the abdomen to the tidal volume increases slightly and the end-expiratory volume of the abdomen decreases below the resting level relatively more than that of the rib cage (Fig. 21; Grimby, Bunn and Mead, 1968).

Konno and Mead's analysis is based on and has shown the high degree of volume dependence between the rib cage and the abdomen under iso-lung volume conditions. On the other hand it shows also the degree of volume independence of these two parts when the lung volume is made to change, i.e. when the system is no longer closed and a second degree of freedom is added. This independence is illustrated by the size of the area enclosed by the dashed line encompassing the extremes of the isopleths in Fig. 20. The maximum width of this area indicates the maximum volume displacement of the abdomen under the condition imposed (see p. 37):

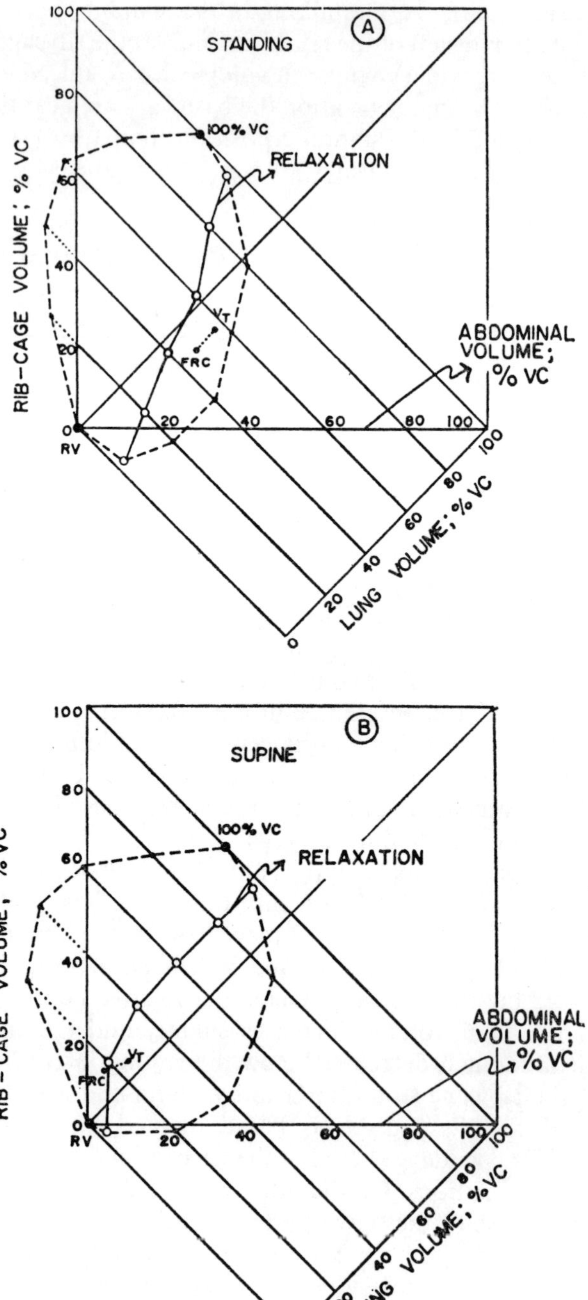

STANDING
A
100% VC
RELAXATION
RIB-CAGE VOLUME; % VC
ABDOMINAL VOLUME; % VC
LUNG VOLUME; % VC
$V_T$
FRC
RV
SUPINE
B
100% VC
RELAXATION
RIB-CAGE VOLUME; % VC
ABDOMINAL VOLUME; % VC
LUNG VOLUME; % VC
$V_T$
FRC
RV


FIG. 20.

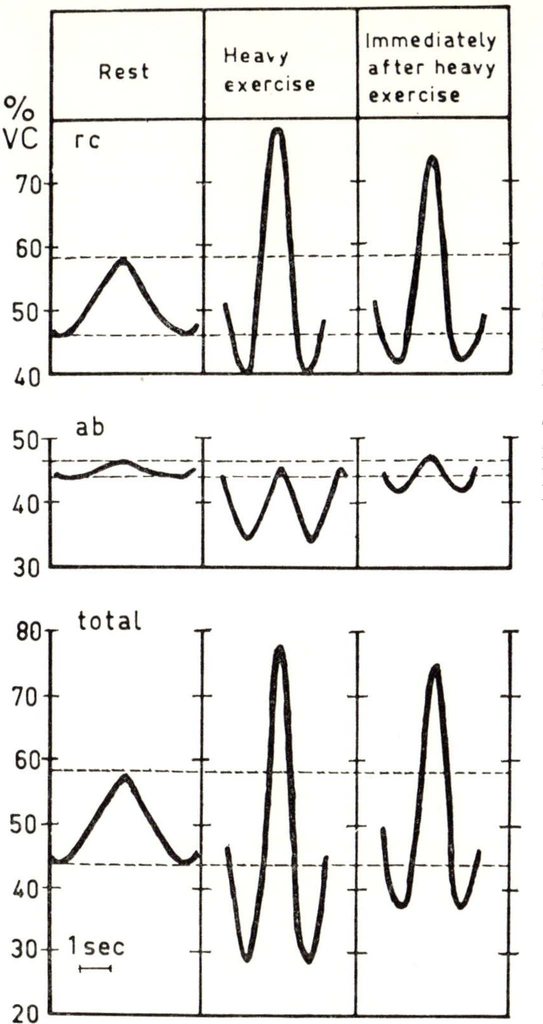


FIG. 21.—TIDAL VOLUMES OF THE RIB CAGE AND OF THE ABDOMEN (ACCORDING TO KONNO AND MEAD'S ANALYSIS) AND TIDAL VOLUME OF THE LUNG AT REST, DURING AND IMMEDIATELY AFTER EXERCISE

The broken lines indicate the end-inspiratory and end-expiratory volume at rest. (From Grimby, Bunn and Mead, 1968.)

---

FIG. 20 (SEE OPPOSITE).—VOLUME DISPLACEMENT OF THE RIB CAGE AND OF THE ABDOMEN

The volume changes are expressed as per cent of the *VC* and are obtained from the diagrams of Fig. 19. The closed circles indicate the extremes of lung volumes with airways open. The continuous line with open circles refers to the relaxed state and the short dotted line indicates the relation during quiet breathing. The FRC point in the standing posture does not fall on the relaxation line because the points of this line were obtained after full inspiration whereas those during quiet breathing were not: long-term effects of the force of gravity probably account for the different displacement of the abdomen. The outermost dashed line indicates the range of possible configurations of the chest wall that can be produced voluntarily while maintaining mouth pressure within $\pm 20$ cm $H_2O$. The 45-degree line indicates equal volume displacements of rib cage and abdomen. The volume displacement of the rib cage in this analysis includes the fraction of the volume moved by the diaphragm not affecting the abdomen motion because of the vertical movement of the rib cage. (From Konno and Mead, 1967.)

it greatly exceeds the abdominal displacement over the VC range. The maximum height of the area indicates the maximum volume displacement of the rib cage: it is about equal to that occurring over the VC range (Konno and Mead, 1967). The sum of the maximum volume displacements of each part is about 25 per cent larger than the VC: hence Konno and Mead conclude that "the volume reserve of the respiratory system is greater than has hitherto been appreciated". In patients with chronic airways obstruction the TLC may be 50 per cent greater than the TLC of normal subjects (Grimby, Takishima, Graham, Macklem and Mead, 1968).

From measurements of changes in the rib cage circumference and of movements of the dome of the diaphragm relative to the rib cage over the IC and the ERV, Wade (1954) expressed the lung volume change as a function of the sum of the two motions measured. Letting:

$\triangle V$ = change of lung volume = $\triangle Vrc + \triangle Vdi$
$\triangle c$ = change of rib cage circumference
$\triangle q$ = change of position of the diaphragmatic dome relative to the rib cage
$x = (\triangle Vrc)/\triangle c$
$y = (\triangle Vdi)/\triangle q$
then $\triangle V = x\triangle c + y\triangle q$

From the data obtained in the standing and supine position Wade has solved a system of two equations (standing and supine) with two unknown quantities (x and y) for the IC and a similar system for the ERV. This approach is similar in principle to that of Konno and Mead since it is also postulates a system with two moving parts (rib cage and diaphragm, instead of abdomen) having a linear volume-motion relationship. It involves the assumption that the unknown quantities of the system (x and y) be constant in both postures within the IC or the ERV. It appears from geometrical consideration that the volume displaced by the rib cage per unit change of circumference (i.e. x) varies with the degree of expansion of the rib cage and that both x and y will vary with posture because of the different shape acquired by the rib cage and the diaphragm in the two positions. It is, however, difficult to estimate how much x and y vary and, therefore, the error introduced by assuming them to be constant.

According to Wade the displacement of the rib cage in the standing and supine position contributes, respectively, 31·8 and 34 per cent of the VC, 14·8 and 33·5 per cent of the ERV, 40·8 and 34·1 per cent of the IC.

Agostoni *et al.* (1965) used a geometric approach to estimate roughly the volume change of the rib cage and obtained that of the diaphragm by subtraction from the lung volume change. This approach is based on measurements of rib cage circumference and diameters, integrated by some measurements on X-ray pictures. The geometric approach offers the advantage of determining the volume displaced by the rib cage and by the diaphragm as a continuous function of the lung volume change over the VC range, but, of course, it involves some questionable assumptions, particularly on the movement of the upper part of the rib cage and of the motion of the boundary between rib cage and abdomen. Probably the error involved by these assumptions is not negligible, and again it is difficult to estimate it theoretically. The data obtained by this approach have been, however, confirmed on the same subjects by two different methods on a narrow volume range near the mid-lung volume (Agostoni *et al.*, 1965;

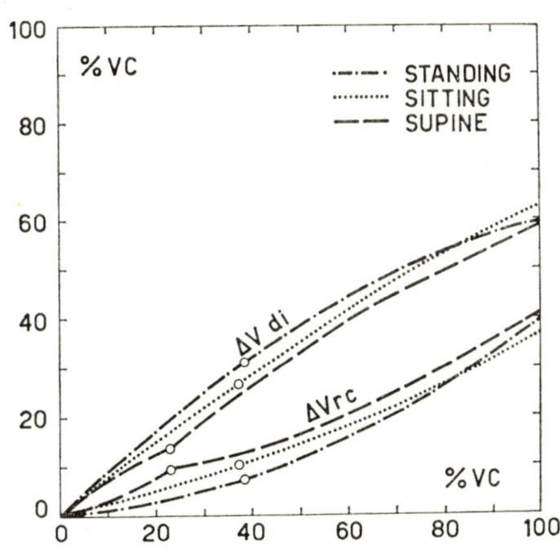


FIG. 22.—ROUGH ESTIMATE OF THE VOLUME DISPLACED BY THE RIB CAGE AND BY THE DIAPHRAGM AS A FUNCTION OF THE LUNG VOLUME CHANGE IN THE STANDING, SITTING AND SUPINE POSITIONS

Both data on the ordinate and on the abscissa are expressed as per cent of the VC in the corresponding positions. Circles indicate the resting volume. Mean of nine subjects. (From Agostoni *et al.*, 1965.)

Agostoni and Mognoni, 1966). This suggests that the net error resulting from the assumptions should be relatively small at least within the tidal volume range. The volumes displaced by the rib cage and by the diaphragm as functions of the change in lung volume in the sitting, standing and supine positions are shown in Fig. 22. The volume contributed by the diaphragm for a small change of lung volume is generally greater than that contributed by the rib cage, except in few situations, as, for instance, at large lung volume in the standing posture. The displacement of the rib cage in the standing, sitting and supine positions contributes, respectively, 39·6, 37·1 and 41·1 per cent of the VC, 18·9, 27·8 and 40·7 per cent of the ERV, 52·5, 42·6 and 41·2 per cent of the IC, and about 38, 35 and 25 per cent of the resting tidal volume. The small tidal displacement of the rib cage in the supine position as compared to the sitting or standing posture has been already discussed in dealing with the changes of the rib cage circumference and of the dorso-ventral motion of the rib cage and abdomen (see pp. 29 and 39).

In three subjects with spinal anaesthesia sufficient to produce motor block up to about $T_1$ (having therefore only the diaphragm, the sterno-cleidomastoids and part of the scaleni active) the inspiratory capacity in the supine position was reduced to 58·5, 57·8 and 19·8 per cent of the values obtained just before the injection of the anaesthetic drug (Eisele *et al.* 1968). In one case the value was so low as to suggest that the end-expiratory level before and after anaesthesia was not the same, and/or the motility of the diaphragm was also impaired, and/or the subject was no longer exerting a maximum effort. In the other two cases the values are similar to that estimated for the diaphragm alone by Agostoni *et al.* (1965). In a man with spinal cord injury at $C_7$ the TLC was 80 per cent of predicted (Stone and Keltz, 1963).

The volume displacement of the abdomen within the tidal volume range has been determined by Bergofsky (1964) by means of partitional plethysmography. In this apparently simple method it is particularly difficult to take into account any motion of the boundary between rib cage and abdomen. The volume displacement of the abdomen obtained by Bergofsky over the tidal volume is much smaller than that obtained by Konno and Mead (1967), which is probably due to the experimental error of partitional plethysmography in contrast to the simple undistorted measurements of Konno and Mead; furthermore the error caused by the volume of chest wall moving out from the plethysmograph when the rib cage is raised has not been taken into account (Agostoni *et al.*, 1965).

In conclusion, it is gratifying to note that notwithstanding the different approaches and assumptions the data of Wade (1954) differ only by 20 per cent from those of Agostoni *et al.* (1965) and that both groups of data in general fit with those of Konno and Mead (1967). Averaging the values of Wade and Agostoni *et al.*, both in the standing and supine positions, the volume contribution of the rib cage over the VC is about 36 per cent and that of the diaphragm 64 per cent. The vertical motion of the rib cage over the VC is such (see p. 34) that about half of the volume displacement of the diaphragm does not produce a displacement of the abdomen, hence the volume displaced by the abdomen over the VC, according to the data of Wade and Agostoni *et al.* is about 32 per cent of the VC, which is equal to the value found by Konno and Mead (average of the standing and supine position). Conversely, if from the average volume displacement of the rib cage according to Konno and Mead (68 per cent of the VC) is subtracted the fraction of the volume displaced by the diaphragm not shared by the abdomen (32 per cent of the VC, from data of Wade and Agostoni *et al.*) the pure rib cage contribution is obtained, which is 36 per cent of the VC, like the average value of Wade and Agostoni *et al.*

Finally, since during quiet breathing in the supine position the tidal vertical movement of the rib cage is small (2–3 mm at the front) the displacement of the abdomen shares almost completely that of the diaphragm, hence the data of Konno and Mead on the volume displacement of the abdomen should be almost as great as those on the displacement of the diaphragm of Agostoni *et al.* In the supine posture the contribution of the abdomen to the tidal volume is about 70 per cent (see p. 39) according to Konno and Mead, and the diaphragm contribution is about 75 per cent (see Fig. 22) according to Agostoni *et al.*

Recently Josenhans and Wang (1970) developed a ballistic method to determine the diaphragmatic contribution to the tidal volume in supine subjects. They found that the percentage of tidal volume contributed by the diaphragm was 51·6 ± 7·2 (S.D.), 63·5 ± 10·5, 66·3 ± 9·6 at the ages of 9 to 16, 18 to 35 and 41 to 75 years, respectively. In females the diaphragm contribution was slightly smaller than in males. In a group of professional singers and woodwind players the diaphragmatic contribution was 67 ± 3·2 per cent. (Josenhans and Wang, 1970).

The contribution of the diaphragm to the tidal volume determined in rabbits by measuring the immediate fall of tidal volume after blocking the conduction in the phrenic nerves is dealt with in the chapter on the Diaphragm.

## Summary

The dorso-ventral diameter of the relaxed rib cage at the xiphoid level changes almost linearly with lung volume: its change over the vital capacity range is about 3 cm, twice that of the lateral diameter.

The change of the rib cage circumference over the vital capacity is about 8 cm at all postures. Over the resting tidal volume it is about 1 cm when standing or sitting, and about 0·6 cm when supine.

At the functional residual capacity the cross sectional area of the rib cage is smaller when standing than when sitting although the lung volume is essentially the same. When supine it is about the same as when sitting, although the lung volume is smaller. Hence because of the effect of gravity, the length of the muscles of the rib cage at FRC may be different at similar lung volume (standing–sitting) or equal at different lung volumes (sitting–supine).

During inspiratory efforts the horizontal section of the rib cage becomes more elliptical, whereas during expiratory efforts it becomes more circular. During hyperventilation or breathing through resistance the change of the dorso-ventral diameter lags behind that of the lateral one. The phase shift increases as the load increases. Hence the forces of the respiratory muscles acting at right angle to the surface of the chest wall operate predominantly on the lateral parts of the rib cage, the frontal part being mainly driven by the pressure acting across it and by the movement of the lateral parts. The deformation occurring under load implies that: (a) some muscles lengthen instead of shortening and vice versa; (b) information from muscle, joint, and lung receptors is out of phase; and (c) the work of breathing is slightly larger than that calculated on the volume-pressure diagram.

The displacement of the dome of the diaphragm over the resting tidal volume is about 1·5 cm when standing and 1·7 cm when supine. Over the vital capacity range the average displacement is about 9·5 cm. About half of this displacement is not transmitted to the abdomen because of the upward movement of the rib cage. At high ventilation the dorso-ventral diameter of the upper abdomen is smaller than during quiet breathing because of the rise of the rib cage.

During moderate contraction of the respiratory muscles the deformation of the rib cage and of the abdomen is negligible. If the lung volume and the degree of extension of the vertebral column are kept constant: (a) the volume change of the rib cage is equal and opposite to that of the abdomen, and (b) there is a unique relation between the motion of a point of the rib cage, or of the abdomen, and the volume displaced by either part. The relation between dorso-ventral motion and volume change is nearly linear for both parts.

About one-third of the vital capacity is contributed by the rib cage and two-thirds by the diaphragm. Only half of the volume displaced by the diaphragm is shared by the abdomen because of the elevation of the rib cage. The rib cage contributes about one-third of the resting tidal volume when standing or sitting, and only one-quarter when supine. This difference depends upon the lower compliance of the rib cage and the higher compliance of the abdomen in the supine position within this volume range.

## REFERENCES

AGOSTONI, E., and MOGNONI, P. (1966). Deformation of the chest wall during breathing efforts. *J. appl. Physiol.*, **21**, 1827–1832.

AGOSTONI, E., MOGNONI, P., TORRI, G., and MISEROCCHI, G. (1966). Forces deforming the rib cage. *Resp. Physiol.*, **2**, 105–117.

AGOSTONI, E., MOGNONI, P., TORRI, G., and SARACINO, F. (1965). Relation between changes of rib cage circumference and lung volume. *J. appl. Physiol.*, **20**, 1179–1186.

AGOSTONI, E., and TORRI, G. (1967). An analysis of the chest wall motions at high values of ventilation. *Resp. Physiol.*, **3**, 318–332.

BERGOFSKY, E. H. (1964). Relative contributions of the rib cage and the diaphragm to ventilation in man. *J. appl. Physiol.*, **19**, 698–706.

DALLY, J. F. H. (1908). An inquiry into the physiological mechanism of respiration, with especial reference to the movements of the vertebral column and diaphragm. *J. Anat.* (*Lond.*), **43**, 93–114.

EISELE, J., TRENCHARD, D., BURKI, N., and GUZ, A. (1968). The effect of chest wall block on respiratory sensation and control in man. *Clin. Sci.*, **35**, 23–33.

GRIMBY, G., BUNN, J., and MEAD, J. (1968). Relative contribution of rib cage and abdomen to ventilation during exercise. *J. appl. Physiol.*, **24**, 159–166.

GRIMBY, G., TAKISHIMA, T., GRAHAM, W., MACKLEM, P., and MEAD, J. (1968). Frequency dependence of flow resistance in patients with obstructive lung disease. *J. clin. Invest.*, **47**, 1455–1465.

Hasselwander, A. (1949). Über die Gestald des Zwerchfells und die Lage des Herzens. *Z. Anat. Entwickl.-Gesch.*, **114**, 375–398.

Heayk, H. von (1960). *The Human Lung*, translated by V. E. Krahl. New York: Hafner.

Herxheimer, H. (1948). The influence of costal and abdominal pressure on the action of the diaphragm in normal and emphysematous subjects. *Thorax*, **3**, 122–126.

Herxheimer, H. (1949). Some observations on the coordination of diaphragmatic and rib movement in respiration. *Thorax*, **4**. 65–72.

Hitzenberger, K. (1927). *Das Zwerchfell in gesunden und kranken Zustand.* Vienna: Springer.

Jordanoglou, J. (1969). Rib movement in health, kyphoscoliosis and ankylosing spondylitys. *Thorax*, **24**, 407–414.

Jordanoglou, J. (1970). Vector analysis of rib movement. *Resp. Physiol.* (in press).

Josenhans, W. T., and Wang, C. S. (1970). A modified ballistic method for measuring axial mass displacement caused by breathing. *J. appl. Physiol.* (In press).

Konno, K., and Mead, J. (1967). Measurement of the separate volume changes of rib cage and abdomen during breathing. *J. appl. Physiol.*, **22**, 407–422.

Krahl, V. E. (1964). Anatomy of the mammalian lung. In: *Handbook of Physiology.* Section 3. Respiration. Vol. 1, pp. 213–284. Eds. Fenn, W. O., and Rahn, H. Washington, D.C.: American Physiological Soc.

Luciani, L. (1911). *Human Physiology*, Vol. I, p. 408. Translated by F. A. Welby. London: Macmillan.

Mains, R. C., Zechman, F. W., and Musgrave, F. S. (1968). Pulmonary mechanics with a simulated postural blood redistribution. *Resp. Physiol.*, **5**, 288–301.

Mead, J., Peterson, N., Grimby, G., and Mead, J. (1967). Pulmonary ventilation measured from body surface movements. *Science*, **156**, 1383–1384.

Polgar, F. (1949). Studies on respiratory mechanics. *Amer. J. Roentgenol.*, **61**, 637–657.

Stone, D. J., and Keltz, H. (1963). The effect of respiratory muscle dysfunction on pulmonary function. *Amer. Rev. resp. Dis.*, **88**, 621–629.

Verzàr, F. (1946). Regulation of the lung volume and its disturbances. *Schweiz. med. Wschr.*, **76**, 932–936.

Wade, O. L. (1954). Movements of the thoracic cage and diaphragm in respiration. *J. Physiol.* (*Lond.*), **124**, 193–212.

Wade, O. L., and Gilson, J. C. (1951). The effect of posture on diaphragmatic movement and vital capacity in normal subjects with a note on spirometry as an aid in determining radiological chest volumes. *Thorax*, **6**, 103–126.

# Chapter III

# STATICS*

E. Agostoni

## VOLUME-PRESSURE RELATIONS OF THE RESPIRATORY SYSTEM DURING RELAXATION†

### Respiratory System

During relaxation of the respiratory muscles under static conditions the pressure exerted by the respiratory system depends upon elastic, surface and gravitational forces operating in the lung and the chest wall. The pressure exerted by the respiratory system is given by the difference between alveolar pressure and body surface pressure; in static conditions alveolar pressure equals the pressure at the airways opening and it is therefore measured in the mouth or in a nostril, when the other openings are closed. To determine the relation between the lung volume and the relaxation pressure of the respiratory system (Rohrer, 1916; Rahn *et al.*, 1946) the subject inspires or expires a volume of air and then relaxes against an obstructed airway and the pressure across the obstruction is measured. Then the subject breathes in or out maximally from a spirometer so that the volume at which the pressure was measured can be related to one extreme of the vital capacity. This volume must be corrected for the compression or expansion of the gas in the respiratory system due to the change of pressure during relaxation. When these measurements are made at various volumes the volume-pressure curve during relaxation is obtained. This method may not be applied to all subjects and some training is necessary.

The volume-pressure curve of the relaxed respiratory system with the trunk erect is shown in Fig. 23 along with the spirogram showing the pulmonary subdivisions. The volume at zero pressure is the resting volume of the respiratory system; it corresponds to the end of a spontaneous expiration during quiet breathing. In the mid-volume range the relation is almost straight, the volume change per 1 cm $H_2O$ being about 2 per cent of the vital capacity. The slope of the static volume-pressure curve is called compliance (C), and usually expressed in litres/cm $H_2O$ or ml/cm $H_2O$. The horizontal distance from the curve to the ordinate at zero pressure indicates the pressure exerted by the passive structure of the system at a given lung volume. Conversely, this distance indicates the pressure that the respiratory muscles must exert to maintain that lung volume with open airways. This

* This chapter is based on "Statics of the respiratory system" by Agostoni and Mead (1964).

† Table IV (p. 141) lists approximate values for a normal adult.

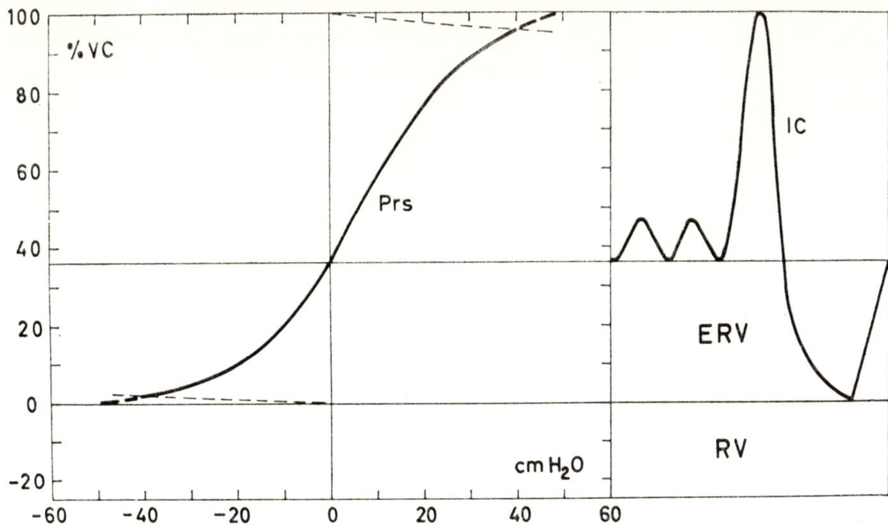


FIG. 23.—STATIC VOLUME-PRESSURE CURVE (PRS) OF THE TOTAL RESPIRATORY SYSTEM DURING RELAXATION IN THE SITTING POSTURE, WITH A SPIROGRAM SHOWING THE SUBDIVISIONS OF LUNG VOLUME

The slanting broken lines indicate the volume change during relaxation against an obstruction due to gas compression at TLC and expansion at RV. The curve was extended to include the full vital capacity range by means of externally applied pressures. (From Rohrer, 1916, as modified by Agostoni and Mead, 1964.)

is strictly true only if the shape of the chest wall is the same when the respiratory muscles are relaxed as when they are active at the same lung volume. In fact the energy of the passive structures of the system is minimum for the configuration occurring during relaxation; whenever, at the same lung volume, this configuration is changed the energy of the passive structure is increased (Agostoni *et al.*, 1965). As contraction of the respiratory muscles deforms the chest wall (see Chapter II), the force exerted by the muscles is slightly larger than that indicated by the volume-pressure diagram during relaxation. The scanty information available on deformation forces is reviewed at the end of this chapter.

Another method of estimating the volume-pressure relation of the relaxed respiratory system in the tidal volume range has been proposed by Heaf and Prime (1956) to overcome the difficulty of voluntary relaxation. With the subject breathing spontaneously, the pressure at the airway opening is raised above that acting on the body surface (positive pressure breathing): if the subject is relaxed at the end of expiration, the relation is obtained by measuring the end-expiratory lung volume and the applied pressure. The relation obtained by this method over the tidal volume range has usually been found similar to that obtained during voluntary relaxation (Heaf and Prime, 1956; Naimark and Cherniack, 1960; Johnson and Mead, 1963; Cherniack and Brown, 1965; Mognoni *et al.*, 1965). On the other hand the volume-pressure curve obtained by Rahn *et al.* (1946) from the end-expiratory values during positive and negative pressure breathing was different from that obtained during voluntary relaxation. This difference seemed related to the fact that in their experiments during PPB and NPB head and neck were not exposed to the same pressure as the rest of the body surface (Rahn *et al.*, 1946; Johnson and Mead, 1963). As, however, the volume-pressure curve obtained during voluntary relaxation with closed glottis is equal to that with open glottis the pressure across the cheeks and the neck does not seem essential (Mognoni *et al.*, 1965). In fact, in some subjects the volume-pressure curve determined from the end-expiratory values during PPB and NPB was different from that obtained during voluntary relaxation even if in the former case head and neck were exposed to the same pressure as the rest of the body (Agostoni, 1962; Mognoni *et al.*, 1965). The different results may in part relate to the period of PPB because the activity of the expiratory muscles seem to appear only after some minutes of PPB (Milic-Emili, personal communication, see Chapter XIII).

In anaesthetised subjects paralysed by pharmacological means the compliance of the respiratory system was found smaller than in conscious subjects in the same posture (Nims *et al.*, 1955; Butler and Smith, 1957; Howell and Peckett, 1957). Howell and Peckett found this difference was due to a decrease of lung compliance during anaesthesia and the compliance of the chest wall was similar to that during voluntary relaxation. Van Lith and his co-workers (1967) found that the compliance of the respiratory system, measured by the method of Heaf and Prime, in conscious subjects was lower than that measured in the same subjects during anaesthesia and neuromuscular blockade. Since the compliance of the lung was the same in both conditions, the expiratory muscles were probably active at the end of expiration during PPB (this was applied only for about 1 min).

**Chest Wall**

The chest wall and the lung are placed in series and therefore the algebraic sum of the pressure exerted by the two parts gives the pressure of the respiratory system ($Pw + Pl = Prs$), whereas the change of volume of each part must be equal and equal to that of the respiratory system ($\triangle Vw = \triangle Vl = \triangle Vrs$). *Prs* is the difference between alveolar pressure and body surface pressure, hence when this latter is atmospheric: $Palv = Pw + Pl$. Since *Pw* is generally taken to indicate the pressure exerted by the relaxed chest wall, when the respiratory muscles are active: $Palv = Pw + Pl + Pmus$.

The pressure exerted by the lung is the difference between alveolar pressure and pleural surface pressure ($Pl = Palv - Ppl$). The pressure exerted by the chest wall is the difference between pleural surface pressure and body surface pressure ($Pw = Ppl - Pbs$). The chest wall pressure may be obtained either indirectly by subtracting the pressure exerted by the lung from that exerted by the respiratory system, or directly from *Ppl* measurements. In fact, when the subject is relaxed (with closed airways to keep a static condition): *Palv* = *Pl* + *Pw*, since *Pl* = *Palv* − *Ppl*, substituting, *Pw* = *Ppl* (Howell and Peckett, 1957; Knowles *et al.*, 1959). Stated in other words (Bouhuys *et al.*, 1966): the pressure between two points must equal the sum of the pressure differences across all intervening structures, the structures between the pleural surface and the surfaces of the body are the lung in one direction and the chest wall in the other.

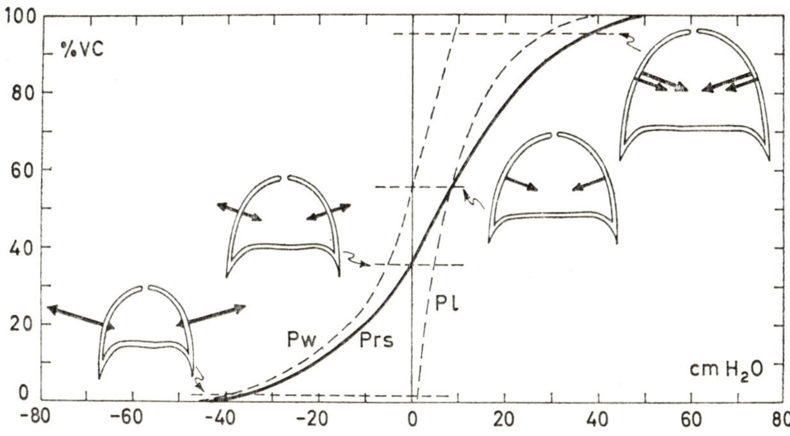


FIG. 24.—STATIC VOLUME-PRESSURE CURVES OF THE LUNG (Pl), CHEST WALL (Pw), AND TOTAL RESPIRATORY SYSTEM (PRS), DURING RELAXATION IN THE SITTING POSTURE

The static forces of the lung and of the chest wall are pictured by the arrows in the side drawings. The dimensions of the arrows are not to scale; the volume corresponding to each drawing is indicated by the horizontal broken lines. (From Rahn *et al.*, 1946; and Knowles *et al.*, 1959, as modified by Agostoni and Mead, 1964.)

Accordingly: *Ppl* = *Palv* − *Pl*, and *Ppl* = *Pmus* + *Pw*. *Palv* may be made equal to zero holding a given lung volume with open airways, then *Ppl* = −*Pl*. *Pmus* may be made equal to zero by relaxing the muscles with the airways closed, then *Ppl* = *Pw*.

The static volume-pressure curves of the relaxed chest wall and of the lung, with trunk erect, are shown in Fig. 24. At the resting volume of the respiratory system the chest wall recoils outwards with a pressure equal to that by which the lung recoils inwards. The resting volume of the lung is below zero per cent of the VC. That of the chest wall is about 55 per cent of the VC. Above this volume both the chest wall and the lung recoil inwards, whereas below this volume the chest wall recoils outwards, hence the lung and the chest wall behave like two opposing springs. According to Turner *et al.* (1968) the compliance of the chest wall decreases slightly above 80–90 per cent VC and the resting volume of the chest wall is higher than indicated above. In the tidal volume range the compliance of the chest wall and that of the lung are about the same. Since the chest wall and the lung are placed in series the sum of the reciprocals of the compliance of the lung and of the chest wall equals the reciprocal of the compliance of the respiratory system: $1/Cl + 1/Cw = 1/Crs$.

If in a relaxed subject both pleural spaces were opened to the ambient air, the lung would collapse and the chest wall would expand to its resting volume. A volume of air of about 60 per cent of the VC would then be sucked into the pleural spaces. If a smaller amount of air were introduced into the pleural spaces the lung would shrink and the chest wall would expand until a new equilibrium is reached between the two opposing recoils at a value of *Pl* between 0 and −5 cm $H_2O$ (see Fig. 24) depending on the volume of air introduced. The partitioning of this volume between the chest wall and the lung depends upon the features of the volume-pressure curve of each part (Fenn, 1954).

For analytical purpose pleural surface pressure has been and will be dealt with as a single value; this, however, occurs only when a sufficient amount of gas is present in the pleural space. Under physiological conditions pleural surface pressure varies at different sites because of the effects of gravity on the lung and the chest wall and of the different shape of these two structures (Mead, 1961; Krueger *et al.*, 1961; Milic-Emili, Mead and Turner, 1964; Milic-Emili *et al.*, 1966; Kaneko *et al.*, 1966; Bryan *et al.*, 1966; Glazier *et al.*, 1967; Proctor *et al.*, 1968; Agostoni and D'Angelo, 1968; Hogg and Nepszy, 1969; Hoppin *et al.*, 1969; McMahon *et al.*, 1969; D'Angelo *et al.*, 1970; Agostoni *et al.*, 1970). It is therefore important to keep in mind that the balance between the lung and the chest wall under physiological conditions results from a wide distribution of pressures (see p. 57). The value of pleural surface pressure used in man for the analysis in terms of volume-pressure curve is obtained from oesophageal pressure measurements. This value must not be considered as an average of pleural surface pressure in different parts, but only as an index of the average pleural surface pressure. The principle underlying this technique as well as the references of practical interest are given by Mead and Milic-Emili (1964). The concepts of pleural surface pressure and pleural liquid pressure are dealt with by Agostoni and Mead (1964, see also Agostoni, 1969).

The static volume-pressure relations have been represented as single lines, implying that a unique pressure corresponds to a given volume. Actually the pressure for a given volume may be slightly different according to the previous volume history of the respiratory system: the pressure being lower on deflation than on inflation. Hence the static volume-pressure relations obtained in steps from minimum to maximum lung volume and vice versa are not superimposed but form a static hysteresis loop. This depends upon viscoelasticity, such as stress adaptation, i.e. a rate-dependent phenomenon, and upon plasticity, i.e. a rate-independent phenomenon. Hysteresis occurs both in the lung (Von Neergard, 1929; Mead *et al.*, 1957; Butler, 1957; Clements *et al.*, 1958; Brown *et al.*, 1959; Hughes *et al.*, 1959; Marshall and Widdicombe, 1960; Cavagna *et al.*, 1962; Van de Woestijne, 1969; Sharp *et al.*, 1967; Bachofen, 1968; Saibene and Mead, 1969) and in the chest wall (Butler, 1957; Van de Woestijne, 1967; Sharp *et al.*, 1967). In the lung it is mainly due to surface properties, whereas in the chest wall it seems mainly related to muscles and ligaments because both skeletal muscles (Buchtal and Rosenfalck, 1957) and elastic fibres (Remington, 1955) exhibit hysteresis. The hysteresis of the respiratory system for a volume change of 3 litres above FRC and expressed as ratio of loop width to loop height is about 0·08; that of the chest wall is about the same (Sharp *et al.*, 1967). For volume changes such as those occurring during quiet breathing hysteresis is negligible.

**Rib Cage and Abdomen-diaphragm**

The rib cage and the abdomen-diaphragm may be considered to operate approximately in parallel (this would be strictly true only if pleural surface pressure were uniform), hence they exert the same pressure: $Prc = P(ab + di) = Pw$, and the volume displaced by the rib cage plus that displaced by the abdomen-diaphragm gives the volume change of the chest wall: $\triangle Vrc + \triangle V(ab + di) = \triangle Vw$. It is thus possible to build an approximate volume-pressure curve of the rib cage and of the abdomen-diaphragm if the volume contributed by these two parts is known (see Chapter II, p. 37) (Agostoni and Mead, 1964). These relations are shown in Fig. 25; they have been calculated without taking into account the different volume distribution between rib cage and abdomen-diaphragm occurring during relaxation as compared to that occurring when a given lung volume is actively held with open airway. This difference in fact varies between subjects and, though important to illustrate the effect of deformation, it cannot be estimated with sufficient accuracy to be dealt with here (Agostoni *et al.*, 1965; Konno and Mead, 1968). In analysing these curves it must be remembered that part of the volume displacement of the abdomen-diaphragm can take place because of the vertical movements of the rib cage (see Chapter II).

Since the abdomen (wall and content) and the diaphragm are placed in series the pressure of the abdomen-diaphragm is the sum of the pressure exerted by its component parts, namely $Pab$ and $Pdi$ (Fig. 25). These are obtained by estimating the pressure on the abdominal side of the diaphragm from gastric pressure measurement. As the volume of gas in the abdomen is normally negligible, the volume of the abdomen can be considered constant when its pressure is changed. Hence the values of the abdominal pressure at different lung volume may profitably be represented

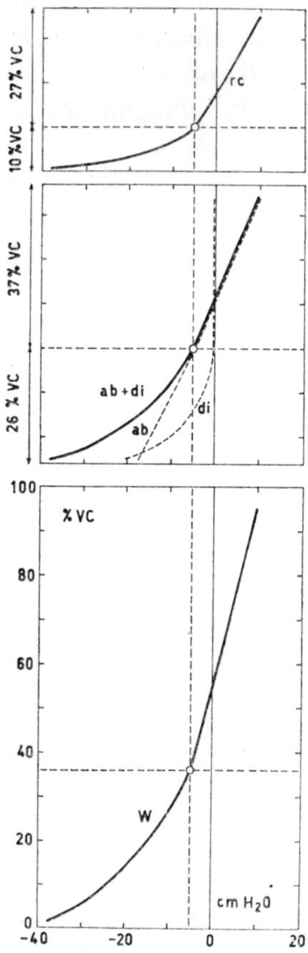


FIG. 25.—STATIC VOLUME-PRESSURE CURVES OF THE RIB CAGE (UPPER DIAGRAM) AND OF THE ABDOMEN-DIAPHRAGM (MIDDLE DIAGRAM) DURING RELAXATION IN THE SITTING POSTURE

The circles refer to the resting volume of the respiratory system (end of spontaneous expiration). The lower diagram shows the volume-pressure relation of the relaxed chest wall. Since the rib cage and the abdomen-diaphragm are approximately placed in parallel they are subjected to the same pressure (*Pw*) and $\Delta Vrc + \Delta V(ab + di) = \Delta Vw = \Delta Vrs$. The broken lines in the middle diagram show the pressure contributed by the abdomen and by the diaphragm in series. In rabbits and dogs Pdi at large lung volumes becomes positive and hence Pab lower than Pw. (Modified from Agostoni and Mead, 1964.)

in the volume-pressure diagram of the respiratory system (Agostoni and Rahn, 1960; Agostoni and Mead, 1964).

The diagrams of Fig. 25 demonstrate several important features: the extremely low compliance of the relaxed rib cage below FRC, the nearly constant compliance of the relaxed abdomen over the VC range, the lack of any transdiaphragmatic pressure above FRC and the smoothly increasing stiffness of the relaxed diaphragm below FRC. The importance of these features may be better understood by analysing the effects of gravity and posture.

## Effects of Gravity and Posture

The volume-pressure curves considered so far refer to the erect trunk. They change with posture mainly because of the effect of gravity on the abdomen. The average pressure contributed by the various parts of the respiratory system in the sitting and supine postures are shown in Fig. 26. For the purpose of this analysis pleural surface pressure is assumed to be uniform. The upper diagrams show the volume-pressure curves of the chest wall along with the pressure contributed by the diaphragm and by the abdomen at different lung volumes. The lower diagrams show the volume-pressure relations of the chest wall, of the lung and of the whole respiratory system.

From the mechanical point of view the abdomen, during relaxation, can be likened to a liquid-filled container in which part of its lateral wall is distensible (Rohrer, 1925; Duomarco and Rimini, 1947; Agostoni and Rahn, 1960; Agostoni and Mead, 1964). From the data of Duomarco and Rimini (1947), it can be estimated that at the end of a spontaneous

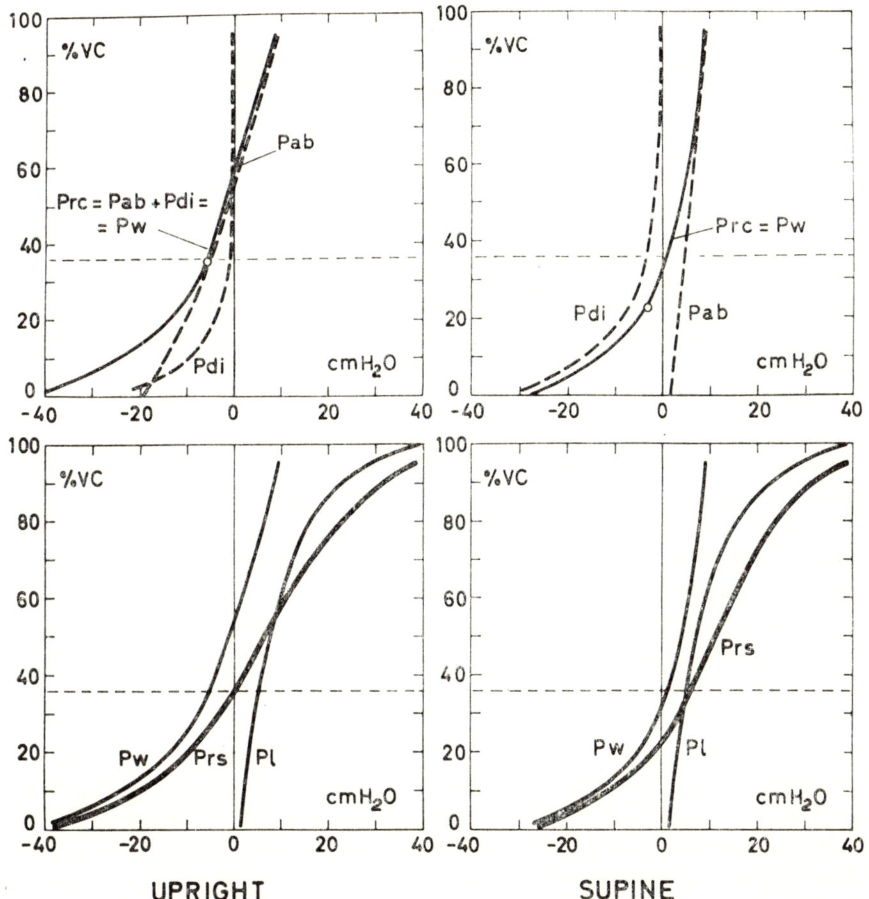


FIG. 26.—PRESSURES CONTRIBUTED BY THE VARIOUS PARTS OF THE RESPIRATORY SYSTEM IN THE SITTING AND SUPINE POSTURE

The upper diagrams show the volume-pressure relationships of the chest wall and the pressures contributed by the rib cage, diaphragm, and abdomen. The circles indicate the resting volume of the respiratory system. The lower diagrams show the volume-pressure relationships of the chest wall, lung and total respiratory system. (Modified from Agostoni and Mead, 1964.)

expiration in the upright posture the level at which the abdominal pressure is equal to the ambient pressure, or "zero level", is 3–4 cm below the diaphragmatic dome. Hence the pressure on the abdominal side of the diaphragm is almost equal to that one would expect to find on its other side: i.e. the diaphragm is essentially relaxed and undistended.

When the subject is supine the "zero level" at the end of spontaneous expiration corresponds to the ventral wall of the abdomen (Duomarco and Rimini, 1947): the diaphragm is distended and balances with its own elasticity the weight of the abdomen. The pressure across the central part of the diaphragm becomes approximately nil only at full inspiration (Agostoni and Mead, 1964). The "zero level" in the prone position corresponds to the dorsal wall of the abdomen, and when the subject is lying on one side, it is midway between the two sides (Duomarco and Rimini, 1947). The "zero level" shifts less with lung volume in the supine than in the upright position, as can be inferred from the values of *Pab* in Fig. 26.

In the upright posture gravity acts in the inspiratory direction on the abdomen–diaphragm and in the expiratory direction on the rib cage. The effect of the abdomen is greater at small than at large volumes, because at large volumes the height of the abdomen is less and its wall stiffer. When the subject, at FRC, moves from the sitting to the standing position the abdominal wall, because of the abdominal weight, pulls the ribs down and in (see Chapter II) and therefore gravity, acting through the abdomen, greatly enhances its small direct expiratory effect on the rib cage. When the subject moves to the standing position his lung volume is slightly increased and oesophageal pressure accordingly decreased, whereas the cross section of his rib cage is reduced to approximately the same extent as in the seated relaxed subject whose lung volume is reduced from FRC to about 16 per cent VC (see Chapter II) (Agostoni, Mognoni, Torri and Miserocchi, 1966). Hence the end-expiratory volume of the rib cage is markedly smaller in the standing than in the sitting position: this volume difference, plus the concomitant small increase of lung volume (about 2 per cent VC) being taken up by the displacement of the abdomen–diaphragm. This change of shape of the chest wall implies an equal change of shape of the lung: as the relationship between oesophageal pressure and change of lung volume is not appreciably affected by posture, this could suggest that the recoil of the lung does not change appreciably in spite of a change of its shape, but probably small changes of pleural surface pressure may occur in different parts without affecting the oesophageal pressure. On the other hand the forces changing the size of the rib cage when the subject moves to the standing position must be marked, because of the small compliance of the rib cage below FRC. These considerations show the difficulty of measuring the pure elastic recoil of the rib cage in a gravitational field (Agostoni, Mognoni, Torri and Miserocchi, 1966).

In the supine posture, the gravitational effect changes little with volume and its action is expiratory both on the abdomen and the rib cage. As a consequence of gravity, the compliance of the chest wall, and hence that of the total respiratory system, in the mid-volume range increases as one changes from the upright to the supine posture. The reduction of the resting volume between upright and supine posture is entirely abdominal; actually, the abdominal shift is somewhat larger than the total, since the rib cage volume increases slightly (see Chapter II) (Agostoni and Mead, 1964). When a subatmospheric pressure of 55 cm $H_2O$ is applied below

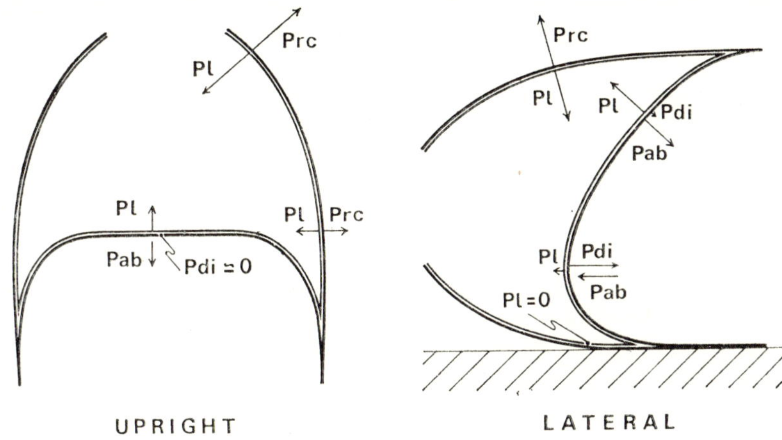


FIG. 27.—SCHEME OF THE PROBABLE DISTRIBUTION OF PRESSURES IN THE RESPIRATORY SYSTEM AT THE END OF A SPONTANEOUS EXPIRATION IN THE UPRIGHT AND LATERAL POSTURES

This scheme takes into account the effect of gravity both on the chest wall and on the lung, and is based on direct and indirect data on man. (Duomarco and Rimini, 1947; Agostoni and Rahn, 1960; Milic-Emili, Mead and Turner, 1964; Milic-Emili *et al.*, 1966; Kaneko *et al.*, 1966) and on direct data on dog and rabbit. (Duomarco and Rimini, 1947; Agostoni and D'Angelo, 1968; D'Angelo *et al.*, 1970.)

the iliac crest of a supine subject the volume-pressure curve of his respiratory system during relaxation becomes almost equal to that in the sitting posture (Zechman *et al.*, 1967): i.e. the change of the abdominal pressure thus produced is such as to counterbalance the gravitational effect of the abdomen in the supine position and to simulate that in the standing position.

In order to undertake the foregoing analysis of the effect of gravity and posture in terms of volume-pressure diagrams pleural surface pressure has been considered uniform. A scheme of the probable distribution of the pressure on the pleural surface at the resting volume of the respiratory system in the upright and lateral postures is given in Fig. 27, taking into account the effect of gravity on both the lung and the chest wall. In the upright posture the lung facing the diaphragmatic dome recoils inwards with a pressure that is probably 3–4 cm $H_2O$, i.e. nearly equal to the pull of the abdominal weight (Duomarco and Rimini, 1947; Agostoni and Rahn, 1960; Milic-Emili *et al.*, 1966; D'Angelo *et al.*, 1970). The superior part of the lung recoils inwards with a pressure that is probably about 10 cm $H_2O$ (Milic-Emili *et al.*, 1966; D'Angelo *et al.*, 1970); the superior part of the rib cage balances the lung recoil with an equal outwards recoil. In the lateral position the recoil of the lung in the lowermost part is probably nil (at least this is the case in dogs, cats and rabbits, Agostoni and D'Angelo, 1968; D'Angelo *et al.*, 1970), and the lowermost part of the chest wall does not pull outwards being fixed by the weight of the chest on the supporting surface. The upper part of the lung recoils inwards with a pressure that should be about 6 cm $H_2O$ (Kaneko *et al.*, 1966; D'Angelo *et al.*, 1970); this recoil is balanced by the opposite one of the superior part of the rib cage. The relaxed diaphragm in the upper part is pulled cranially by the lung recoil and caudally by the abdominal pressure that is subatmospheric. In the lower part both the lung and the abdomen distend the diaphragm cranially, the latter progressively more than the former.

Fig. 28*a* shows values for the major pulmonary subdivisions in various postures; most of the effect of posture depends on the influence of gravity. The largest variation is in the resting end-expiratory volume (FRC). The increase of FRC in the seated subject with the arms supported reflects the influence of the weight and position of the shoulder girdle (Craig, 1960). As the subject leans forward the "zero level" of the abdomen moves downwards, because also the lower part of the abdomen becomes flexible. This effect is further increased in the knee-elbow position, when the "zero level" is presumably near the ventral wall of a relaxed abdomen. When the prone posture is assumed, the ventral abdominal wall is supported from the outside; the "zero level" moves up toward the dorsal wall of the abdomen and the FRC decreases (Agostoni and Mead, 1964).

During gradual tilting (Fig. 28*b*), from standing to supine the FRC decreases almost linearly with the angle of tilt from the vertical to the horizontal. Beyond the horizontal, the volume change is reduced (Colville *et al.*, 1956). Indeed, in subjects suspended on a smooth flat surface by their ankles, beyond about −10° the FRC increases (Fig. 28*b*, broken line). Two events are involved here: the hydrostatic pressure, which continues to increase as the head is tilted downward, is increasingly opposed by the elastic recoil of the diaphragm; at the same time the action of gravity on the rib cage, shoulder girdle and associated structures, which is expiratory in the upright posture, becomes inspiratory in the head-down positions, countering the abdominal hydrostatic pressure and eventually superseding when the diaphragm becomes taut (Agostoni and Mead, 1964).

Fig. 28*c* shows that tall individuals have larger FRC than short

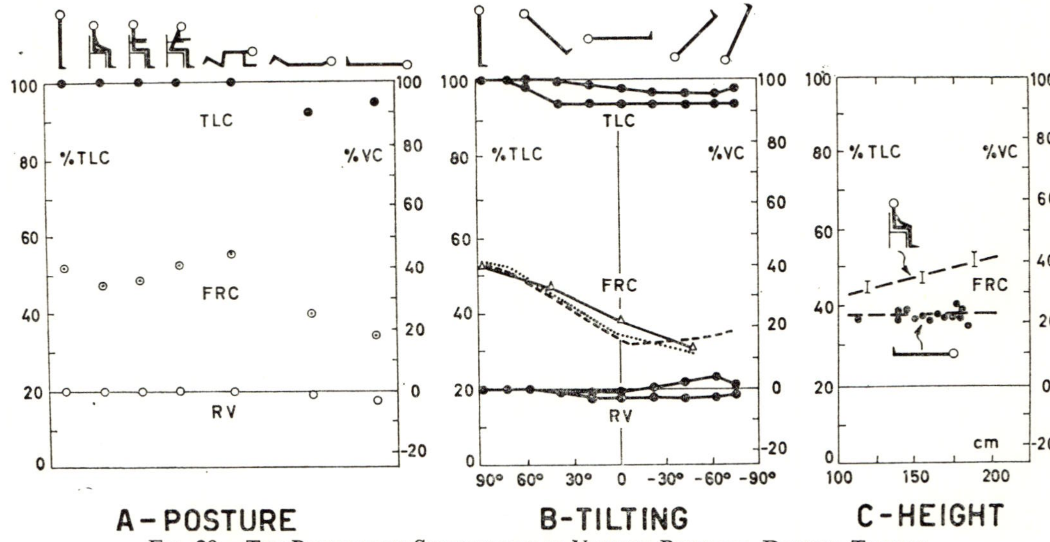


FIG. 28.—THE PULMONARY SUBDIVISIONS IN VARIOUS POSTURES, DURING TILTING AND AS A FUNCTION OF BODY HEIGHT

In instances where RV was not determined it was assumed to be 20% of TLC in the upright posture. The data were obtained from the following sources: A: *Standing* (Wade and Gilson, 1951; J. Mead, unpublished observation); *seated erect* (Craig, 1960; Hurtado and Fray, 1933; Withfield *et al.*, 1950); *seated erect, arms supported* (Craig, 1960); *seated leaning forward, arms supported* (Craig, 1960); *on hands and knees* (Craig, 1960); *prone* (Moreno and Lyons, 1961); *supine* (Craig, 1960; Dittmer and Grebe, 1958; Hurtado and Fray, 1933; Moreno and Lyons, 1961; Withfield *et al.*, 1950). B: *Triangles* (Wade and Gilson, 1951); all the remaining points were obtained by J. Mead. The dotted (supported at shoulders) and broken lines (supported by ankles) are average values for five subjects. The TLC and RV values are for individual subjects. C: *Seated* (Cook and Hamann, 1961); the ranges are the standard error for groups of 10 subjects; the broken line is drawn by eye; *supine*. (Morse *et al.*, 1952; Robinson, 1938). (From Agostoni and Mead, 1964.)

individuals in the upright posture, but similar FRC in recumbency, as is to be expected on hydrostatic grounds (Agostoni and Mead, 1964).

The only significant changes in TLC and RV occur between the upright posture and recumbency: these changes are less than 10 per cent (Agostoni and Mead, 1964). They occur within minutes and the following facts suggest that they are due mainly to shifts in blood between the thoracic cavity and the rest of the body: (*a*) the differences between upright and supine VC are smaller after applications of tourniquets to the extremities (Amussen *et al.*, 1939; Dow, 1939; Hamilton and Morgan, 1932; Mills, 1949; Sjöstrand, 1941) or immersion in water (Hamilton and Mayo, 1944): i.e. in conditions that minimise the differences of thoracic blood volume between the two postures; (*b*) VC changes with ambient temperature, according to the expected movements of blood from the thorax to the extremities and vice versa (Glaser, 1949; Rahn *et al.*, 1949); (*c*) the VC, in the supine posture, increases 5·6 per cent if the alveolar pressure is maintained at 55 cm $H_2O$ for 5 sec by a Valsalva manoeuvre near maximal inspiration immediately before a VC measurement; the same manoeuvre in the upright posture produces only half this increase (Bateman, 1950; Fowler *et al.*, 1951).

At mid-lung volume an increase in intrathoracic, extrapulmonary blood volume causes a decrease in lung volume, a simultaneous increase in chest wall volume and an associated change in pleural pressure, which under static conditions and with the airways obstructed become less subatmospheric (Mains *et al.*, 1968). As far as the volume extremes are concerned, an increase in intrathoracic, extrapulmonary blood volume would be expected to reduce both TLC and RV, and are therefore in accord with the changes observed in passing from the upright to the supine posture. The intrathoracic, extrapulmonary blood volume should change with pleural pressure, increasing as pleural pressure becomes more subatmospheric. Accordingly, the vital capacity of the chest wall should exceed that of the lungs by an amount equal to this blood volume change. This discrepancy can be reduced by manoeuvres which displace blood out of the chest wall, as previously mentioned (Agostoni and Mead, 1964).

The displacement of the volume-pressure curve of the relaxed chest wall occurring when a sitting subject is submerged up to the xiphoid or to the shoulders is shown in Fig. 29. The changes occurring during submersion to the xiphoid reflect the lack of the downward pull produced by the gravitational field on the abdomen. Hence the broken curve of Fig. 29 should approximate the volume pressure curve of the relaxed chest wall when gravity free; it should be displaced about 1 cm $H_2O$ to the left because of the lack of shoulder weight (Craig, 1960). Accordingly the expiratory reserve volume when gravity free should decrease by about 8 per cent of the VC (Agostoni, Gurtner, Torri and Rahn, 1966).

During submersion to the shoulders the end-expiratory volume is reduced to about 11 per cent of the VC (Fig. 29). The volume decrease produced by this submersion is almost entirely (5/6) due to the displacement of the abdomen–diaphragm, because of the small compliance of the rib cage at small volume and because of the higher hydrostatic pressure acting on the abdomen. The diaphragm at the end of expiration is displaced craniad nearly as far as at full expiration under normal condition and is essentially relaxed. Only part of the ambient pressure is transmitted through the abdominal wall, which is stretched toward the abdominal cavity and recoils outwards, as could be anticipated by the middle diagram of Fig. 25. The average hydrostatic pressure acting on the submerged rib cage is

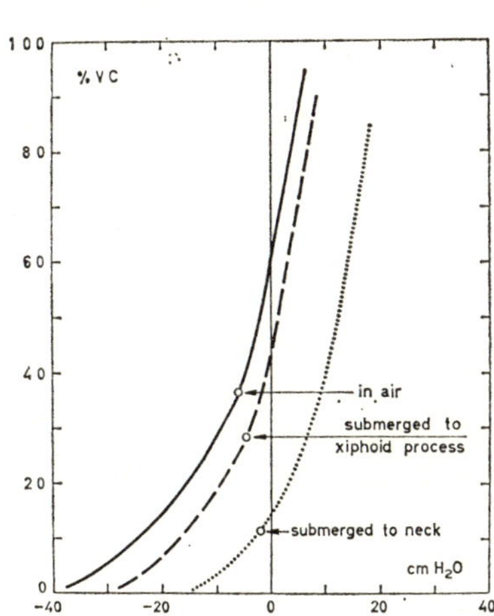


FIG. 29.—VOLUME-PRESSURE CURVES OF THE RELAXED CHEST WALL WHEN THE SUBJECT IS SITTING IN AIR (CONTINUOUS LINE) AND DURING SUBMERSION UP TO THE XIPHOID PROCESS (BROKEN LINE) AND UP TO THE NECK (DOTTED LINE)
The circles indicate the points corresponding to the end of spontaneous expirations; the pressure volume curve of the lung is not appreciably changed during submersion and therefore the volume-pressure curve of the relaxed respiratory system during submersion undergoes the same shift as that of the chest wall. The volume differences at the upper end of the curves are due, partly to the larger compression of the gas, partly to the reduction of the upper limit of the VC occurring during submersion. (From Agostoni, Gurtner, Torri and Rahn, 1966.)

about 17 cm $H_2O$. The average hydrostatic pressure acting on the chest wall is about 20 cm $H_2O$ (Jarrett, 1965; Agostoni, Gurtner, Torri and Rahn, 1966). A subject completely immersed in the head-up posture chooses, however, to breathe at a pressure lower than this to avoid the discomfort produced by the stretching of the pharynx and of the cheeks (Thompson and McCally, 1967). In the supine subject submerged to the ventral wall the average hydrostatic pressure acting on the chest wall is 6–7 cm $H_2O$ (Hong *et al.*, 1960; Jarrett, 1965).

In the subject submersed to the shoulders the vital capacity is reduced by about 9 per cent (Fig. 29): two-thirds of this reduction are probably accounted for by a decrease of the upper volume extreme caused by the hydrostatic pressure counter-acting that exerted by the inspiratory muscles (see p. 63), and one-third by the shift of blood into the chest cavity (Agostoni, Gurtner, Torri and Rahn, 1966).

**Changes throughout the Life Span**

The static features of the respiratory system change throughout life. In the mature foetus the resting volume of the chest wall corresponds to that of the nonaerated lung and the pressure on the pleural surface is not lower than that on the body surface (Bernstein, 1882; Agostoni *et al.*, 1958; Agostoni, 1959; Avery and Cook, 1961). Above the resting volume the chest wall of the mature foetus does not recoil appreciably over a considerable volume range (Agostoni, 1959; Avery and Cook, 1961), whereas below resting volume it resists further compression (Avery and Cook, 1961). In the newborn the specific compliance (i.e. the compliance corrected for size, by relating it to the FRC or to lung weight) of the chest wall is particularly high (Agostoni, 1959; Avery and Cook, 1961; Richards and Bachman, 1961), that of the lung is slightly lower than in the adult (Avery and Cook, 1961). The main static change of the respiratory system during the growth is the increasing outward recoil of the chest wall; it is not clear how much of this is due to changes in the mechanical properties of the chest wall and how much to a disproportionate growth of the chest wall relative to that of the lung.

From young adulthood the vital capacity decreases almost linearly with age; at 70 it is about three quarters of that at 30 (Robinson, 1938; Baldwin *et al.*, 1948; Withfield *et al.*, 1950; Needham *et al.*, 1954; Pemberton and Flanagan, 1956; Briscoe, 1965). This decrease is due to an increase of the residual volume (Kaltreider *et al.*, 1938; Greifenstein *et al.*, 1952; Needham *et al.*, 1954; Norris *et al.*, 1956; Pierce and Ebert, 1958; Briscoe, 1965; Turner *et al.*, 1968). In fact the total lung capacity does not decrease (Norris *et al.*, 1956; Pierce and Ebert, 1958; Turner *et al.*, 1968); in the cases in which a small decrease was found (Needham *et al.*, 1954; Permutt and Martin, 1960; Mittman *et al.*, 1965) this was probably due to the fact that the older subjects were smaller, as pointed out by Turner *et al.* (1968). The recoil of the lung decreases with age particularly at high lung volume (Frank *et al.*, 1957; Pierce and Ebert, 1958; Permutt and Martin, 1960; Cohn and Donoso, 1963; Mittman *et al.*, 1965; Turner *et al.*, 1968). In the region of spontaneous breathing, the compliance of the lung increases with age while that of the chest wall decreases. In the 20-year-old subject the lungs are less compliant than the chest wall whereas in the 60-year-old the reverse is true and the overall compliance is somewhat less (Turner *et al.*, 1968). At low lung volume the outwards recoil of the chest wall increases with age. On the other hand the resting volume of the chest wall seems to decrease in the old subjects, hence the volume-pressure curve of the chest wall should become less steep with age, pivoting around a point at about mid-lung volume, where its recoil remains the same (Turner *et al.*, 1968). The increase of FRC with age found by most investigators (Greifenstein *et al.*, 1952; Kaltreider *et al.*, 1938; Needham *et al.*, 1954; Norris *et al.*, 1956; Frank *et al.*, 1957; Pierce and Ebert, 1958) is therefore mainly due to the decrease of lung recoil and is less marked than that of RV (Turner *et al.*, 1968).

## VOLUME-PRESSURE RELATIONS OF THE RESPIRATORY SYSTEM DURING STATIC MUSCULAR EFFORTS

### Alveolar Pressure

The alveolar pressures during maximum static inspiratory and expiratory efforts exerted for 1–2 sec at different lung volumes are shown in Fig. 30. The outer curves (solid lines) of this diagram represent therefore the volume pressure relationship of the system during maximum inspiratory and expiratory efforts. The horizontal distance between these curves and the relaxation pressure curve gives the net pressure exerted by the contraction of the respiratory muscles (broken lines). Similar curves are obtained, if, instead of performing the effort against the obstructed airways, the subject starting from full inspiration or expiration breathes into or out of containers of different capacity (Cook *et al.*, 1964).

The volume-pressure relations of the respiratory system during maximum efforts in groups of males of different ages are shown in Fig. 31. The pressures exerted by children are relatively high as compared with those of men. This is probably related to the smaller radius of curvature of the rib cage, of the diaphragm and of the abdominal wall in the children (Agostoni and Mead, 1964; Cook *et al.*, 1964). In fact, according to Laplace law, the pressure exerted in a cylinder or a sphere for a given tension of the wall is inversely proportional to the radius. Hence, if the radius of curvature of the wall is small, thin muscles may exert pressures as great as those exerted by thick muscles of a wall with a greater radius. Human newborn may lower the intrathoracic pressure to $-70$ cm $H_2O$ to overcome the high resistance to the first breath (Karlberg, 1957).

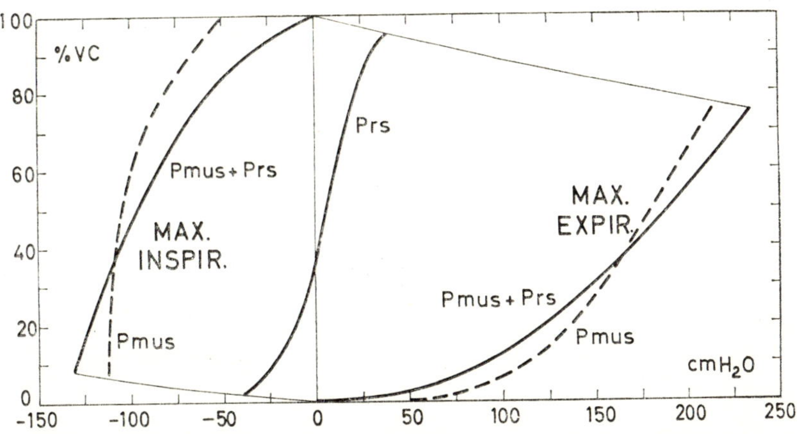


FIG. 30.—LUNG VOLUME AGAINST ALVEOLAR PRESSURE DURING MAXIMUM STATIC INSPIRATORY AND EXPIRATORY EFFORTS AND DURING RELAXATION IN THE SITTING POSTURE.

The broken lines indicate the pressure contributed by the muscles. (From Rohrer, 1916, as modified by Agostoni and Mead, 1964.)

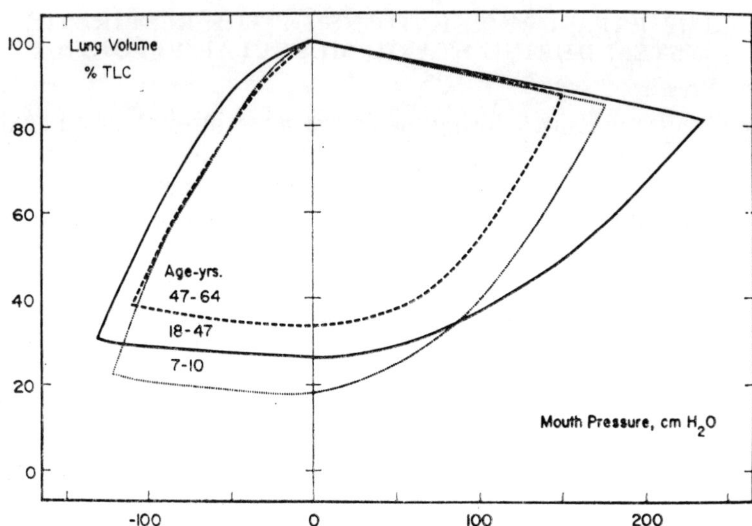


FIG. 31.—VOLUME-PRESSURE DIAGRAMS OF THE RESPIRATORY SYSTEM DURING MAXIMUM STATIC INSPIRATORY AND EXPIRATORY EFFORTS MADE BY MALES OF DIFFERENT AGES

(From Cook *et al.*, 1964.)

The volume-pressure relation of the respiratory system during maximum efforts in young adult women as compared with men of similar age is shown in Fig. 32. A great part of the difference in the maximum pressures

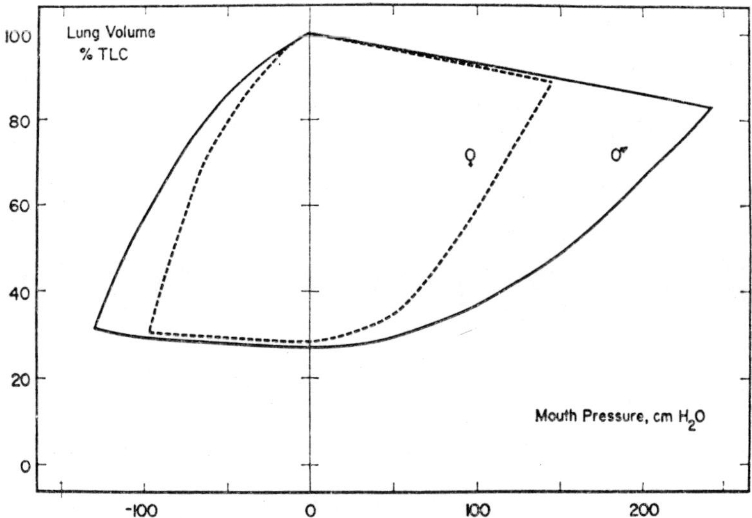


FIG. 32.—VOLUME-PRESSURE DIAGRAMS OF THE RESPIRATORY SYSTEM DURING MAXIMUM STATIC INSPIRATORY AND EXPIRATORY EFFORTS MADE BY YOUNG ADULT MALES AND FEMALES

(From Cook *et al.*, 1964.)

seems due to the difference in the strength of the accessory muscles (Ringqvist, 1966). Maximum inspiratory pressures in the Korean diving women are significantly greater than in a control group, whereas the maximum expiratory pressures are about the same. The greater development of inspiratory muscles is probably related to the condition of negative pressure breathing that these women undergo daily while in water between dives (Song *et al.*, 1963).

The decrease of the maximum respiratory pressures with age, their difference between sexes and their scattering among subjects parallel those observed for the maximum strength of other groups of muscles in the body (Ringqvist, 1966).

The values given in Figs. 30–32 do not necessarily represent the maximum that the inspiratory and expiratory muscles can exert, because, particularly in the expiratory efforts, antagonist muscles may be active (see below, transdiaphragmatic pressure). In this case the pressure exerted by the agonists could be higher than that recorded in term of alveolar pressure. Moreover, pressure measurements do not give information about the muscle forces deforming the chest wall (see below).

These values of maximum pressure are obtained by voluntary efforts in the laboratory and they do not necessarily represent the effect of a maximum contraction of the muscles involved. The time course of the pressure change during the quickest and maximum effort suggests that the degree of excitation can be maximum as during tetanic stimulation (see Chapter IV). It is not known, however, whether all muscle fibres are excited.

The maximum values of pressure found in recent years (Agostoni and Fenn, 1960; Agostoni and Rahn, 1960; Cook *et al.*, 1964; Milic-Emili, Orzalesi, Cook and Turner, 1964; Ringqvist, 1966) are higher than many other given in the older literature (for a complete survey of data available in the literature concerning maximum respiratory pressure see Ringqvist, 1966). These differences are probably related to: (*a*) the type of mouthpiece used (Cook *et al.*, 1964); (*b*) the inhibition elicited by the pain at the tympanic membrane (Agostoni and Fenn, 1960) when measurements are made through the nostrils; (*c*) the inhibition possibly elicited by the interference with circulation if the subject is asked to maintain the pressure for more than 1 - 2 seconds (Mosler and Balsamoff, 1924; Milic-Emili, Orzalesi, Cook and Turner, 1964).

The expiratory pressures are larger when the chest is inflated, whereas the inspiratory pressures are larger when it is deflated. This behaviour, besides being influenced by the mechanical features of the passive structures involved, by the action of the antagonist muscles (see below) and by a hypothetical inhibition of the efforts, depends upon the force-length relation of the muscles (see Chapter I). The length of the expiratory muscles should increase roughly with the cube root of the lung volume and vice versa for the inspiratory muscles. It must be recalled in this connection that the length of the respiratory muscles is greatly influenced also by posture (see Chapter II).

During slight neuromuscular blockade either with a non-depolarising agent such as tubocurarine (Johansen *et al.*, 1964) or with a depolarising agent such as decamethonium (Jørgensen *et al.*, 1966) the maximum inspiratory and expiratory pressures are reduced comparatively less than the maximum force of other muscles, as those involved in hand grip and head lift.

When the pressure in the lung is changed, the lung volume changes owing to the compression or expansion of the gas, as shown in Figs. 30–32: as a consequence, the pressure attained is smaller than it would be if the system had been filled with liquid. This effect may become very large if the same efforts are performed at low ambient pressure. At high altitude therefore the whole diagram is very much curtailed, although, of course, the actual mechanics of the chest are not changed (Rahn *et al.*, 1946).

**Abdominal and Thoracic Pressures**

Alveolar or oesophageal pressure measurements give only the resultant of the action of agonist and antagonist groups of muscles when they act simultaneously. A separation of the contribution of inspiratory and expiratory muscles may be done only at the abdominal boundary of the respiratory system through transdiaphragmatic and transabdominal pressure measurements (Agostoni and Rahn, 1960). The pressure on the thoracic and abdominal side of the diaphragm at different lung volumes during inspiratory, expiratory and expulsive efforts are shown in Fig. 33.

During maximum inspiratory efforts, the abdominal pressure, in the majority of subjects, remains roughly as during relaxation (Fig. 33). Up to about 60 per cent VC, the transdiaphragmatic pressure remains about the same or decreases slightly as the volume increases; above this volume the transdiaphragmatic pressure decreases progressively. In a minority of subjects the abdominal pressure increases at large lung volumes because of the contraction of the abdominal muscles (Mills, 1950; Campbell 1952; Campbell and Green, 1953; Delhez *et al.*, 1959). In these subjects the transdiaphragmatic pressure decreases only slightly above 60 per cent VC; hence at large volume it becomes progressively higher than the pressure across the chest wall, whereas at small volumes it is somewhat smaller than the chest wall pressure in all subjects. The different behaviour of *Pdi* at large lung volumes suggests that, at these volumes, the diaphragm exerts more pressure if the abdominal muscles are contracted. This view is supported by the high value of *Pdi* that can be reached during expulsive efforts at large lung volume (see below). It is not clear why in some subjects the muscles of the abdominal wall contract at large lung volume. In such subjects the contraction of the muscles of the abdominal wall is elicited before the balance between inspiratory muscle force and opposing passive

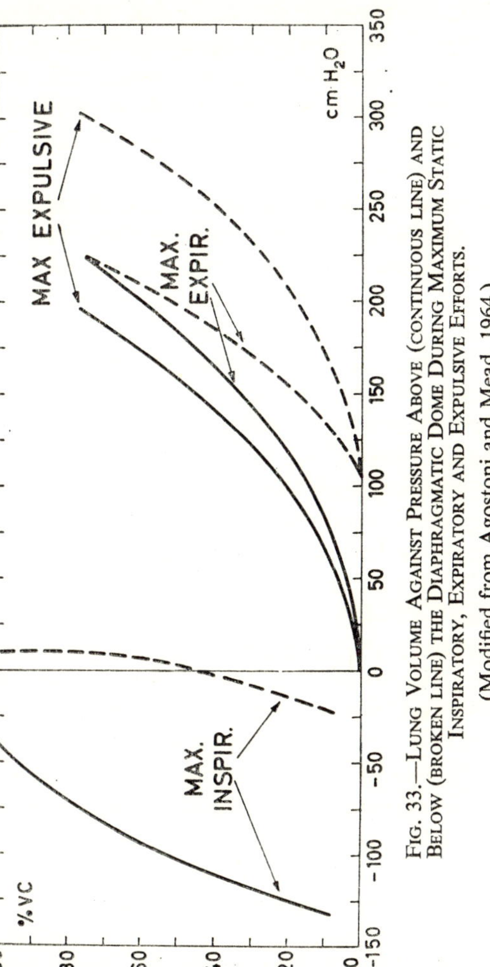


FIG. 33.—LUNG VOLUME AGAINST PRESSURE ABOVE (CONTINUOUS LINE) AND BELOW (BROKEN LINE) THE DIAPHRAGMATIC DOME DURING MAXIMUM STATIC INSPIRATORY, EXPIRATORY AND EXPULSIVE EFFORTS.

(Modified from Agostoni and Mead, 1964.)

force is reached, while in the majority of subjects such contraction is not elicited or it is inhibited by training (Agostoni and Mead, 1964).

During a moderate expiratory effort above the resting volume the transdiaphragmatic pressure is nil (Agostoni and Rahn, 1960), but during maximum effort there is an abdominal thoracic pressure difference (Fig. 33). Since at these volumes one would not expect the diaphragm to be passively distended, this pressure should be due to diaphragmatic contraction, as was confirmed by electromyographic findings (Agostoni *et al.*, 1960). At small lung volumes the transdiaphragmatic pressure increases progressively as lung volume decreases; it is greater than at relaxation because of diaphragmatic contraction (Agostoni and Torri, 1962). This suggests that the abdominal muscles contribute more than the other expiratory muscles and that the contraction of the diaphragm balances this difference. This mechanism becomes more evident when expulsive efforts are made (Fig. 33). In this manoeuvre the muscles of the abdominal wall contract more vigorously than during maximum expiratory efforts and the diaphragm also increases its activity. *Pab* reaches its maximum, and *Ppl* becomes somewhat smaller than during maximum expiratory efforts, suggesting that in this instance the action of the muscles of the abdominal wall on the rib cage is more than balanced by diaphragmatic contraction (Agostoni and Mead, 1964). At low lung volume the diaphragm can develop a greater pressure than the abdominal muscles and it is strongly but not maximally contracted (Milic-Emili, Orzalesi, Cook and Turner, 1964). The maximum static value of *Pab* during 2 sec expulsive efforts is about 400 cm $H_2O$, and the maximum value of *Pdi* is about 200 cm $H_2O$ (Agostoni and Mead, 1964).

At large lung volume the transdiaphragmatic pressure is larger during expulsive than during inspiratory efforts; i.e. the diaphragmatic contraction yields larger abdominal-thoracic pressure difference when the lower ribs are squeezed by the contraction of the abdominal muscles. As shown in the previous chapter, the lateral diameter of the rib cage, relative to its iso-lung volume value during relaxation, is reduced by the activity of the expiratory muscles and increased by that of the inspiratory ones. The marked reduction of the radius of curvature of the diaphragm along the lateral diameter could explain why the transdiaphragmatic pressure is greater during expulsive efforts.

The resting length of the diaphragm in the sitting posture is about at FRC (see p. 54). To obtain the pressure exerted by the diaphragmatic contraction below FRC the transdiaphragmatic pressure during relaxation must therefore be subtracted from that exerted during the effort at the same lung volume. It then appears that the pressure exerted by the diaphragm contraction is maximum at FRC and only slightly less at RV. Two opposite factors are here involved. At low lung volumes the diaphragm is above its resting length and therefore, according to the length-force relation of the muscle, its force should decrease as the lung volume decreases. On the other hand the radius of curvature of the diaphragm decreases with the lung volume and therefore, according to Laplace law, the pressure exerted, for a given force of the muscle, should be greater the smaller the radius.

It is difficult to assess the lung volume at which the abdominal muscles are at their resting length because the abdominal wall is always distended by the weight of the abdomen and because the distance between rib cage and pelvis may be changed markedly by flexion and extension of the vertebral column. It could be, however, that the weight of the abdomen is supported by the thick superficial fascia of the abdomen; if this hypothesis is correct the abdominal muscles could be below their resting length even at large lung volumes. Some facts and considerations suggest that this could be the case. (*a*) The pressure exerted by the abdominal muscles increases with the lung volume and is still increasing at the largest volume at which the effort may be done. Because of the force-length relation of the muscle, this suggests that the abdominal muscles are still below their resting length. It is possible, of course, that some muscles are distended at large volume if the vertebral column is greatly distended, but in performing an expiratory or expulsive effort the vertebral column is flexed and the distension disappears. (*b*) The slope of the relationship between lung volume and abdominal pressure during relaxation does not decrease at large volumes as would be expected if the muscles were distended beyond their resting length.

## FACTORS LIMITING THE VOLUME EXTREMES

The following mechanisms have been proposed as factors limiting the volume extremes: (*a*) obstruction of the airways, (*b*) decreasing force of the agonist muscles because of mechanical disadvantage or reflex inhibition, and (*c*) action of antagonist muscles.

### Upper Volume Extreme

If, as proposed by Mills (1950), glottis closure is the main factor limiting the expansion of the lung, oesophageal pressure at the end of a maximum inspiration should be much more negative than is generally found. In fact, Mead and his colleagues (1963) found that only one in six normal subjects closed the upper airways at the end of maximum inspiration.

A balance between the passive opposing force, which rises markedly at both volume extremes (see Fig. 30) and the driving force of the agonist muscles, which decreases at both extremes, is the factor setting the upper volume limit in those subjects in whom at the end of maximum inspiration the abdominal muscles do not contract (Mead *et al.*, 1963). Many subjects, however, contract the abdominal muscles at full inspiration, as shown by their marked electrical activity (Campbell, 1952; Campbell and Green,

1953; Delhez *et al.*, 1959; Mills, 1950). This contraction may lead to a marked increase of the abdominal pressure (Campbell and Green, 1953; Mills, 1950) and hence of the transdiaphragmatic pressure (Agostoni and Rahn, 1960), depending upon the geometry of the chest wall at full expiration (Mead, personal communication). In the former case the increase of the abdominal pressure antagonises the action of the diaphragm; in the latter case the contraction of the abdominal muscles could antagonise the action of the inspiratory muscles on the rib cage. Hence, when there is a contraction of the abdominal muscles, the limit to the upper volume extreme seems set by the contraction of the antagonist muscles. It is difficult to assess whether the decrease of the force of the agonist muscles at large lung volume depends only on mechanical disadvantage or also on a reflex inhibition of the effort.

**Lower Volume Extreme**

The oesophageal pressure at full expiration may be found to be 1–2 cm $H_2O$ below atmospheric (Mead and Milic-Emili, 1964): hence generalised closure of the airways cannot be considered the factor limiting the lower volume extreme. Below 20 per cent of the VC, airways in the inferior part of the lung close (Milic-Emili *et al.*, 1966; Burger and Macklem, 1968): pleural surface pressure on the lowermost part of the lung at RV is probably positive even in the upright subject. The collapse of some airways could conceivably elicit some reflex mechanism preventing further collapse of the lung, but no experimental evidence supports this hypothesis.

Leith and Mead (1967) found that when an external pressure assisting expiration was suddenly applied at or near RV the spirograms of young individuals showed a sudden decrement, while those of some old subjects were not influenced. This indicates that the lower volume extreme in the young subjects is set by a static balance of opposing forces operating in the chest wall, whereas in most old subjects it is set by a dynamic balance operating in the airways through a flow-limiting mechanism. In fact in young subjects: (*a*) the maximum expiratory flow near RV is effort dependent (see Chapter IV), (*b*) the maximum expiration is performed quickly, (*c*) at full expiration a condition of no flow may be kept for several seconds. In contrast, in most old subjects, because of decreased lung recoil and airway conductance, the maximum expiratory flow near RV is not effort dependent and the expiration at low lung volume proceeds so slowly that it is ended by an abrupt inspiration before the expiratory flow has ceased. It may be that in old subjects the collapsed or minimal air volume of the lung increases so greatly as to exceed the volume to which the respiratory muscles could compress the chest wall, if the flow-limiting mechanism did not set the minimum volume of the system (Leith and Mead, 1967).

At the end of maximum expiration, in young subjects, the diaphragm is contracted as shown both by the transdiaphragmatic pressure higher than during relaxation (Agostoni and Rahn, 1960) and by the electrical activity of the diaphragm (Agostoni and Torri, 1962; Delhez *et al.*, 1964). It seems therefore that contraction of antagonist muscles contributes to the static balance of forces limiting the lower volume extreme, at least at the abdominal boundary of the chest wall. The mechanism leading to the marked contraction of the diaphragm seems related to the simultaneous activity of the abdominal muscles rather than to a reflex, because this contraction is irrelevant when the lung volume is reduced to the same extent by an external force, as during submersion or breathing from a tank in which the pressure is kept subatmospheric (Agostoni, Gurtner, Torri and Rahn, 1966). A stretch reflex in the diaphragm seems unlikely because of the paucity of muscle spindles (see Chapter VI).

## FORCES DEFORMING THE RIB CAGE

As mentioned above (see p. 49) the volume-pressure diagram cannot provide full information on the forces of the muscles and of the passive structures of the chest wall when the shape of the chest wall at a given lung

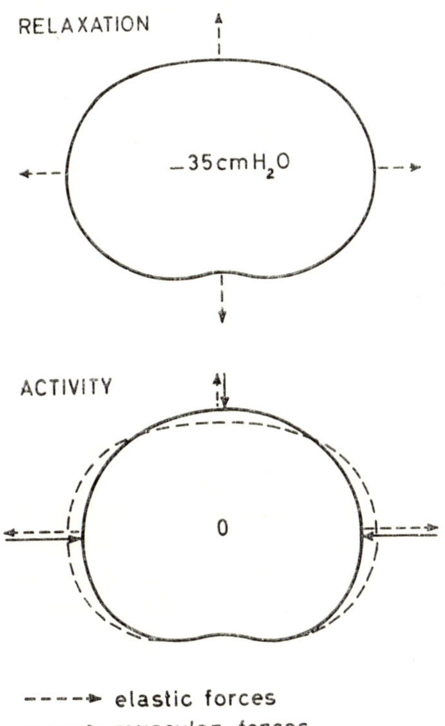


FIG. 34.—SCHEMA OF THE FORCES ACTING AT RIGHT ANGLE TO THE SURFACE OF THE RIB CAGE ALONG THE LATERAL AND DORSO-VENTRAL DIAMETER DURING RELAXATION AND ACTIVITY OF THE RESPIRATORY MUSCLES AT RV.

The broken line in the lower drawing gives the profile during relaxation in order to show the change of shape of the cross section of the rib cage caused by the action of the muscles. For explanation see text.

volume changes. Information on the force of deformation is however scanty.

The changes of shape of the cross section of the rib cage produced by the action of the respiratory muscles show that the forces of the respiratory muscles acting at right angle to the surface of the chest wall are not evenly distributed (see Chapter II). A comparison between the situations during relaxation and during activity of the respiratory muscles at RV, illustrated by Fig. 34, may help in understanding the forces responsible for the changes of shape (Agostoni and Mognoni, 1966). During relaxation the pressure inside the rib cage is about 35 cm $H_2O$ below atmospheric: the equilibrium on the lateral diameter is set by a force, corresponding to the transmural pressure, pushing in and an equal elastic force arising from the stretched wall pushing out (broken arrows, Fig. 34). Similar forces are operating on the dorso-ventral and the other diameters and are responsible for the shape of the rib cage under these conditions. When the subject contracts his muscles to hold that volume with open airways the pressure inside is almost equal to atmospheric. If the muscle forces were equally distributed the shape would not change and only a slight decrease of volume would occur. If the muscle forces are not evenly distributed and that acting on the lateral diameter is greater than the opposite force exerted by the passive structures, the lateral diameter decreases until the elastic force arising from the stretched wall becomes great enough to balance the muscle force as shown by the schema of Fig. 34. This implies that the muscle force acting on the dorso-ventral diameter is smaller than the elastic one, and therefore this diameter increases until the elastic force is decreased enough for a new equilibrium, as shown by the schema of Fig. 34. This is the change of shape actually occurring (see Chapter II). The same analysis may be applied to other lung volumes and manoeuvres; this case was chosen only because the transmural pressure during activity is nil and therefore the analysis is simpler.

In order to get some information on the magnitude of the force necessary to change the lateral diameter as during expiratory efforts the relaxed rib cage has been squeezed along the lateral diameter and the applied force as well as the change of diameters have been recorded (Agostoni, Mognoni, Torri and Miserocchi, 1966). For instance during maximum static expiratory effort at mid-lung volume the lateral diameter decreases by about 22 mm; in order to produce this change a force of 10–12 kg is required. The knowledge of these forces is of little interest *per se*, because they cannot be directly compared with the volume-pressure diagram, but it permits calculation of at least part of the work of deformation which may be directly compared with the work calculated on the volume-pressure diagram (see Chapter V).

## REFERENCES

AGOSTONI, E. (1959). Volume-pressure relationships of the thorax and lung in the newborn. *J. appl. Physiol.*, **14**, 909–913.

AGOSTONI, E. (1962). Diaphragm activity and thoracoabdominal mechanics during positive pressure breathing. *J. appl. Physiol.*, **17**, 215–220.

AGOSTONI, E. (1969). The thickness and the pressure of the pleural liquid. In: *The Pulmonary Circulation and Interstitial Space*, pp. 65–77. Ed. A. P. Fishman. Chicago Univ. Press.

AGOSTONI, E., and D'ANGELO, E. (1968). The recoil of the most dependent part of the lung. *Resp. Physiol.*, **5**, 379–384.

AGOSTONI, E., D'ANGELO, E., and BONANNI, M. V. (1970). The effect of the abdomen on the vertical gradient of pleural surface pressure. *Resp. Physiol.*, (**8**, 332-346).

AGOSTONI, E., and FENN, W. O. (1960). Velocity of muscle shortening as a limiting factor in respiratory air flow. *J. appl. Physiol.*, **15**, 349–353.

AGOSTONI, E., GURTNER, G., TORRI, G., and RAHN, H. (1966). Respiratory mechanics during submersion and negative-pressure breathing. *J. appl. Physiol.*, **21**, 251–258.

AGOSTONI, E., and MEAD, J. (1964). Statics of the respiratory system. In: *Handbook of Physiology*. Section 3, Respiration. Vol. 1, pp. 387–409. Eds. W. O. Fenn and H. Rahn. Washington, D.C.: American Physiological Soc.

AGOSTONI, E., and MOGNONI, P. (1966). Deformation of the chest wall during breathing efforts. *J. appl. Physiol.*, **21**, 1827–1832.

AGOSTONI, E., MOGNONI, P., TORRI, G., and FERRARIO-AGOSTONI, A. (1965). Static features of the passive rib cage and abdomen-diaphragm. *J. appl. Physiol.*, **20**, 1187–1193.

AGOSTONI, E., MOGNONI, P., TORRI, G., and MISEROCCHI, G. (1966). Forces deforming the rib cage. *Resp. Physiol.*, **2**, 105–117.

AGOSTONI, E., and RAHN, H. (1960). Abdominal and thoracic pressures at different lung volumes. *J. appl. Physiol.*, **15**, 1087–1092.

AGOSTONI, E., SANT'AMBROGIO, G. and DEL PORTILLO CARRASCO, H. (1960). Electromyography of the diaphragm in man and transdiaphragmatic pressure. *J. appl. Physiol.*, **15**, 1093–1097.

AGOSTONI, E., TAGLIETTI, A., FERRARIO-AGOSTONI, A., and SETNIKAR, I. (1958). Mechanical aspects of the first breath. *J. appl. Physiol.*, **13**, 344–348.

AGOSTONI, E., and TORRI, G. (1962). Diaphragm contraction as a limiting factor to maximum expiration. *J. appl. Physiol.*, **17**, 427–428.

ASMUSSEN, E., CHRISTENSEN, E. H., and SJÖSTRAND, T. (1939). Über die Abhängigkeit der Lungenvolumen von der Blutverteilung. *Skand. Arch. Physiol.*, **82**, 193–200.

AVERY, M. E., and COOK, C. D. (1961). Volume-pressure relationships of lungs and thorax in fetal, newborn and adult goats. *J. appl. Physiol.*, **16**, 1034–1038.

BACHOFEN, H. (1968). Lung tissue resistance and pulmonary resistance. *J. appl. Physiol.*, **24**, 296–301.

BALDWIN, E., COURNAND, A., and RICHARDS, D. W., JR. (1948). Pulmonary insufficiency. I. Physiological classification, clinical methods of analysis, standard values in normal subjects. *Medicine* (*Baltimore*), **27**, 243–278.

BATEMAN, J. B. (1950). Studies of lung capacities and intrapulmonary mixing: normal lung capacities. *J. appl. Physiol.*, **3**, 133–142.

BERNSTEIN, J. (1882). Zur Entstehung der Aspiration der Thorax bei der Geburt. *Pflügers Arch. ges. Physiol.*, **28**, 229–242.

BOUHUYS, A., PROCTOR, D. F., and MEAD, J. (1966). Kinetic aspects of singing. *J. appl. Physiol.*, **21**, 483–496.

BRISCOE, W. A. (1965). Lung volumes. In: *Handbook of Physiology*. Section 3, Respiration. Vol. 2, pp. 1345–1379. Eds. W. O. Fenn and H. Rahn. Washington, D.C.: American Physiological Soc.

BROWN, E. S., JOHNSON, R. P., and CLEMENTS, J. A. (1959). Pulmonary surface tension. *J. appl. Physiol.*, **14**, 717–720.

BRYAN, A. C., MILIC-EMILI, J., and PENGELLY, D. (1966). Effect of gravity on the distribution of pulmonary ventilation. *J. appl. Physiol.*, **21,** 778–784.

BUCHTAL, F., and ROSENFALCK, P. (1957). Elastic properties of striated muscle. In: *Tissue Elasticity*, pp. 73–93. Ed. J. W. Remington. Washington, D.C.: American Physiological Soc.

BURGER, E. J., JR., and MACKLEM, P. (1968). Airway closure: demonstration by breathing 100 per cent $O_2$ at low lung volumes and by $N_2$ washout. *J. appl. Physiol.*, **25**, 139–148.

BUTLER, J. (1957). The adaptation of the relaxed lungs and chest wall to changes in volume. *Clin. Sci.*, **16**, 421–433.

BUTLER, J., and SMITH, B. H. (1957). Pressure-volume relationships of the chest in the completely relaxed anesthetized patient. *Clin. Sci.*, **16**, 125–146.

CAMPBELL, E. J. M. (1952). An electromyographic study of the role of the abdominal muscles in breathing. *J. Physiol.* (*Lond.*), **117**, 222–233.

CAMPBELL, E. J. M., and GREEN, J. H. (1953). The variations in intra-abdominal pressure and the activity of the abdominal muscles during breathing; a study in man. *J. Physiol.* (*Lond.*), **122**, 282–290.

CAVAGNA, G., BRANDI, G., SAIBENE, F., and TORELLI, G. (1962). Pulmonary hysteresis. *J. appl. Physiol.*, **17,** 51–53.

CHERNIACK, R. M., and BROWN, E. (1965). A simple method for measuring total respiratory compliance: normal values for males. *J. appl. Physiol.*, **20**, 87–91.

CLEMENTS, J. A., BROWN, E. S., and JOHNSON, R. P. (1958). Pulmonary surface tension and the mucus lining of the lungs: some theoretical considerations. *J. appl. Physiol.*, **12**, 262–268.

COHN, J. E., and DONOSO, H. D. (1963). Mechanical properties of lung in normal men over 60 years old. *J. clin. Invest.*, **42**, 1406–1410.

COLVILLE, P., SHUGG, C., and FERRIS, B. G., JR. (1956). Effects of body tilting on respiratory mechanics. *J. appl. Physiol.*, **9**, 19–24.

COOK, C. D., and HAMANN, J. F. (1961). Relation of lung volumes to height in healthy persons between the ages of 5 and 38 years. *J. Pediat.*, **59**, 710–714.

COOK, C. D., MEAD, J., and ORZALESI, M. M. (1964). Static volume-pressure characteristics of the respiratory system during maximal efforts. *J. appl. Physiol.*, **19**, 1016–1022.

CRAIG, A. B., JR. (1960). Effects of position on expiratory reserve volume of the lungs. *J. appl. Physiol.*, **15**, 59–61.

D'ANGELO, E., BONANNI, M. V., MICHELINI, S., and AGOSTONI, E. (1970). Topography of the pleural surface pressure in rabbits and dogs. *Resp. Physiol.* **8,** 204-229.

DELHEZ, L., PETIT, J.-M., and MILIC-EMILI, G. (1959). Influence des muscles expirateurs dans la limitation de l'inspiration. (Étude electromyographique chez l'homme). *Rev. franç. Étud. clin. biol.*, **4**, 815–818.

DELHEZ, L., TROQUET, J., DAMOISEAU, J., PIRNAY, F. DEROANNE, R., and PETIT, J.-M. (1964). Influence des modalités d'exécution des maneuvres d'expiration forcée et d'hyperpression thoraco-abdominale sur l'activité électrique du diaphragme. *Arch. int. Physiol. Biochem.*, **72**, 76–94.

DITTMER, D. S., and GREBE, R. M. Eds. (1958). *Handbook of Respiration*, pp. 28–40. Philadelphia: Saunders.

DOW, P. (1939). The venous return as a factor affecting the vital capacity. *Amer. J. Physiol.*, **127**, 793–795.

DUOMARCO, J. L., and RIMINI, R. (1947). *La Presion Intraabdominal en el Hombre*. Buenos Aires: El Ateneo.

FENN, W. O. (1954). The pressure-volume diagram of the breathing mechanism. In: *Handbook of Respiratory Physiology*, pp. 19–27. Ed. M. W. Boothby. Texas: USAF School of Aviation Medicine, Randolph AFB.

FOWLER, R. C., GUILLET, M., and RAHN, H. (1951). Lung volume changes with positive and negative pulmonary pressures, pp. 522–528. Ohio: Techn. Rept. 6528, Wright-Patterson AFB.

FRANK, N. R., MEAD, J., and FERRIS, B. G., JR. (1957). The mechanical behaviour of the lungs in healthy elderly persons. *J. clin. Invest.*, **36,** 1680–1687.

GLASER, E. M. (1949). The effect of cooling and warming on the vital capacity, forearm and hand volume, and skin temperature of man. *J. Physiol.* (*Lond.*), **109**, 421–429.

GLAZIER, J. B., HUGHES, J. M. B., MALONEY, J. E., and WEST, J. B. (1967). Vertical gradient of alveolar size in lungs of dogs frozen intact. *J. appl. Physiol.*, **23**, 694–705.

GREIFENSTEIN, F. E., KING, R. M., LATCH, S. S., and COMROE, J. H., JR. (1952). Pulmonary function studies in healthy men and women 50 years and older. *J. appl. Physiol.*, **4**, 641–648.

HAMILTON, W. F., and MAYO, J. P. (1944). Changes in the vital capacity when the body is immersed in water. *Amer. J. Physiol.*, **141**, 51–53.

HAMILTON, W. F., and MORGAN, A. B. (1932). Mechanism of the postural reduction in vital capacity in relation to orthopnea and storage of blood in the lungs. *Amer. J. Physiol.*, **99**, 526–533.

HEAF, P. J. D., and PRIME, F. J. (1956). The compliance of the thorax in normal human subjects. *Clin. Sci.*, **15**, 319–327.

HOGG, J. C., and NEPSZY, S. (1969). Regional lung volume and pleural pressure gradient estimated from lung density in dogs. *J. appl. Physiol.*, **27**, 198–203.

HONG, S. K., TING, E. Y., and RAHN, H. (1960). Lung volumes at different depths of submersion. *J. appl. Physiol.*, **15**, 550–553.

HOPPIN, F. G., JR., GREEN, I. D., and MEAD, J. (1969). Distribution of pleural surface pressure in dogs. *J. appl. Physiol.*, **27,** 863–873.

HOWELL, J. B. L., and PECKETT, B. W. (1957). Studies of the elastic properties of the thorax of supine anesthetized paralyzed human subjects. *J. Physiol.* (*Lond.*), **136**, 1–19.

HUGHES, R., MAY, A. J., and WIDDICOMBE, J. G. (1959). Stress relaxation in rabbits' lungs. *J. Physiol.* (*Lond.*), **146**, 85–97.

HURTADO, A., and FRAY, W. W. (1933). Studies of total pulmonary capacity and its subdivisions. III. Changes with body posture. *J. clin. Invest.*, **12**, 825–832.

JARRETT, A. S. (1965). Effect of immersion on intrapulmonary pressure. *J. appl. Physiol.*, **20**, 1261–1266.

JOHANSEN, S. H., JØRGENSEN, M., and MOLBECH, S. (1964). Effect of tubocurarine on respiratory and nonrespiratory muscle power in man. *J. appl. Physiol.* **19**, 990–994.

JOHNSON, L. F., JR., and MEAD, J. (1963). Volume-pressure relationships during pressure breathing and voluntary relaxation. *J. appl. Physiol.*, **18**, 505–508.

JØRGENSEN, M., MOLBECH, S., and JOHANSEN, S. H. (1966). Effect of decamethonium on head lift, hand grip, and respiratory muscle power in man. *J. appl. Physiol.*, **21**, 509–512.

KALTREIDER, N. L., FRAY, W. W., and HYDE, H. van Z. (1938). The effect of age on the total pulmonary capacity and its subdivisions. *Amer. Rev. Tuberc.*, **37**, 662–689.

KANEKO, K., MILIC-EMILI, J. DOLOVICH, M. B., DAWSON, A., and BATES, D. V. (1966). Regional distribution of ventilation and perfusion as a function of body position. *J. appl. Physiol.*, **21**, 767–777.

KARLBERG, P. (1957). Breathing and its control in premature infant. In: *Physiology of Prematurity*, pp. 77–150. Ed. J. T. Lanman. New York: Macy.

KNOWLES, J. H., HONG, S. K., and RAHN, H. (1959). Possible errors using esophageal balloon in determination of pressure-volume characteristics of the lung and thoracic cage. *J. appl. Physiol.*, **14**, 525–530.

KONNO, K., and MEAD, J. (1968). Static volume-pressure characteristics of the rib cage and abdomen. *J. appl. Physiol.*, **24**, 544–548.

KRUEGER, J. J., BAIN, T., and PATTERSON, J. L., JR. (1961). Elevation gradient of intrathoracic pressure. *J. appl. Physiol.*, **16**, 465–468.

LEITH, D. E., and MEAD, J. (1967). Mechanisms determining residual volume of the lungs in normal subjects. *J. appl. Physiol.*, **23**, 221–227.

MAINS, R. C., ZECHMAN, F. W., and MUSGRAVE, F. S. (1968). Pulmonary mechanics with a simulated postural blood redistribution. *Resp. Physiol.*, **5**, 288–301.

MARSHALL, R., and WIDDICOMBE, J. G. (1960). Stress relaxation of the human lung. *Clin. Sci.*, **20**, 19–31.

MCMAHON, S. M., PROCTOR, D. F., and PERMUTT, S. (1969). Pleural surface pressure in dogs. *J. appl. Physiol.*, **27**, 881–885.

MEAD, J. (1961). Mechanical properties of lungs. *Physiol. Rev.*, **41**, 281–330.

MEAD, J., and MILIC-EMILI, J. (1964). Theory and methodology in respiratory mechanics with glossary of symbols. In: *Handbook of Physiology*. Section 3, Respiration. Vol. 1, pp. 363–376. Eds. W. O. Fenn and H. Rahn. Washington, D.C.: American Physiological Soc.

MEAD, J., MILIC-EMILI, J., and TURNER, J. M. (1963). Factors limiting depth of a maximal inspiration in human subjects. *J. appl. Physiol.*, **18**, 295–296.

MEAD, J., WHITTENBERGER, J. L., and RADFORD, E. P., JR. (1957). Surface tension as a factor in pulmonary volume-pressure hysteresis. *J. appl. Physiol.*, **10**, 191–196.

MILIC-EMILI, J., HENDERSON, J. A. M., DOLOVICH, M. B., TROP, D., and KANEKO, K. (1966). Regional distribution of inspired gas in the lung. *J. appl. Physiol.*, **21**, 749–759.

MILIC-EMILI, J., MEAD, J., and TURNER, J. M. (1964). Topography of esophageal pressure as a function of posture in man. *J. appl. Physiol.*, **19**, 212–216.

MILIC-EMILI, J., ORZALESI, M. M., COOK, C. D., and TURNER, J. M. (1964). Respiratory thoraco-abdominal mechanics in man. *J. appl. Physiol.*, **19**, 217–223.

MILLS, J. N. (1949). The influence upon the vital capacity of procedures calculated to alter the volume of blood in the lungs. *J. Physiol.* (*Lond.*), **110**, 207–216.

MILLS, J. N. (1950). The nature of the limitation of maximal inspiratory and expiratory efforts. *J. Physiol.* (*Lond.*), **111**, 376–381.

MITTMAN, C., EDELMAN, N. H., NORRIS, A. H., and SHOCK, N. W. (1965). Relationship between chest wall and pulmonary compliance and age. *J. appl. Physiol.*, **20**, 1211–1216.

MOGNONI, P., TORRI, G., and AGOSTONI, E. (1965). Confronto della relazione volume-pressione del sistema respiratorio ottenuta con diversi procedimenti. *Atti Accad. Naz. Lincei*, **38**, 925–928.

MORENO, F., and LYONS, H. A. (1961). Effect of body posture on lung volumes, *J. appl. Physiol.*, **16**, 27–29.

MORSE, M., SCHULTZ, F. W., and CASSELS, D. E. (1952). The lung volume and its subdivisions in normal boys 10–17 years of age. *J. clin. Invest.*, **31**, 380–391.

MOSLER, E., and BALSAMOFF, S. (1924). Uber den Valsalva-versuch. *Klin. Wschr.* **3**, 491–495.

NAIMARK, A., and CHERNIACK, R. M. (1960). Compliance of the respiratory system and its components in health and obesity. *J. appl. Physiol.*, **15**, 377–382.

NEEDHAM, C. D., ROGAN, M. C., and MCDONALD, I. (1954). Normal standards for lung volumes, intrapulmonary gas-mixing, and maximum breathing capacity. *Thorax*, **9**, 313–325.

NEERGAARD, K. VON (1929). Neue Auffassungen über einen Grundbegriff der Atemmechanik. Die Retraktionskraft der Lunge, abhängig von der Oberflachenspannung in den Alveolen. *Z. ges. exp. Med.*, **66**, 373–394.

NIMS, R. G., CONNER, E. H., and COMROE, J. H., JR. (1955). The compliance of the human thorax in anesthetized patients. *J. clin. Invest.*, **34**, 744–750.

NORRIS, A. H., SHOCK, N. W., LANDOWNE, M. and FALZONE, J. A. (1956). Pulmonary function studies: Age differences in lung volumes and bellows functions. *Gerontologia* (*Basel*), **11**, 379–387.

PEMBERTON, J., and FLANAGAN, E. G. (1956). Vital capacity and timed vital capacity in normal men over forty. *J. appl. Physiol.*, **9**, 291–296.

PERMUTT, S., and MARTIN, H. B. (1960). Static pressure-volume characteristics of lungs in normal males. *J. appl. Physiol.*, **15**, 819–825.

PIERCE, J. A., and EBERT, R. V. (1958). The elastic properties of the lungs in the aged. *J. Lab. clin. Med.*, **51**, 63–71.

PROCTOR, D. F., CALDINI, P., and PERMUTT, S. (1968). The pressure surrounding the lungs. *Resp. Physiol.*, **5**, 130–144.

RAHN, H., FENN, W. O., and OTIS, A. B. (1949). Daily variations of vital capacity, residual air and expiratory reserve including a study of the residual air method. *J. appl. Physiol.*, **1**, 725–736.

RAHN, H., OTIS, A. B., CHADWICK, L. E., and FENN, W. O. (1946). The pressure-volume diagram of the thorax and lung. *Amer. J. Physiol.*, **146**, 161–178.

REMINGTON, J. W. (1955). Hysteresis loop behavior of the aorta and other extensible tissues. *Amer. J. Physiol.*, **180**, 83–95.

RICHARDS, C. C., and BACHMAN, L. (1961). Lung and chest wall compliance of apneic paralyzed infants. *J. clin. Invest.*, **40**, 273–278.

RINGQVIST, T. (1966). The ventilatory capacity in healthy subjects. *Scand. J. clin. Lab. Invest.*, **18**, Suppl. 88.

ROBINSON, S. (1938). Experimental studies of physical fitness in relation to age. *Arbeitsphysiologie*, **10**, 251–323.

ROHRER, F. (1916). Der Zusammenhang der Atemkräfte und ihre Abhängigkeit vom Dehnungszustand der Atmungsorgane. *Pflügers Arch. ges. Physiol.*, **165**, 419–444.

ROHRER, F. (1925). Physiologie der Atembewegung. In: *Handbuch der normalen und pathologischen Physiologie.* Vol. II, pp. 70–127. Eds. Bethe, A., Bergmann, G. V., Embden, G., and Ellinger, A. Berlin: Springer.

SAIBENE, F., and MEAD, J. (1969). Frequency dependence of pulmonary quasi-static hysteresis. *J. appl. Physiol.*, **26**, 732–737.

SHARP, J. T., JOHNSON, F. N., GOLDBERG, N. B., and VAN LITH, P. (1967). Hysteresis and stress adaptation in the human respiratory system. *J. appl. Physiol.*, **23**, 487–497.

SJÖSTRAND, T. (1941). Über die Bedeutung der Lungen als Blutdepot beim Menschen. *Acta physiol. scand.*, **2**, 231–248.

SONG, S. H., KANG, D. H., KANG, B. S., and HONG, S. K. (1963). Lung volumes and ventilatory responses to high $CO_2$ and low $O_2$ in the ama. *J. appl. Physiol.*, **18**, 466–470.

THOMPSON, L. J., and MCCALLY, M. (1967). Role of transpharyngeal pressure gradients in determining intrapulmonary pressure during immersion. *Aerospace Med.*, **38**, 931–935.

TURNER, J. M., MEAD, J., and WOHL, M. E. (1968). Elasticity of human lungs in relation to age. *J. appl. Physiol.*, **25**, 664–671.

VAN DE WOESTIJNE, K. P. (1967). Influence of forced inflations on the creep of lungs and thorax in the dog. *Resp. Physiol.*, **3**, 78–89.

VAN LITH, P., JOHNSON, F. N., and SHARP, J. T. (1967). Respiratory elastances in relaxed and paralyzed states in normal and abnormal men. *J. appl. Physiol.*, **23**, 475–486.

WADE, O. L., and GILSON, J. C. (1951). The effect of posture on diaphragmatic movement and vital capacity in normal subjects with a note on spirometry as an aid in determining radiological chest volumes. *Thorax*, **6**, 103–126.

WITHFIELD, A. G. W., WATERHOUSE, J. A. H., and ARNOTT, W. M. (1950). The total lung volume and its subdivisions. A study in physiological norms. I. Basic data. *Brit. J. soc. Med.*, **4**, 1–25.

ZECHMAN, F. W., MUSGRAVE, F. S., MAINS, R. C., and COHN, J. E. (1967). Respiratory mechanics and pulmonary diffusing capacity with lower body negative pressure. *J. appl. Physiol.*, **22**, 247–250.

# Chapter IV

# DYNAMICS*

E. Agostoni

## DYNAMIC PROPERTIES OF THE RELAXED RESPIRATORY SYSTEM†

The force applied to a mechanical system is met by an equal opposing force developed by the system (Newton's third law of motion). The opposing force is the sum of the forces depending on the instantaneous displacement, velocity and acceleration of the system. The mechanical properties of the system related to these physical quantities are the elasticity, the friction and the inertia.

If the motion of any given point of the system is uniquely related to the overall motion, the system has only one degree of freedom, i.e. it has only one way in which to move. The number of degrees of freedom is therefore defined by the number of independent variables.

As long as the system has only one degree of freedom and its physical properties are constants (i.e. the system is linear) the balance between applied and opposing force may be simply expressed by the equation of motion:

$$Fapp = E\,s + R\,\mathrm{d}s/\mathrm{d}t + M\,\mathrm{d}^2s/\mathrm{d}t^2$$

where $Fapp$ is the applied force, $s$ the displacement, $t$ the time, and $E$, $R$ and $M$ constants expressing, respectively, the modulus of elasticity, the frictional resistance and the mass. The first term on the right side of the equation is the static component and the other two are the dynamic components. Whereas electrical systems are linear, mechanical systems are generally nonlinear, i.e. their properties are not expressed by single constants. With these systems, such as for the respiratory one, the equation of motion may be only approximate, or it must be left in a more general form, and graphic means may then offer the best analytical tool (Mead and Milic-Emili, 1964).

In a three dimensional (or acoustical) system, such as the respiratory one, the applied pressure is opposed by pressures related to its volume, flow and volume acceleration; its mechanical properties are the compliance ($C$), the flow-resistance ($R$), and the inertance ($I$). These are analogous, in electrical systems, to the capacitance, the resistance and the inductance (Olson, 1958).

Dynamic pressures are related to flow and volume acceleration. The

* Parts of this chapter are based on "Dynamics of breathing" by Mead and Agostoni (1964).
† Table IV (p. 141) lists approximate values for a normal adult.

pressure related to volume acceleration is separated from that related to flow by measuring the pressure when flow is zero and substracting the static component. Volume acceleration is determined from the simultaneous slope of the flow tracing. At ordinary breathing frequencies the pressure producing acceleration is negligible and therefore the pressure measured, corrected for the static component, is related only to flow (Mead and Agostoni, 1964).

**Flow-resistance**

Measurements of the flow-resistance of the relaxed chest wall, and hence of the relaxed respiratory system, face the following difficulties: (*a*) voluntary relaxation in dynamic conditions; (*b*) apportioning the volume change between the rib cage and the abdomen (see Kinematics chapter). In practice the second difficulty has been neglected and the dynamic properties of the chest wall have been dealt with as if the chest wall had only one degree of freedom. This is a considerable simplification, because the chest wall has a variety of ways in which to move depending on the motion of the rib cage and abdomen. The dynamic properties of the chest wall have been studied relatively little and the data are less reliable than those for the lung (Mead and Agostoni, 1964).

The flow-resistance of the relaxed respiratory system may be determined during relaxed expiration or inspiration. In fact after a fraction of a second from the beginning of the manoeuvre the accelerative pressure becomes essentially nil and all the static pressure of the relaxed system is related to flow. The driving pressure can then be estimated from measurements of volume if the static volume-pressure relation is known. Hence from simultaneous measurements of volume and flow during a relaxed expiration the flow-resistance of the whole respiratory system may be determined within the volume range covered by the manoeuvre (Comroe *et al.*, 1954; Brody, 1954; McIlroy *et al.*, 1963). If the flow-resistance of the lung is also determined (see below), that of the chest wall can be determined by subtraction from the flow-resistance of the whole system, being the lung and the chest wall placed in series. The flow-resistance of the chest wall may be obtained directly if pleural pressure is measured simultaneously with the flow during a relaxed expiration. The flow-resistive pressure of the relaxed chest wall is in faet given, for each lung volume, by the pressure measured in dynamic conditions minus that measured in static conditions during relaxation.

The flow-resistance of the respiratory system can also be obtained, by driving the relaxed system with a pump: the flow produced and the pressure applied are measured, and a correction is applied for the static component of pressure (Otis *et al.*, 1950). Another approach, based on externally applied pressures, is that of measuring the mechanical impedance of the system while it is driven at high frequencies. (Du Bois *et al.*, 1956). This approach stems from the following. When a sine wave airflow is applied to the airway openings the resultant pressure changes across the respiratory system are related to the resistive impedance and the reactive impedance; the latter in turn is the sum of the reactance related to the inertance and of that related to the compliance of the respiratory system. The reactance related to inertance is directly proportional to frequency, while that related to compliance is inversely proportional to frequency. Hence the reactance related to inertance, that is negligible at normal breathing frequency, becomes important at higher frequencies. Since the two reactances are 180° out of phase there is a frequency at which they are of equal magnitude and opposite in sign. At this frequency, called the natural or resonant frequency of the system, pressure and flow are in phase and the pressure change depends only upon the flow-resistance and the rate of flow. This, however, is not strictly true for the complexity of the respiratory system. Indeed, the compliance, the flow-resistance and the inertance of the chest wall change appreciably over a wide range of frequency and the flow-resistance measured with this method is smaller than that during relaxed expiration (Mead and Agostoni, 1964). Brody *et al.* (1960) provided values for the flow-resistance of the respiratory system during forced oscillation of the respiratory system at 10 cycles/sec in various postures. Since Mead (1960) showed that these measurements could be made during spontaneous breathing by superimposing the forced oscillations on the breathing pattern, this method has been widely used also in patients and recently improved (Ferris *et al.*, 1964; Sharp *et al.*, 1964*a*; Fisher *et al.*, 1968; Grimby *et al.*, 1968; Goldman *et al.*, 1970). Grimby *et al.* (1968) showed that artificial resonance can be achieved at different frequencies by adding to the pressure signals ones proportional either to instantaneous volume or to volume acceleration. This made it possible to measure the flow-resistance at various high frequencies. Moreover, Goldman *et al.* (1970) eliminated the need to achieve or simulate resonance by utilizing two specific instants in the induced cycle where the components of the applied pressure difference due to the reactances related to compliance and inertance are nil.

The flow-resistance of the lung (gas plus tissues) is obtained by measuring the flow and the pleural pressure, and correcting this latter for the simultaneous static component.

The relation between pressure and flow in the respiratory system is alinear because of the turbulence of the gas. This relation is expressed by Rohrer's (1915) empirical equation: $Pres = K_1\dot{V} + K_2\dot{V}^2$, though it may also be fitted by the equation: $Pres = A\dot{V}^B$ (Ainsworth and Eveleigh, 1952), where $B$ is about 1·3. According to the Rohrer equation the resistance ($R = P/\dot{V}$) increases linearly with flow, being: $R = K_1 + K_2\dot{V}$. The constant $K_1$ is related to the laminar flow, and, during mouth breathing, is about 2·4 cm $H_2O$/l/sec; $K_2$ is related to the turbulent flow, and,

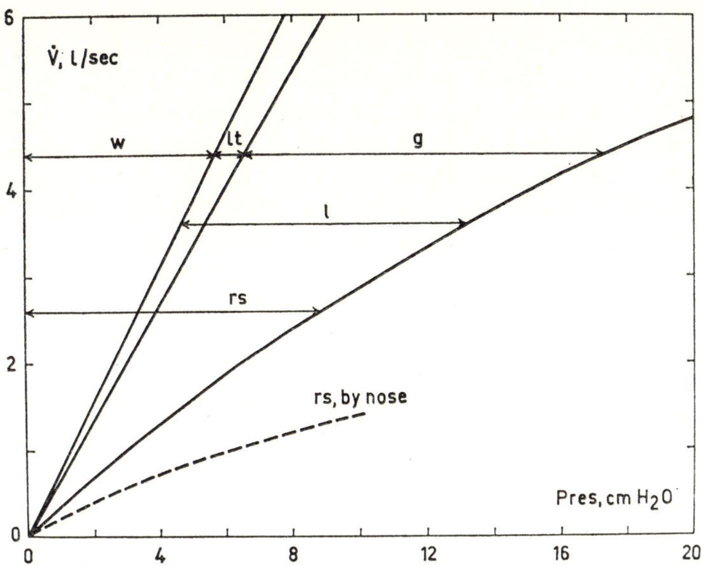


FIG. 35.—FLOW-PRESSURE RELATIONSHIP FOR AN ADULT MAN (VC = 5 LITRES) IN THE MIDRANGE OF LUNG VOLUME

The respiratory system (rs) has been subdivided into chest wall (w) and lungs (l), and the latter has been further subdivided into gas (g) and tissue (lt). (Modified from Mead and Agostoni, 1964.)

during mouth breathing, is about 0·3 cm $H_2O/(l/sec)^2$ (Mead and Agostoni, 1964). The significance of these constants however is not well defined. A more adequate physical definition of the flow pattern has been recently provided by Jaeger and Matthys (1968) for the unbranched part of the airways. For the branched part the following relationship should hold according to Pedley *et al.* (1970): $P = k(\rho\mu)^{1/2}\ \dot{V}^{3/2}$, where $\rho$ and $\mu$ are the density and the viscosity of the gas, respectively, and $k$ a factor related to morphological features.

The flow-pressure relationship of the relaxed respiratory system and of its various parts at mid-lung volume for an adult man with a vital capacity of about 5 litres is shown in Fig. 35. During quiet breathing through the nose about 50 per cent of the flow-resistance of the respiratory system is due to the nose (Mead, 1960; Butler, 1960; Ferris *et al.*, 1964). The flow-resistance of the relaxed chest wall for a flow of 1 l/sec is about 75 per cent of that of the lung (tissue and gas) during mouth breathing and about 30 per cent during nose breathing. As the flow rate increases the flow-resistance of the lung becomes a progressively increasing portion of the resistance of the whole system (Mead and Agostoni, 1964, and unpublished observations). During quiet breathing the expiratory flow-resistance of the airways is a little larger than the inspiratory, mainly because of the laryngeal narrowing; this narrowing disappears when ventilation increases (Ferris *et al.*, 1964). The flow-resistance of the extra-thoracic airways contributes 35–45 per cent of the total airways flow-resistance during mouth-breathing (Hyatt and Wilcox, 1961; Ferris *et al.*, 1964). The conductance ($1/R$) of the airways increases linearly with lung volume because the lumen of the airways is widened by the greater recoil of the lung (Briscoe and DuBois, 1958). Airway flow-resistance may change markedly when the viscosity and the density of the gas breathed are changed. Through its effect on gas density a decrease of ambient pressure decreases the flow-resistance (Otis and Bembower, 1949), whereas an increase increases the flow-resistance (Marshall *et al.*, 1956; Bühlmann, 1963). The flow-resistance of the chest wall should decrease with increasing lung volume because the pressure associated with tissue viscous resistance is proportional to the linear velocity of tissues and for a given volume change the linear movement of tissue decreases as the volume increases (Grimby *et al.*, 1968). The conductance of the respiratory system also increases linearly with lung volume (Fisher *et al.*, 1968).

### Inertance, Time Constant, Damping and Natural Frequency

The inertance of the respiratory system is apparently linear and amounts to only 0·01 cm $H_2O/l/sec^2$ (DuBois *et al.*, 1956; Mead, 1956; Sharp *et al.*, 1964*b*); it is mainly contributed by the gas in the airways, the part of the system with the smallest mass but with the greatest linear velocities and hence accelerations (Mead, 1956). In obese subjects the inertance is mainly contributed by the tissues (Sharp *et al.*, 1964*b*).

As the inertance is negligible in most conditions the respiratory system may be considered a first order system (a resistance-capacitance system in electrical terms); but this can only be an approximation as the resistance and compliance of the respiratory system vary with the lung volume.

The time constant* of the respiratory system within the tidal volume range is about 0·35 sec: it is about the same for the lung (during mouth breathing) and for the chest wall (McIlroy *et al.*, 1963).

The lung is highly overdamped, whereas the chest wall is underdamped and may therefore oscillate (Zechman *et al.*, 1965). The whole respiratory system is more than critically damped (Brody *et al.*, 1956; Hull and Long, 1961). The natural frequency, $f_n = [1/(2\pi)]\sqrt{1/(IC)}$, of the human respiratory system is about 5 cycles/sec (DuBois *et al.*, 1956; Coerman *et al.*,

* The time constant is given by the resistance ($R$) times the compliance ($C$). During relaxed expiration, in subjects who are able to relax their respiratory muscles (McIlroy *et al.*, 1963), as a first approximation the volume decreases exponentially according to the formula: $V = V_0e^{-t/RC}$, where $V_0$ is the volume at the beginning of the relaxed expiration. When $t = RC$, i.e. in a period equal to the time constant, 63 per cent of the decay is accomplished. Since flow is the time derivative of volume and since the derivative of an exponential function is still an exponential function, the flow also should decrease exponentially; more particularly: $\dot{V} = -(1/RC)V_0e^{-t/RC}$. In fact the flow during a relaxed expiration rapidly reaches a peak, because of the negligible inertia, and then decreases approximately in an exponential way.

1960; Zechman *et al.*, 1965; Fisher *et al.*, 1968). It appears from this formula that when the natural frequency and the compliance of the respiratory system are known, its inertance can be calculated.

## DYNAMICS OF THE MAXIMUM CONTRACTION OF THE RESPIRATORY MUSCLES

The pressure exerted by the respiratory muscles (*Pmus*) is opposed by pressures related to the volume, flow and acceleration of the respiratory system. The first arises from static properties (*Pst*), the second from flow-resistive properties (*Pres*), and the third from inertial properties (*Pin*). Considering for simplicity and for analytical purpose the respiratory system (*rs*) as a linear system with only one degree of freedom, its equation of motion is: $-Pmus = Pst(rs) + Pres(rs) + Pin(rs)$. *Pmus* cannot be measured directly, but it can be calculated if the values on the right side of this equation are known. They can be estimated from measurements of the volume events and from previous determination of the static, resistive and inertial features of the relaxed respiratory system. Since the pressure of the respiratory system is the sum of the pressures contributed by the lung and by the chest wall, this procedure in practice is followed only for studies of the chest wall, because the pressures contributed by the lung are obtained directly from oesophageal pressure measurements (see p. 82) (Mead and Agostoni, 1964).

As pointed out by Fenn (1963) the essential difference between the limb movements and the act of breathing from the mechanical point of view is that in the limb movements most of the muscle force is balanced by inertial forces, whereas in the breathing act it is balanced by elastic and frictional forces.

The basic properties of the skeletal muscles, extensively studied in the limb, are shared by the respiratory muscles and are relevant to the dynamics of breathing.

### Time Course of the Pressure Exerted during the Quickest and Maximum Respiratory Efforts

As mentioned in Chapter I the force developed in an isometric contraction requires an appreciable time to reach its final value, mainly because of the series elastic element within the muscle. During an isometric contraction of a mammalian limb muscle tetanically stimulated through its motor nerve the force rises exponentially with a half time ranging from 30 to 50 msec. A similar time course, except for the first 20 msec, has been found during a maximum voluntary activation (Wilkie, 1949; Merton, 1954). The slow rise during the first 20 msec occurs probably because excitation of all the neurons involved is not synchronous and/or because there may be differences in the velocity of propagation of the impulses. After this short period, however, as shown by Wilkie (1949) for the flexors of the forearm, the degree of excitation is constant throughout all the effort, and therefore the time course of the force developed depends upon the contractile properties of the muscle.

This should also be the case for the respiratory muscles, but because of the complexity of the respiratory system the analysis can be only approximate and it is convenient to discuss first the peculiar conditions of the system in this respect.

The time course of the pressure should differ from that of the force exerted by the respiratory muscles because of the deformation of the chest

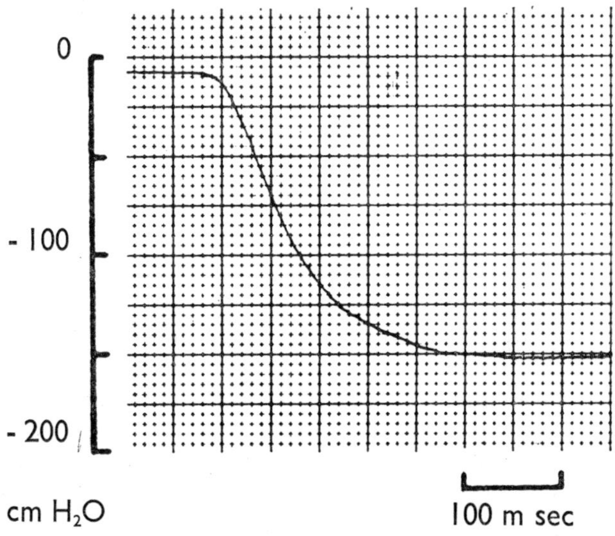


FIG. 36.—CHANGE OF THE OESOPHAGEAL PRESSURE AGAINST TIME DURING THE QUICKEST AND MAXIMUM STATIC INSPIRATORY EFFORT PERFORMED AT FRC (From Mognoni *et al.*, 1968.)

wall (see Chapters II and III) and because of the compliance of the gas. A muscle exerting a force on the wall of a hollow system produces a pressure within it which depends upon the geometrical relationship between the muscle and the system. In so far as this coupling remains constant during the contraction the time course of the force applied and that of the pressure exerted are the same. Because of the deformation, however, this coupling could change even during a static contraction: then the time course of the pressure may be appreciably different from that of the force. This difference is difficult to assess. On the other hand the lag introduced by the gas phase, calculated from the time constant of the passive respiratory system with closed airways (about 5 msec), turns out to be negligible (Mognoni *et al.*, 1968).

During the maximum and quickest static efforts the oesophageal and gastric pressures change approximately as a simple exponential function of time after the first 30–50 msec (Fig. 36). The values of half times recorded are of the same order of magnitude as those: (*a*) of the force exerted by the flexors of the forearm during maximum voluntary effort (Wilkie, 1949), (*b*) of the force exerted by the adductor pollicis during maximum voluntary effort or tetanic stimulation (Merton, 1954), (*c*) of the oesophageal or transdiaphragmatic pressure of cats and dogs in which the diaphragm was tetanically stimulated through the phrenic nerves (Mognoni *et al.*, 1968). This similarity suggests that: (*a*) the time course of the pressure change during the respiratory effort is not greatly affected by the deformation of the chest wall and it is therefore representative of the time course of the force of the respiratory muscles, (*b*) the degree of excitation of the respiratory muscles during the maximum and quickest efforts could be maximum and constant as during tetanic stimulation, except, as for every voluntary effort, at the onset of the contraction. The time course of the phenomenon during the quickest efforts seem therefore to have almost reached the limit set by the mechanical properties of the muscles (Mognoni *et al.*, 1968). There is no evidence that all the fibres of the muscles involved are excited as during tetanic stimulation: the experiments of Merton (1954) show that this is possible in the limbs, but the differences between the limbs and the respiratory system have to be considered. The maximum values of pressure are similar during both the quickest and the relatively slow efforts (Mognoni *et al.*, 1968).

The average value of half time for the oesophageal pressure change during the quickest inspiratory efforts at FRC ($50 \pm 1{\cdot}4$ msec) is not significantly different from that of the gastric pressure change during the quickest expiratory efforts at FRC ($46 \pm 4{\cdot}1$ msec). On the other hand both these values are significantly lower than that of the oesophageal pressure change during the quickest expiratory efforts at FRC ($61{\cdot}2 \pm 2{\cdot}9$ msec). During these efforts the oesophageal pressure is the resultant of the action of agonist and antagonist groups of muscles (Agostoni and Rahn, 1960; Agostoni and Mead, 1964) and probably the change of pressure is slowed down by the action of the antagonist muscles (Mognoni *et al.*, 1968).

The time course of the pressure exerted by the respiratory muscles at the beginning of the dynamic efforts (open airways) is almost the same as that during static efforts. This suggests that during the quickest and maximum dynamic efforts the contraction is not slowed down by neural control (Mognoni *et al.*, 1968).

### Muscular Factors limiting the Maximum Frequency

The time necessary to reach the peak muscle pressure during the quickest and maximum dynamic effort at FRC is about 0·2 sec (Mognoni *et al.*, 1968). The shortest time for complete relaxation seems to be at least

0·2 sec, but its determination in man is difficult and has not yet been studied systematically. The major part of these times seems related to muscle properties. One can then speculate on the muscular limit to the maximum frequency. At a frequency of 3–4 cycles/sec in man the activity of one group of muscles should start to increase before that of the previous cycle has ended: this would reduce the pressure change and hence the tidal volume because the antagonist group of muscles should be opposed even when it is developing its maximal pressure for that condition. As the frequency is increased still further this process increases and the tidal volume becomes eventually nil.

Otis and Guyatt (1968) studied the relationship between the tidal volume and the period of the breathing cycle during maximal voluntary ventilation. The relationship can be expressed by the following formula: $t = t_0 + bV_T$, i.e. the shortest period of the breathing cycle increases linearly with the tidal volume and for a period equal to $t_0$ the tidal volume is nil. If $t$ is expressed in sec and $V_T$ as a fraction of the vital capacity: $t_0 = 0{\cdot}14$ sec and $b = 1{\cdot}5$ sec/VC. Hence $V_T$ becomes nil when the breathing frequency is about 7 cycles/sec; this is also the maximal frequency at which the pressure in the closed respiratory system may be made to change, and is only a little above the natural frequency of the respiratory system. The relationship between the tidal volume and the shortest period of the breathing cycle found by Otis and Guyatt (1968) is similar to that found by Fenn (1938) between the amplitude of the movement of a limb or digit and the shortest period of oscillation.

## Muscular Factors limiting the Maximum Flow

The data on the time course of the pressure exerted during the quickest static respiratory effort indicate that about 0·2 sec are necessary to reach the maximum pressure achievable at a given lung volume. During this time an appreciable amount of air may be moved in or out of the lung when the airways are open: hence the rate at which the pressure can be developed limits the dynamic pressure and hence the flow at the beginning of the effort. As the time course of the pressure during the quickest static effort seems mainly related to the series elastic elements of the muscles (see above), the initial flow should be mainly limited by this mechanical feature of the muscle.

Another basic property of the muscle, that described by the force-velocity relationship (see Chapter I), sets a more general limit to the maximum flow. By performing the maximum and quickest respiratory efforts through graded resistances* it has been shown that the maximum pressure exerted by the respiratory muscles, at a given lung volume,

* Either by adding external resistances or by changing the density of the gas and therefore also the intrathoracic flow-resistance.

decreases as the air* flow increases (Agostoni and Fenn, 1960; Schilder *et al.*, 1963; Hyatt and Flath, 1966). As the flow increases the rate of muscle shortening increases and hence the force exerted, measured in terms of pressure, decreases (Otis, 1954; Agostoni and Fenn, 1960). If the muscles could develop their full force independently of the speed of shortening, the pressure, and therefore the flow, should reach higher values than those actually recorded. This explanation, however, holds only if the muscles are maximally excited regardless of the resistance encountered: a proprioceptive feedback could in fact diminish the stimulus when the resistance is low. As is the case with the limbs (Fenn and Marsh, 1935; Wilkie, 1950), this does not occur in the respiratory system because the iso-volume pressure-flow relationship obtained in cats during tetanic stimulation of the phrenic nerves (i.e. in absence of proprioceptive feedback) shows the same features as those obtained in man during voluntary efforts (Mognoni *et al.*, 1968). As in man, when no resistance is added the pressure is about half that obtained in isometric condition. It seems therefore that the maximum air flow, like the maximum velocity of limbs movements, is limited by the rate with which the muscles are able to transform chemical into mechanical energy (Agostoni and Fenn, 1960).

**Iso-volume Flow-pressure Relationships**

Maximum iso-volume $\dot{V}$-$P$ relationships at FRC are shown by the lines A-E and D-F in Fig. 37. These lines are obtained by making maximum inspiratory and expiratory efforts through different resistances. Points B and C indicate conditions in the absence of added resistance, between B-A and C-D external resistances are added up to a complete closure of the airways (A and D). Points E and F are obtained while breathing 20 per cent $O_2$ in He, i.e. by lowering airways resistance (Agostoni and Fenn, 1960; Hyatt and Flath, 1966). The maximal iso-volume $\dot{V}$-$P$ relationships are almost straight except for the expiratory lines at large lung volume which are concave upwards (Hyatt and Flath, 1966).

The line B-C indicates the normal iso-volume $\dot{V}$-$P$ relationship. On the expiratory side the flow reaches a maximum before the line intersects the maximum $\dot{V}$-$P$ relationship and then levels off or may even decrease (Hyatt *et al.*, 1958; Fry and Hyatt, 1960; Hyatt and Flath, 1966; Mead *et al.*, 1967; Pride *et al.*, 1967). This further limit to flow is due to the collapse of a segment of the intrathoracic airways (Einthoven, 1892; Dayman, 1951; Campbell *et al.*, 1957; Fry, 1958; Hyatt *et al.*, 1958; Fry and Hyatt, 1960; Macklem and Wilson, 1965; Mead *et al.*, 1967; Pride *et al.*, 1967; Macklem and Mead, 1968). When the maximum expiratory flow,

* Because of the compressibility of the air, the pressure exerted by the respiratory muscles should be related to the rate of change of the chest wall volume (or chest wall flow) and not to the air flow at the airways opening. The difference however is immaterial for the purpose of this analysis (Hyatt and Flath, 1966) and the term air flow has been therefore used for simplicity.

for a given volume, is reached, the pressure at the inner wall of the intrathoracic airway is greater than that at the outer surface, at the alveolar end, but it is less at the tracheal end. It follows that at some intermediate point along the airways the transmural pressure is nil. The intrathoracic airways may then be divided into two segments: one running from the alveoli to the points where the pressure across the wall of the airways is nil (equal pressure points), and the other downstream (Macklem and Wilson, 1965; Mead *et al.*, 1967). Since the two segments are in series, the flow is known if the driving pressure and the resistance of either of them is known. In the downstream segment the driving pressure is pleural pressure and the resistance is variable because of the dynamic compression of the airways: the pressure and the resistance are effort dependent. In the upstream segment, the driving pressure is the static recoil of the lung, which, like the resistance, is effort independent and constant for a given volume. Under these conditions the maximum expiratory flow is no longer effort dependent, but is directly related to the static recoil of the lung, and inversely related to the flow-resistance of the upstream segment (Mead *et al.*, 1967). In other words flow depends upon the difference between alveolar pressure and the pressure surrounding the airways and not upon the total pressure drop between the alveoli and the atmosphere. This condition is analogous to that of a Starling resistor or of a waterfall; actually, Pride *et al.* (1967) considered this as the mechanism limiting flow, and data of Macklem and

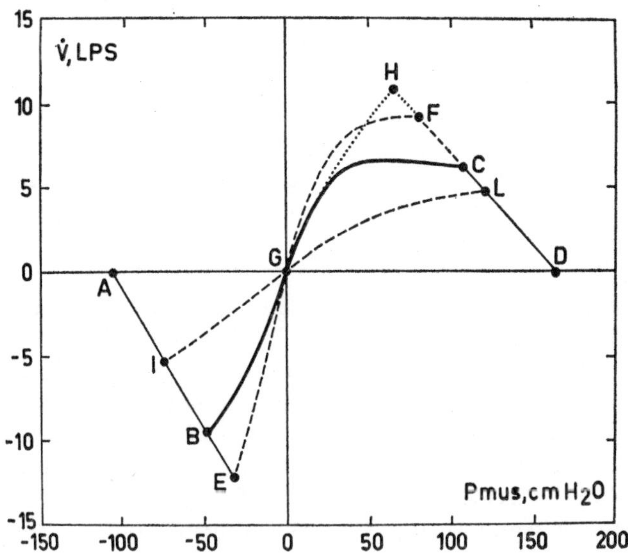


FIG. 37.—FLOW-PRESSURE CURVES AT MID-LUNG VOLUME SUMMARISING THE FACTORS LIMITING MAXIMUM FLOW

For explanation see text. (Modelled after Fenn, 1963; Mead and Agostoni, 1964 and Hyatt and Flath, 1966.)

Mead (1968) support this view. Flow would be effort dependent until the transmural pressure at some point along the airways reaches a critical value and forms a constricted orifice that limits flow, independently of the conditions downstream. Then the flow becomes effort independent, being directly related to the difference between the static recoil of the lung and the critical transmural pressure, and inversely related to the resistance of the segment upstream to the orifice (Pride *et al.*, 1967). This latter is probably a few cm downstream to the equal pressure points (Macklem and Mead, 1968). Changes in the upstream segment account for most of the limitation of the expiratory flow seen in disease (Mead *et al.*, 1967).

At mid-lung volume the equal pressure points are just beyond (on the alveolar side) the lobar bronchi; hence the dynamic compression involves the lobar and mainstem bronchi and the intrathoracic trachea (Macklem and Wilson, 1965). The equal pressure points move upstream (towards the alveoli) as the flow increases but when the maximal expiratory flow is reached they become fixed despite further increases in pleural surface pressure. The equal pressure points develop in the intrathoracic trachea at about 75 per cent of the VC and move upstream as the lung volume decreases until near the RV, where they move again downstream and disappear (Macklem and Wilson, 1965; Mead *et al.*, 1967). Hence the maximum expiratory flow is effort dependent only at lung volumes above 75 per cent VC and close to RV. At large volumes in fact the recoil of the lung is great enough to prevent the pleural pressure becoming higher than the lateral pressure in the airways during an expiratory effort. At volumes close to RV, because of the rising opposition of the chest wall to further volume reduction and the decreasing muscle force, pleural pressure does not become greater than the lateral pressure in the airways, at least in young individuals (Mead *et al.*, 1967).

An increase of the expiratory pressure above the value at which the flow levels off is utilised only to increase the kinetic energy of the air stream, and hence to assist expectoration (Mead and Agostoni, 1964; Hyatt and Flath, 1966). Cough is only effective downstream from the equal pressure points: since these move upstream as the lung volume decreases a series of coughs without intervening inspirations would tend to clear progressively the smaller airways (Mead *et al.*, 1967). In the lower part of the lung of the upright man the equal pressure points can reach bronchi with a diameter of about 1 mm (Macklem and Mead, 1968).

The line E-F in Fig. 37 indicates the iso-volume $\dot{V}$-$P$ line obtained while breathing 20 per cent $O_2$ in He. The expiratory flow levels off again, but this occurs at a higher value of flow (Schilder *et al.*, 1963; Hyatt and Flath, 1966). Breathing a gas with a density of one-third that of air at sea level, such as a mixture of 20 per cent $O_2$ in He, the peak inspiratory flow is about 30 per cent higher than normal; on the other hand with a gas having a density 3 times greater than that of air at sea level the peak inspiratory flow is reduced to about 60 per cent (Marshall *et al.*, 1956; Maio and Fahri, 1967).

The line G-H indicates the expiratory iso-volume $\dot{V}$-$P$ relation that would obtain if there were no compression of the airways. This line and the line G-B on the inspiratory side are iso-volume $\dot{V}$-$P$ lines with nearly constant resistance but variable effort, whereas the maximal iso-volume $\dot{V}$-$P$ lines are constant effort but variable resistance lines. The line I-L is an iso-volume $\dot{V}$-$P$ line obtained after adding a high external resistance: the expiratory part intersects the maximum iso-volume $\dot{V}$-$P$ line while still rising, hence the flow is limited by the muscle properties and not by airway collapse.

### Volume-pressure and Volume-flow Relationships

The maximum static pressures exerted by the respiratory muscles are indicated by the outer broken lines of the upper diagram in Fig. 38 (see Fig. 30). They enclose the potential range of pressures available to move air in or out of the system; any point inside the boundary can represent a different pattern of muscular contraction because of the possibility of antagonist muscle activity. The sigmoid shaped line of the upper diagram, —*Pst* (*rs*), defines the muscle pressures necessary to balance the pressure of the relaxed respiratory system at each lung volume: this line defines therefore the condition of no flow. When muscle pressure is to the right of this line, alveolar pressure is above atmospheric and air moves out of the system; when muscle pressure is to the left, alveolar pressure is below atmospheric and air moves in. In the areas between the zero flow line (sigmoid line) and the ordinate at zero pressure, the pressure exerted by the passive structure exceeds muscle pressure and therefore the chest wall moves in an opposite direction to the contraction of the muscles: the contracting muscles are lengthened and perform negative work. In the remaining areas muscle pressure exceeds the pressure exerted by the passive structure and therefore the chest wall moves in the same direction as the contraction of the muscles: the contracting muscles shorten and perform positive work. In either case antagonist muscles may of course be active (Mead and Agostoni, 1964).

Maximum static pressures are never achieved under dynamic conditions, they may be approached only when breathing through resistances. In fact maximum pressure decreases as the velocity of muscle shortening, and therefore the air flow, increases (see p. 88). The maximum pressures exerted by the respiratory muscles during a forced VC are shown by the continuous lines of the upper diagram of Fig. 38. The rounding at the beginning of the manoeuvres reflects the time necessary to develop the force (see p. 88).

The lower diagram of Fig. 38 shows the corresponding flows. During the forced inspiratory VC the maximum flow is reached at mid-lung volume

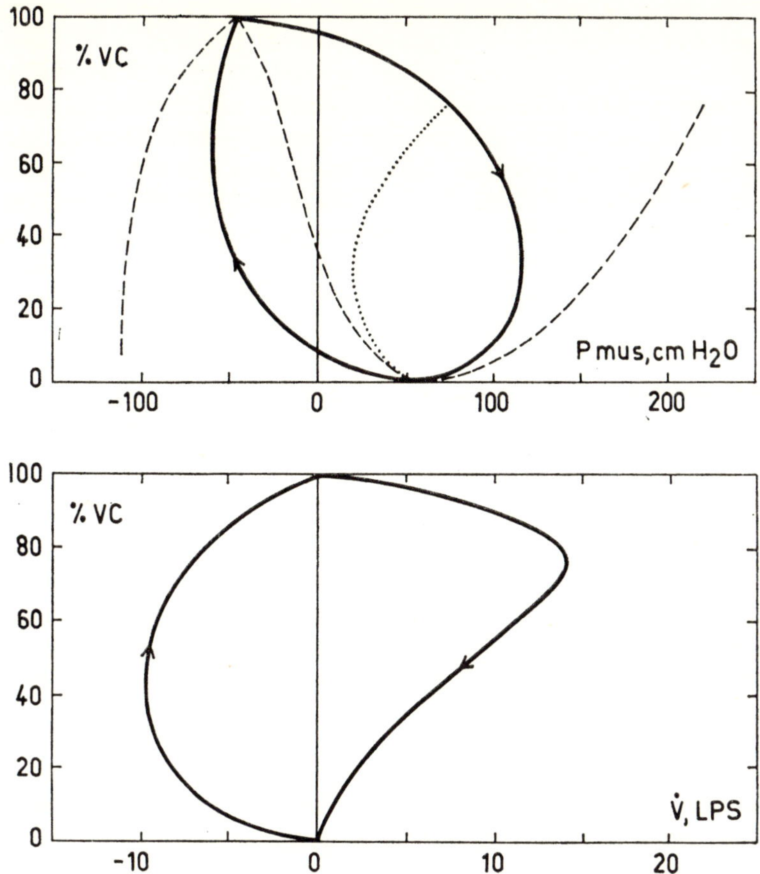


FIG. 38.—VOLUME-PRESSURE AND VOLUME-FLOW RELATIONSHIPS DURING FORCED VITAL CAPACITY

The outer broken lines refer to the maximum static effort (Fig. 30); the inner broken line indicates the muscle pressure necessary to balance the static pressure of the relaxed respiratory system (for further explanation see text). The continuous lines refer to the forced vital capacity and indicate therefore the maximum dynamic pressure exerted by the muscles or the maximum flow. *Pmus* indicates the resultant of the activity of agonist and antagonist muscles when they act simultaneously. The dotted line indicates approximately the limit beyond which a further increase of muscle pressure does not increase the expiratory flow. (Modified from Mead and Agostoni, 1964.)

where the pressure is about half the static one. This difference shows the great effect of muscle properties on the dynamics of breathing (see p. 89). During the forced expiratory VC the maximum flow is reached at about 80 per cent VC, where the pressure is still far from its maximum; the flow then decreases markedly, notwithstanding the increasing pressure, because of the compression of the intrathoracic airways (see p. 89). Higher flows can be reached, except at very low lung volume, by reducing the density of the gas breathed (see Fig. 37) (Schilder *et al.*, 1963; Hyatt and Flath,

1966). Because of the compression of the airways the flow-resistance increases and the velocity of muscle shortening decreases: for this reason the maximum dynamic pressures on expiration are a higher fraction of the corresponding static pressures than on inspiration. Furthermore during forced expiration antagonist muscles are generally active (see p. 68) and therefore the maximum pressure of the expiratory muscles is greater than indicated by *Pmus* in Fig. 38.

The dotted line in the upper diagram of Fig. 38 indicates approximately the limit beyond which any further increase of muscle pressure does not increase the expiratory flow because of the compression of the intrathoracic airways. The compression of the intrathoracic airways is directly related to pleural pressure, but, for consistency with the other values of pressure appearing in this diagram, these limit values have also been expressed as muscle pressure; they have been calculated, taking into account the corresponding static and dynamic pressure of the relaxed chest wall, from the limit values of pleural pressure at various lung volumes found in the literature (Schilder *et al.*, 1963; Macklem and Wilson, 1965; Hyatt and Flath, 1966; Mead *et al.*, 1967; Pride *et al.*, 1967; Olafsson and Hyatt, 1969).

If the expiratory vital capacity is started from a condition of maximum static expiratory effort the flow cannot be higher because, owing to the compression of the gas, the lung volume at the opening of the airways is already in the range where the flow is not effort dependent.

The area enclosed by the continuous lines in the upper diagram represents the maximum potential work of the respiratory muscles under normal conditions. When the system is loaded the maximum potential work increases approaching the area encompassed by the maximum static pressure lines.

The volume-pressure and volume-flow relations during maximum breathing capacity, i.e. the maximum voluntary ventilation that can be endured for 15 sec (Comroe *et al.*, 1962), are shown in Fig. 39 (continuous lines). During maximum breathing capacity the peak flows correspond to those attained during forced vital capacity at the same lung volume. During the expiratory phase the peak pressure is lower than that of the forced vital capacity at the same volume, but overshoots the value at which compression of the airways limits a further increase of the expiratory flow (dotted line). Maximum breathing capacity takes place in a volume range where the limit to maximum flow is set by compression of the airways; the subject perhaps tends to avoid the useless effort, but the lower peak pressure could also in some cases be due to a stronger antagonist activity of the diaphragm (Milic-Emili *et al.*, 1964).

The volume change of the respiratory system during a cycle of maximum breathing capacity (indicated by the height of the continuous loop) is appreciably greater than the volume of air displaced in or out of it (indi-

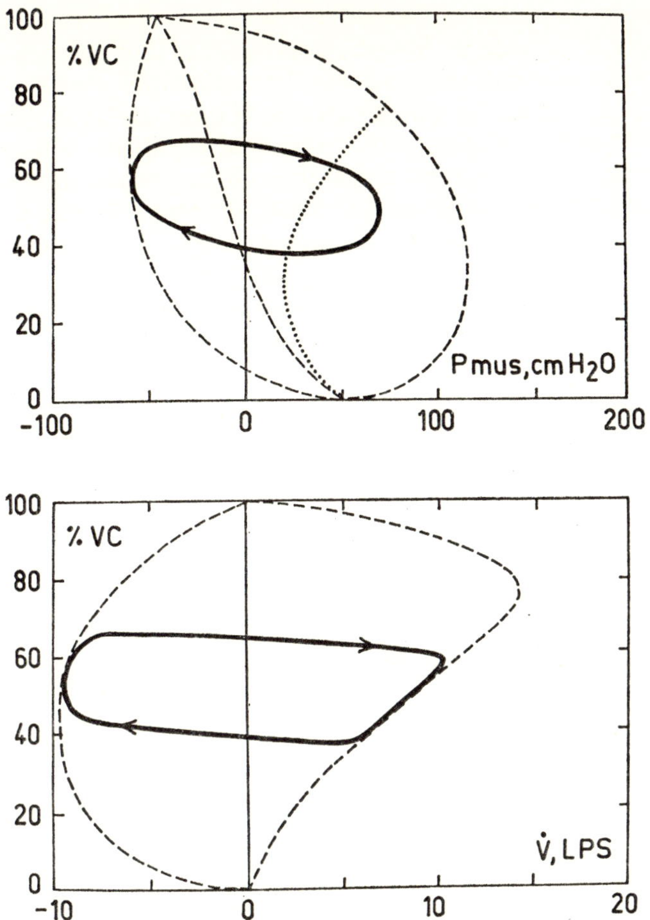


FIG. 39.—VOLUME-PRESSURE AND VOLUME-FLOW RELATIONSHIPS DURING MAXIMUM BREATHING CAPACITY

The broken and dotted lines are repeated from Fig. 38 for comparison. *Pmus* indicates the net pressure contributed by the muscles. (Modified from Mead and Agostoni, 1964.)

cated by the vertical distance between the intercepts of the loop with the sigmoid line) because of the compression and expansion of the gas (Jaeger and Otis, 1964; Milic-Emili *et al.*, 1964).

The lung volume at which the maximum breathing capacity takes place is the one yielding the largest ventilation when, as it happens, the period of inspiration equals that of expiration. Since in these circumstances the peak flow bears about the same relation to mean flow in both phases, peak flows are about the same in both phases (Mead and Agostoni, 1964).

During maximum breathing capacity the volume acceleration becomes high and therefore the pressure necessary to overcome inertia becomes relevant: in the case illustrated by Fig. 39 it corresponds to about 7 per cent of the total pressure swing (neglecting the inertia of the chest wall which has not been calculated in these diagrams) (Mead and Agostoni, 1964).

The maximum ventilation that can be endured for 4 min is 60–70 per cent of the maximum breathing capacity (Freedman, 1966; Tenney and Reese, 1968) and is similar to the maximum ventilation attained during muscular exercise. For periods longer than 10 min the maximum voluntary ventilation is about 50 per cent of MBC (Tenney and Reese, 1968, see Energetics chapter). According to Shephard (1967), however, 70–80 per cent of the MBC can be sustained for 15 min when the subject is exercising at 80 per cent of his aerobic work capacity. The MBC is inversely related to the square root of the gas density (Miles, 1966; see also Cotes, 1954; Wood, 1963 and Maio and Farhi, 1967).

## ANALYSIS OF SPONTANEOUS BREATHING

### Time Course of the Volume, Flow and Pressure

During spontaneous breathing at rest the tidal volume is about 10 per cent of the VC and the breathing frequency about 15 cycles/min. The

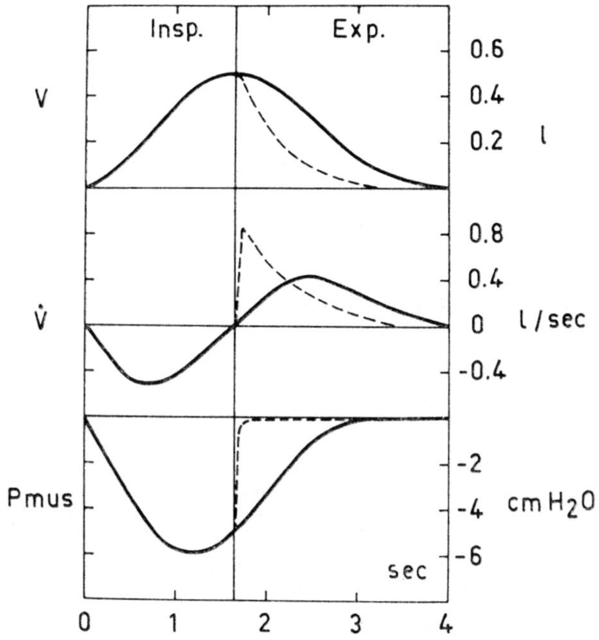


Fig. 40.—Changes of Volume, Flow and Muscle Pressure During a Spontaneous Breathing Act Through the Nose of an Adult Male at Rest

Breathing frequency 15 cycles/min, ventilation 7·5 litres (BTPS)/min. The broken lines indicate the pattern occurring if expiration were completely passive, i.e. not braked by the persistent activity of the inspiratory muscles. The data for pressure are obtained from the flow-pressure diagram of Fig. 35, and taking a value of 0·1 litre/cm $H_2O$ for the compliance of the respiratory system. (Modified from Mead and Agostoni, 1964.)

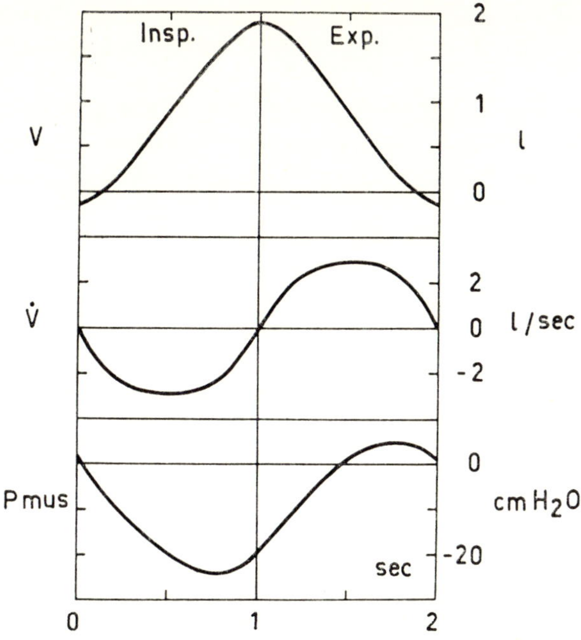


FIG. 41.—CHANGES OF VOLUME, FLOW AND MUSCLE PRESSURE DURING A SPONTANEOUS BREATHING ACT THROUGH THE MOUTH OF AN ADULT MALE PERFORMING AN EXERCISE OF MEDIUM INTENSITY.

Breathing frequency 30 cycles/min, ventilation 60 litres/min (BTPS). The data for pressure are obtained as in Fig. 40: at the large lung volume reached by this breath the flow-resistance of the lung is only 3/4 of that given in Fig. 35, but this is compensated by the decrease of the lung compliance. (Modified from Mead and Agostoni, 1964.)

changes of lung volume, the flow and the pressure contributed by the muscles during a spontaneous breathing cycle through the nose are shown in Fig. 40. Inspiration takes less than half cycle, the flow pattern is dome shaped, the rise being slightly more steep than the fall. The peak pressure contributed by the inspiratory muscles is attained between the peak flow and the peak volume change. During expiration the peak flow is reached approximately at one-third of the phase. It would occur very early in expiration if the expiration were completely passive (broken line), the inertia of the system being so small that little time would be spent in acceleration. If the expiration were completely passive the decrease of the volume and flow would be exponential (broken lines, see p. 100). The pattern of the volume and flow during spontaneous expiration indicates therefore that the activity of the inspiratory muscles subsides slowly during expiration. This is shown both by the pressure contributed (bottom diagram of Fig. 40) and by electromyographic findings (Green and Howell, 1959; Agostoni *et al.*, 1960; Petit *et al.*, 1960; Lourenço and Mueller, 1967). This activity opposes the recoil of the passive structures, thus smoothing the change of volume. The inspiratory muscles are lengthened during this braking process, i.e. they do negative work. The expiratory muscles do not intervene because the recoil developed during inspiration is sufficient to complete expiration (see below).

During exercise of medium intensity (Fig. 41), when the ventilation is increased 8-fold and the breathing frequency 2-fold from the resting values, the volume changes become more symmetrical in the two phases. The end-expiratory lung volume is slightly below the resting level provided that posture is not markedly altered. The activity of the inspiratory muscles persists during expiration but for a shorter period, and during the last part of expiration the expiratory muscles intervene as indicated by the positive values of *Pmus*. In fact the recoil developed during inspiration is no longer sufficient to provide the expiratory flow required by this degree of ventilation. Meanwhile the flow-resistance is about halved because mouth breathing has spontaneously replaced nose breathing.

During maximum exercise ventilation the ventilation increases about 16-fold and the breathing frequency about 3-fold from the resting levels. The tidal volume is about 50 per cent of the VC and the end-expiratory lung volume is below the resting level by about 5–15 per cent of the VC (Asmussen and Christensen, 1939; Mead and Agostoni, 1964; Agostoni and Torri, 1967). Inspiratory and expiratory muscles act simultaneously during an appreciable period at the end of expiration and probably also at the end of inspiration. The study of their separate contribution requires a more complicated analysis that will be dealt with below (see p. 102).

The tidal volume for a given value of ventilation may change according to the type of exercise. During exercise on the treadmill or by bicycle ergometer as well as during the recovery, the tidal volume in general increases linearly with the ventilation up to 60–70 l/min (Milic-Emili and Cajani, 1957; Hey *et al.*, 1966). From these values of ventilation up to the maximum exercise ventilation the tidal volume remains nearly constant (Milic-Emili and Cajani, 1957); when the ventilation is further increased voluntarily the tidal volume decreases progressively.

## Volume-pressure and Volume-flow Relationships

The volume pressure and volume-flow relationships during spontaneous breathing at rest, during maximum aerobic exercise (i.e. the maximum exercise that can be sustained for a long period), and during maximum exercise are shown in Fig. 42 (continuous lines). During maximum exercise the tidal volume spans the steepest portion of the static volume-pressure curve of the relaxed respiratory system. The peak inspiratory pressure is less than that occurring during maximum breathing capacity, the peak expiratory pressure is markedly less and never exceeds the flow-limiting pressure (Olafsson and Hyatt, 1969). It would be interesting to know the mechanism by which the action of expiratory muscles is regulated so efficiently. It is probably part of the more complex mechanism that regulates the ventilatory pattern in such a way as to require the minimum cost (see Chapter V).

The region between the resting and the maximum exercise ventilation represents the performance range during spontaneous breathing. The long-

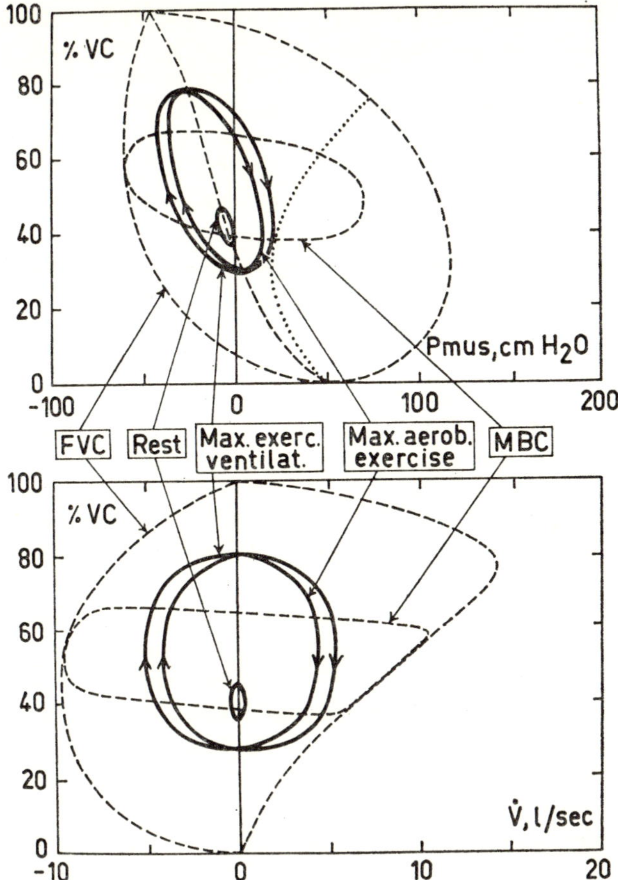


FIG. 42.—VOLUME-PRESSURE AND VOLUME-FLOW RELATIONSHIPS DURING SPONTANEOUS BREATHING AT REST, DURING MAXIMUM AEROBIC EXERCISE AND DURING MAXIMUM EXERCISE VENTILATION

The broken and dotted lines are repeated from Fig. 39 for comparison. *Pmus* indicates the net pressure contributed by the muscles. (Modified from Mead and Agostoni, 1964; see also Fenn, 1951.)

term reserve of the breathing pump corresponds to the region between the resting loop and the maximum aerobic exercise loop. It corresponds to about a 14-fold increase of ventilation and should be limited by the oxygen supply to the respiratory muscles. The short-term reserve is the difference between MBC and maximum aerobic exercise: i.e. about a 2-fold increase of the latter or a 28-fold increase of the resting ventilation. The short-term reserve may also be visualised as the region between the maximum aerobic exercise and the forced vital capacity loops.

### Volume-pressure Relationship in terms of Pleural Pressure

For a direct analysis of the static and dynamic pressure contributed by the various parts of the respiratory system it is convenient to think in terms of pleural pressure (see Chapter III for the simplification required) and hence to represent the volume-pressure curve of the lung with negative sign. This representation, introduced by Heaf and Prime (1954) and developed by Campbell in the first edition of this monograph, is illustrated within the tidal volume range by Fig. 43. When the lung volume is held constant with open airways the mirror image of the static volume-pressure curve of the lung is obtained, i.e. curve *—Pst* (*l*); when the airways are closed and the muscles relaxed the static volume-pressure curve of the relaxed chest wall is obtained, i.e. curve *Pst* (*w*) (see Chapter III). The horizontal distance between *—Pst* (*l*) and *Pst* (*w*) indicates the pressure that the respiratory muscles must contribute to keep the respiratory system at a given volume with open airways, it corresponds therefore to the pressure given by the static volume-pressure curve of the respiratory system during relaxation. The pressure is contributed by the inspiratory muscles when *—Pst* (*l*) is to the left of *Pst* (*w*) and by the expiratory muscles when it is to the right, i.e. below the resting volume of the respiratory system where the two curves intersect.

So far the static conditions have been considered. When the subject breathes some pressure must be added to overcome the flow-resistance. The heavy continuous line of Fig. 43 is the pathway of pleural pressure under dynamic conditions, hence the horizontal distance between it and *—Pst* (*l*) gives the pressure necessary to overcome the flow-resistance of the lung (gas and tissue). The pressure necessary to overcome the flow resistance of the chest wall is indicated by the distance between the dotted line and *Pst* (*w*). The dotted line cannot be obtained directly in the spontaneously breathing subject, but only in a paralysed subject during artificial respiration. The dotted line in Fig. 43 has been calculated from the values of flow after measuring the flow-resistance of the chest wall (see p. 81). The distance between the heavy continuous line and the dotted one gives the pressure exerted by the muscles in dynamic conditions. The pressure is contributed by the inspiratory muscles when the continuous line is to the left of the dotted line and by the expiratory ones when it is to the right. During the expiration illustrated this distance reflects the persistent activity of the inspiratory muscles serving to brake the recoil of the lung and thus provide a smooth transition from inspiration to expiration. If expiration were passive from its onset and if chest wall flow-resistance were equal to that of the lungs, expiration would proceed down a line midway between the

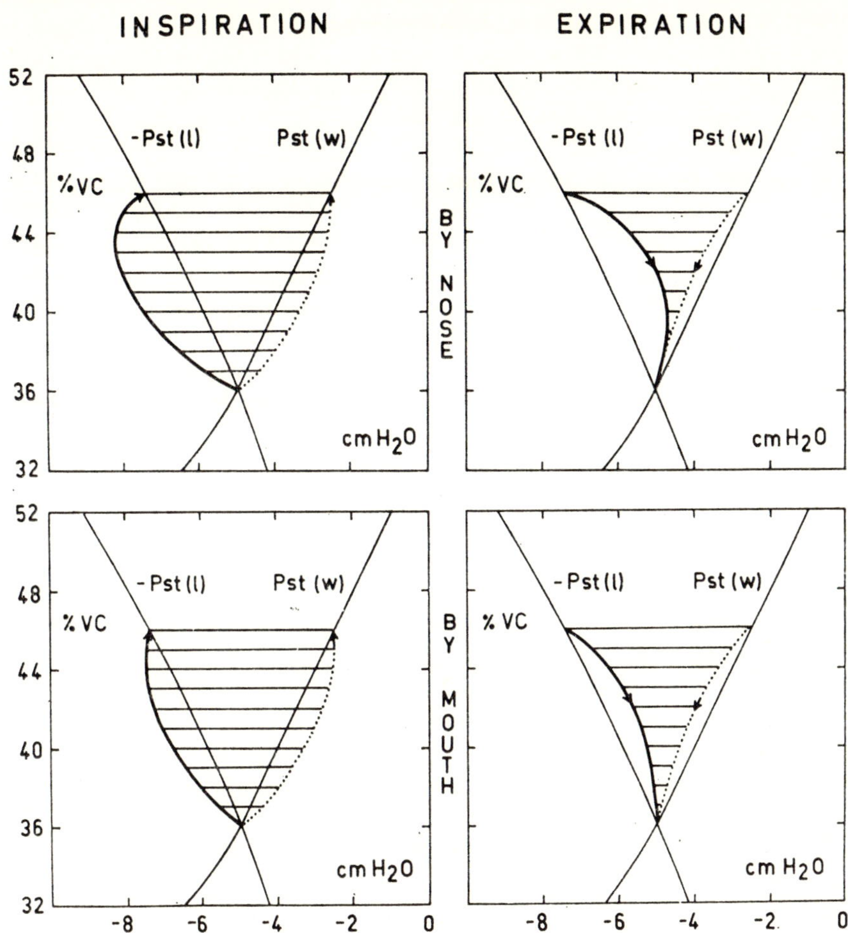


FIG. 43.—VOLUME-PRESSURE DIAGRAMS IN TERMS OF PLEURAL PRESSURE

The upper diagrams refer to the quiet breathing act through the nose illustrated in Fig. 40. the lower diagrams illustrate the pattern that would occur if the same volume events of Fig. 40 were produced during mouth breathing. The curve —Pst(1) indicates the static pressure exerted by the lung on the pleural surface, the curve Pst(w) indicates the static pressure exerted by the relaxed chest wall. The heavy continuous line is the pathway of pleural pressure under dynamic conditions: the horizontal distance between it and —Pst(l) gives the pressure overcoming the flow resistance of the lung. The pressure overcoming the flow resistance of the chest wall is given by the horizontal distance between the dotted line and Pst(w). The heavy continuous and dotted lines are obtained from the volume events of Fig. 40 and the flow-pressure relationships of Fig. 35. The pressure contributed by the inspiratory muscles is given by the horizontal distance between the heavy continuous line and the dotted line. The work done by the inspiratory muscles is given by the hatched area: it is positive during inspiration and negative during expiration. For further explanation see the text.

two static curves, $-Pst\ (l)$ and $Pst\ (w)$. The work done by the inspiratory muscles is given by the hatched area (see Chapter V).

If a subject performs a deep expiration below the resting level the activity of the expiratory muscles persists during the first part of the following inspiration braking the outward recoil of the chest wall (Agostoni, 1961; see also Fig. 63).

It appears from the right upper diagram of Fig. 43 that during nose breathing an increase of ventilation of only 2–3 fold should move the continuous line appreciably to the right of the dotted line, i.e. it should involve the intervention of the expiratory muscles in the last part of expiration. As the shift from nose to mouth breathing seems to occur at a ventilation of about 35 litres/min (Saibene and Mognoni, personal communication), the above consideration implies that during spontaneous breathing the expiratory muscles intervene for relatively small increase of ventilation.

On the other hand, during mouth breathing, as shown by the right lower diagram of Fig. 43, the intervention of the expiratory muscles should occur at a higher value of ventilation. While Campbell (1952, 1955) did not find an electrical activity of the expiratory muscles (abdominal and intercostals) up to 40–50 l/min, Taylor (1960) did find electrical activity of the internal intercostal for lower values of ventilation (see Chapter VII). The volume-pressure diagram of Fig. 44 analyses a breathing cycle at a ventilation of 60 l/min (see Fig. 41) which involves a clear contribution of the expiratory muscles in the last part of the expiration (hatched area).

These diagrams give only the net pressure, or work, contributed by the respiratory muscles, hence if inspiratory and expiratory muscles act simultaneously, the contribution of the antagonist muscles and the corresponding portion of the agonist ones escape this analysis.

## Abdominal Pressure and Diameter

The measurements of the abdominal pressure (see Chapter III) enable the contribution of antagonist muscles to be separately assessed even when they act simultaneously (Agostoni and Rahn, 1960), and provide some basic information for the analysis of the relative motion of the rib cage and of the abdomen-diaphragm (Agostoni, 1961; Mead and Agostoni, 1964).

The pathway of the pressure on the abdominal side of the diaphragm during a spontaneous breathing cycle is shown in Fig. 45. The horizontal distance between the heavy continuous and broken lines corresponds to the transdiaphragmatic pressure. At the beginning of inspiration it is smaller than the pressure contributed by the inspiratory muscles as given by transthoracic pressure measurements; thereafter it becomes progressively larger. This behaviour of transdiaphragmatic pressure is mainly conditioned by the vertical movement of the rib cage (see Chapter II) and is therefore

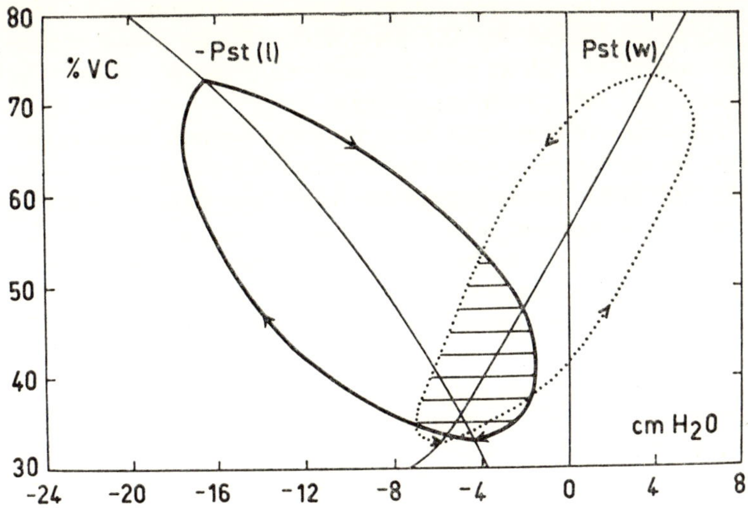


FIG. 44.—VOLUME-PRESSURE DIAGRAM IN TERMS OF PLEURAL PRESSURE

It refers to the breathing act illustrated by Fig. 41. The heavy continuous line is the pathway of the pleural pressure under dynamic conditions: the horizontal distance between it and —Pst(1) gives the pressure overcoming the flow-resistance of the lung. The pressure overcoming the flow-resistance of the chest wall is given by the horizontal distance between the dotted line and Pst(w). The heavy continuous and dotted lines are obtained from the volume events of Fig. 41 and the flow-pressure relationship of Fig. 35, taking into account the decrease of lung flow-resistance with increase of lung volume (Briscoe and DuBois, 1958). The net pressure contributed by the expiratory muscles is given by the horizontal distance between the heavy continuous line and the dotted line when the former is to the right of the latter. The work performed by the expiratory muscles is given by the hatched area.

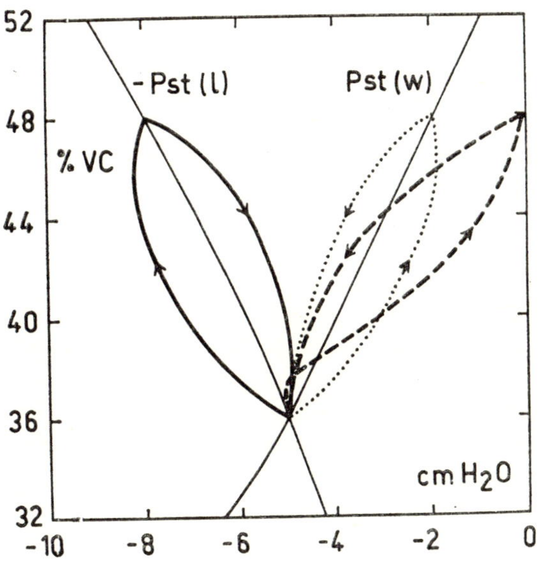


FIG. 45.—VOLUME-PRESSURE DIAGRAM IN TERMS OF PLEURAL PRESSURE INCLUDING THE PRESSURE ON THE ABDOMINAL SIDE OF THE DIAPHRAGM (BROKEN LINE)

Breathing cycle through the mouth. The horizontal distance between the heavy continuous and broken line gives the transdiaphragmatic pressure. (Modified from Agostoni, 1961.)

rather variable. The braking action of the diaphragm in the first part of expiration is shown by the persistence of the transdiaphragmatic pressure, which goes along with the persistence of the electrical activity (Agostoni *et al.*, 1960; Petit *et al.*, 1960).

At high values of ventilation the transdiaphragmatic pressure decreases more quickly during the beginning of expiration but reappears towards the end of it. This latter occurs because the diaphragm starts its activity before the expiration is over and, when the end-expiratory lung volume is below resting level, also because the diaphragm is passively distended. During maximum voluntary ventilation the transdiaphragmatic pressure has been observed to persist throughout the whole expiration (Milic-Emili *et al.*, 1964).

The abdominal pressure during relaxation corresponds to *Pst* (*w*) above the resting volume of the respiratory system, but it is progressively higher at lower lung volume because of the distension of the diaphragm (see Chapter III). The comparison of the abdominal pressure during activity of the respiratory muscles and during relaxation at the same lung volume cannot, however, provide information on the contribution of the abdominal muscles owing to the different volume partitioning between rib cage and abdomen occurring in the two conditions (see Chapter II). In order to interpret, at least qualitatively, the abdominal pressure change it is necessary to compare simultaneous measurements of the changes of the dorso-ventral diameter of the abdomen, of the abdominal pressure and of the lung volume (Agostoni and Torri, 1967). In fact the abdominal diameter and pressure depend essentially upon three factors: (1) the degree of extension of the vertebral column, (2) the position of the diaphragm relative to its insertions (hence the lung volume), (3) the degree of contraction of the abdominal muscles. The first two factors affect the abdominal diameter and pressure in the same direction, whereas the third factor affects the two variables in opposite direction (except during static efforts).

When the ventilation is high (above 100 l/min), the abdominal pressure at the end of inspiration is equal or higher than that at the same lung volume during relaxation, whereas the abdominal diameter is markedly lower (Figs. 46 and 18). This indicates that at high values of ventilation the abdominal muscles contract before the end of inspiration; in fact owing to the decrease of the abdominal diameter, caused by the extension of the vertebral column, the abdominal pressure should be lower and not higher than that during relaxation at the same lung volume if the abdominal muscles were relaxed (Agostoni and Torri, 1967).

Furthermore, at the beginning of inspiration the abdominal pressure decreases in spite of the increase of the abdominal diameter (Fig. 46). This indicates that at the beginning of inspiration when the ventilation is high the abdomen drives the diaphragm: i.e. the gravitational and elastic forces of the abdomen drive it down and forwards more quickly than the action of the diaphragm does. Hence part of the gravitational and elastic work done on the abdomen during expiration is utilised during the next inspiration (Agostoni and Torri, 1967).

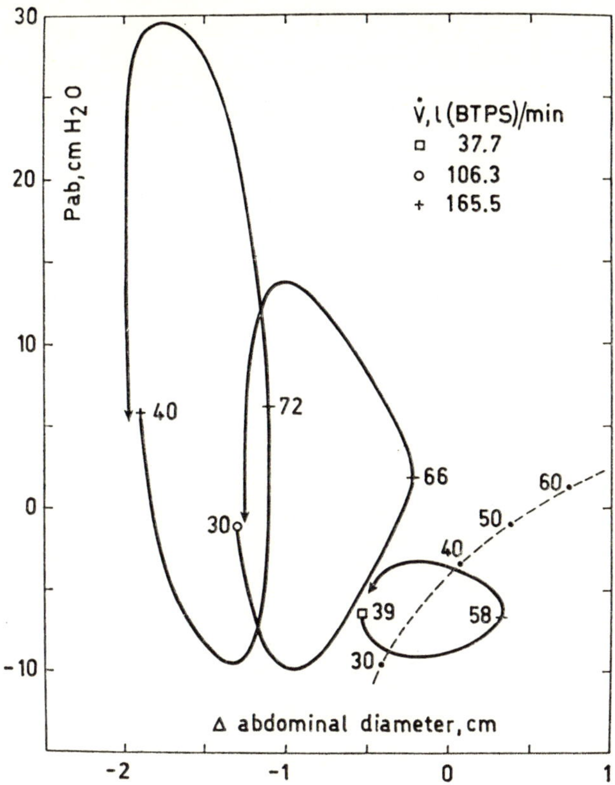


Fig. 46.—Abdominal Pressure Against Change of the Dorso-ventral Diameter of the Upper Abdomen at Various Values of Ventilation

Zero on the abscissa refers to the resting end-expiratory value. The square, the circle and the plus correspond to the beginning of inspiration, while the horizontal bars indicate the end of inspiration. The broken line indicates the relationship during relaxation. The figures along the lines indicate the corresponding lung volume expressed as per cent of the VC. (Modified from Agostoni and Torri, 1967.)

## PHONATION

The breathing pattern during phonation consists of rapid inspirations and prolonged expirations. During the expiration the vocal cords are drawn together by the adductor muscles: the subglottic pressure pushes them apart, while their elastic recoil and the decrease of the lateral pressure due to the increase of kinetic pressure (Bernouilli principle) close them again, thus generating a periodic flow. This produces longitudinal vibrations of the air above the glottis at the frequency of the vocal cord move-

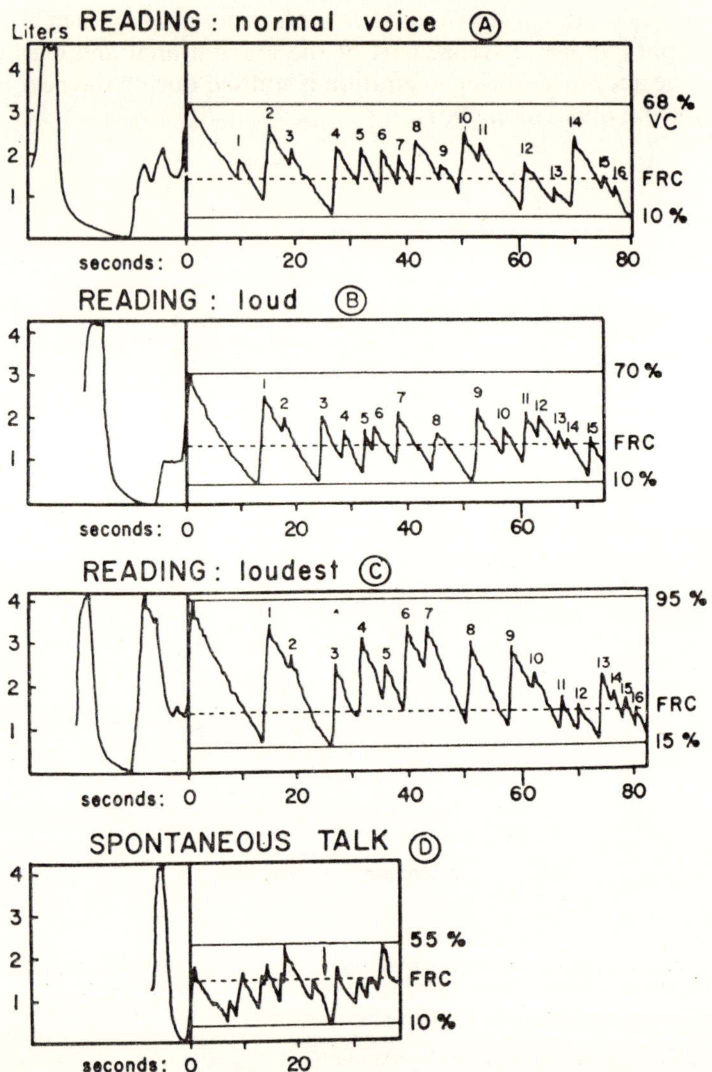


FIG. 47.—VOLUME EVENTS DURING READING OF A WRITTEN ENGLISH TEXT

A: at normal voice; B: at increased loudness, and C: at maximum loudness. D: spontaneous talking, at arrow the subject laughed. (From Bouhuys *et al.*, 1966.)

ments. This frequency is the fundamental tone of the voice. The portion of the cycle through which the glottis is closed decreases as the pitch increases: with a falsetto note a complete closure does not occur.

To produce a tone of constant loudness and pitch the subglottic pressure must remain constant; for a louder tone of the same pitch the subglottic pressure must increase, while the vocal cord tension must decrease in order to keep pitch constant. Phonation requires therefore a fine coordination between the laryngeal and the chest wall muscles (Draper *et al.*, 1959; Cavagna and Margaria, 1965; Bouhuys *et al.*, 1966). A tone of constant pitch and loudness can be produced with different airflows. The minimum airflow necessary to maintain a tone increases with increasing loudness, but it increases less at small than at great loudness (Cavagna and Margaria, 1968). Phonation tends to increase pulmonary ventilation at rest, but to decrease it during muscular exercise. During the hyperpnoea of exercise the intensity of the voice is increased (Otis and Clark, 1968).

During talking (Fig. 47) and singing, up to 90 per cent of the vital capacity may be used without conscious effort. This large volume range is

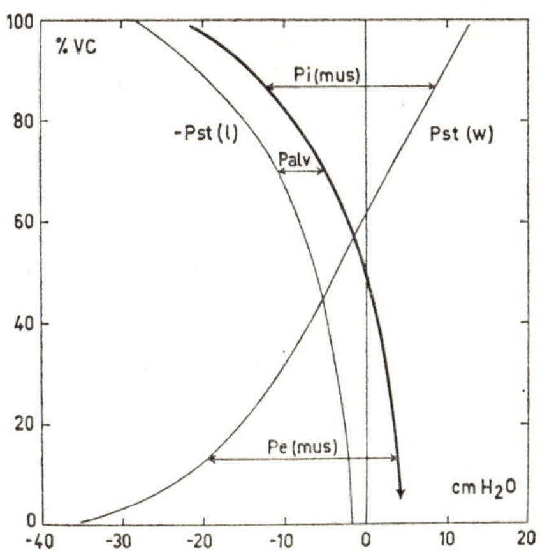


Fig. 48.—Pathway of Pleural Pressure (heavy line) During Singing of a Soft Tone Through Almost the Vital Capacity Range

As the flow-resistance during phonation is essentially given by the glottis, the horizontal distance between the heavy line and the —Pst(1) line gives the alveolar pressure and therefore also the subglottic pressure (Modified from Bouhuys *et al.*, 1966).

possible because small alveolar pressures are required during phonation, whereas during high values of ventilation or wind instrument playing large pressure changes are required (Bouhuys *et al.*, 1966). Nevertheless this feature raises a problem from the point of view of respiratory control. Volume events during singing do not show a clear difference between trained and untrained singers (Bouhuys *et al.*, 1966). Cavagna and Margaria (1965) found that during the production of a soft constant tone the flow decreases with the lung volume; this indicates that not all the flow is utilised to produce the sound at large and mid-lung volumes. This phenomenon, however, does not occur in all subjects according to Bouhuys and his colleagues (1966) (see also Fig. 63).

The expiratory flow is so low during phonation that the resistance to the expiratory flow is essentially given by the glottis: hence the subglottic pressure may be considered equal to the alveolar pressure and calculated from oesophageal pressure measurements if the static volume-pressure relationship of the lung is known (Bouhuys *et al.*, 1966). For a tone of given pitch and loudness the subglottic pressure is essentially constant at different lung volumes as shown by Fig. 48. The heavy line indicates schematically the pleural pressure during singing of a soft tone through almost the full VC range. The horizontal distance between the heavy and the —*Pst* (*l*) line gives the alveolar pressure, and hence the subglottic pressure. For a tone of greater loudness or pitch the heavy line would shift to the right, being still parallel to —*Pst* (*l*). The net pressure contributed by the muscles (*Pmus*) is given by the horizontal distance between the heavy line (pleural pressure) and the *Pst* (*w*) line. When the heavy line is to the left of *Pst* (*w*) the pressure is inspiratory, when it is to the right it is expiratory. It is apparent that a continously changing degree of muscle activity is required to keep an almost steady subglottic pressure as the lung volume diminishes during phonation (Bouhuys *et al.*, 1966).

When the tone is of low intensity and pitch, *Ppl* is smaller than the relaxation value and an inspiratory muscle pressure is recorded from the beginning of expiration to about 60 per cent of the VC. Simultaneous measurements of abdominal pressure have shown that the diaphragm, in the upright posture, contributes less than the inspiratory pressure recorded; it may even be completely relaxed. The low value of *Ppl* is obtained through a stronger contraction of the other inspiratory muscles, which increases the volume of the rib cage: the diaphragm, nearly or completely relaxed, is then sucked up and the abdominal pressure decreases until the hydraulic pull of the abdominal content upon the diaphragm balances *Ppl.* This mechanism cannot operate in the supine position (Bouhuys *et al.*, 1966). This finding of a prevailing action of the inspiratory muscles other than the diaphragm during phonation at large lung volume agrees with the observations of Draper *et al.* (1959) on the electrical activity of the diaphragm and intercostal muscles during speech (see also Chapter VII and Fig. 63) and with those of Konno and Mead (1967) on the dorso-ventral diameter of the rib cage and of the abdomen during singing and talking (see Chapter II). The almost complete lack of diaphragm activity during phonation is probably related to the paucity of proprioceptive control in this muscle (Bouhuys *et al.*, 1966) (see Chapter VI).

## PANTING

Many animals (dogs, cats, rabbits, sheep and cattle) perform a rapid shallow ventilation, called panting, when their body temperature tends to increase. The breathing frequency during panting reaches about 320 cycles/min in the dog (Crawford, 1962), 300 in the sheep (Hales and Webster, 1967), and 200 in the ox (Hales and Findlay, 1968). The tidal volume decreases so that alveolar ventilation remains almost constant, whereas dead space ventilation increases markedly. The ventilation of the dead space may become 90 per cent of total ventilation in the ox, being about 60 per cent under normal conditions (Hales and Findlay, 1968), and about 75 per cent in the sheep, being normally 55 per cent (Hales and Webster, 1967). During exposure to severe heat the breathing frequency reaches slightly lower values while the tidal volume increases so that the alveolar ventilation increases and the alveolar $P_{CO_2}$ decreases. The significance of this kind of panting is not clear: periods of deep panting may alternate with periods of rapid shallow panting (Hemingway, 1938; Albers, 1961; Hales and Webster, 1967; Hales and Findlay, 1968).

The frequencies during panting are probably close to the maximum that the neuromuscular system is able to reach (see p. 87). In the dog the frequency during panting corresponds to the natural frequency of the respiratory system, which in this animal is 5–6 cycles/sec (Crawford, 1962). At the natural or resonant frequency the reactive impedance, i.e. the algebraic sum of the impedance related to inertance and that related to the compliance, is nil. Hence at the resonant frequency the impedance to breathing is minimum, being given only by the resistive component. This implies that at this frequency high ventilation is obtained with minimum energy expenditure and therefore with minimum heat production, which is essential because the purpose of panting is that of increasing heat loss by increasing evaporation from the airways and the lung (see Chapter V).

## REFERENCES


AGOSTONI, E. (1961). A graphical analysis of thoracoabdominal mechanics during the breathing cycle. *J. appl. Physiol.*, **16**, 1055–1059.

AGOSTONI, E., and FENN, W. O. (1960). Velocity of muscle shortening as a limiting factor in respiratory air flow. *J. appl. Physiol.*, **15**, 349–353.

AGOSTONI, E., and MEAD, J. (1964). Statics of the respiratory system. In: *Handbook of Physiology*. Section 3, Respiration. Vol. 1, pp. 387–409. Eds. W. O. Fenn and H. Rahn. Washington, D.C.: American Physiological Soc.

AGOSTONI, E., and RAHN, H. (1960). Abdominal and thoracic pressures at different lung volumes. *J. appl. Physiol.*, **15**, 1087–1092.

AGOSTONI, E., SANT'AMBROGIO, G., and DEL PORTILLO CARRASCO, H. (1960). Electromyography of the diaphragm in man and transdiaphragmatic pressure. *J. appl. Physiol.*, **15**, 1093–1097.

AGOSTONI, E., and TORRI, G. (1967). An analysis of the chest wall motions at high values of ventilation. *Resp. Physiol.*, **3**, 318–332.

AINSWORTH, M., and EVELEIGH, J. W. (1952). A method of estimating lung airway resistance in humans. Ministry of Supply, Chem. Defence Exptl. Estab. Porton Tech. Paper 320.

ALBERS, C. (1961). Der Mechanismus des Wärmehechelns beim Hund. I. Die Ventilation und die arteriellen Blutgase während des Wärmehechelns. *Pflügers Arch. ges. Physiol.*, **274**, 125–147.

ASMUSSEN, E., and CHRISTENSEN, E. H. (1939). Die Mittelkapazität der Lungen bei erhöhtem $O_2$—Bedarf. *Skand. Arch. Physiol.*, **82**, 201–211.

BOUHUYS, A., PROCTOR, D. F., and MEAD, J. (1966). Kinetic aspects of singing. *J. appl. Physiol.*, **21**, 483–496.

BRISCOE, W. A., and DUBOIS, A. B. (1958). The relationship between airway resistance, airway conductance and lung volume in subjects of different age and body size. *J. clin. Invest.*, **37**, 1279–1285.

BRODY, A. W. (1954). Mechanical compliance and resistance of the lung-thorax calculated from the flow recorded during passive expiration. *Amer. J. Physiol.*, **178**, 189–196.

BRODY, A. W., DUBOIS, A. B., NISELL, O. I., and ENGELBERG, J. (1956). Natural frequency, damping factor and inertance of the chest-lung system in cats. *Amer. J. Physiol.*, **186**, 142–148.

BRODY, A. W., O'HALLORAN, P. S., WANDER, H. J., CONNOLLY, J. J., JR., ROLEY, E. E., and KOBOLD, E. (1960). Ventilatory mechanics and strength: long-term re-examinations and position change. *J. appl. Physiol.*, **15**, 561–566.

BUHLMANN, A. A. (1963). Respiratory resistance with hyperbaric gas mixtures. In: *Second Symposium on Underwater Physiology*, pp. 98–107. Eds. C. J. Lambertsen and L. J. Greenbaum, Jr. Washington, D.C.: Nat. Acad. Sci.-Nat. Res. Counc.

BUTLER, J. (1960). Work of breathing through the nose. *Clin. Sci.*, **19**, 55–62.

CAMPBELL, E. J. M. (1952). An electromyographic study of the role of the abdominal muscles in breathing. *J. Physiol.* (*Lond.*), **117**, 222–233.

CAMPBELL, E. J. M. (1955). An electromyographic examination of the role of the intercostal muscles in breathing in man. *J. Physiol.* (*Lond.*), **129**, 12–26.

CAMPBELL, E. J. M., MARTIN, H. B., and RILEY, R. L. (1957). Mechanisms of airway obstruction. *Bull. Johns Hopk. Hosp.*, **101**, 329–343.

CAVAGNA, G. A., and MARGARIA, R. (1965). An analysis of the mechanics of phonation. *J. appl. Physiol.*, **20**, 301–307.

CAVAGNA, G. A., and MARGARIA, R. (1968). Airflow rates and efficiency changes during phonation. *Ann. N.Y. Acad. Sci.*, **155**, 152–164.

COERMAN, R. R., ZIEGENRUECKER, G. H., WITTWER, A. L., and VON GIERKE, E. (1960). The passive dynamic mechanical properties of the human thorax-abdomen system and of the whole body system. *Aerospace Med.*, **31**, 443–455.

COMROE, J. H., JR., FORSTER, R. E., DUBOIS, A. B., BRISCOE, W. A., and CARLSEN, E. (1962). *The Lung*, p. 202. Chicago: Year Book.

COMROE, J. H., JR., NISELL, O. I., and NIMS, R. G. (1954). A simple method for concurrent measurement of compliance and resistance to breathing in anesthetized animals and man. *J. appl. Physiol.*, **7**, 225–228.

COTES, J. E. (1954). Ventilatory capacity at altitude and its relation to mask design. *Proc. roy. Soc. B*, **143**, 32–39.

CRAWFORD, E. C., JR. (1962). Mechanical aspects of panting in dogs. *J. appl. Physiol.*, **17**, 249–251.

DAYMAN, H. (1951). Mechanics of airflow in health and in emphysema. *J. clin. Invest.*, **30**, 1175–1190.

DRAPER, M. H., LADEFOGED, P., and WHITTERIDGE, D. (1959). Respiratory muscles in speech. *J. Speech Res.*, **2**, 16–27.

DUBOIS, A. B., BRODY, A. W., LEWIS, D. H., and BURGESS, B. F., JR. (1956). Oscillation mechanics of lungs and chest in man. *J. appl. Physiol.*, **8**, 587–594.

EINTHOVEN, W. (1892). Ueber die Wirkung der Bronchialmuskeln, nach einer neuen Methode untersucht, und über Asthma nervosum. *Pflügers Arch. ges. Physiol.*, **51**, 367–445.

FENN, W. O. (1938). The mechanics of muscular contraction in man. *J. appl. Physiol.*, **9**, 165–177.

FENN, W. O. (1951). Mechanics of respiration. *Amer. J. Med.*, **10**, 77–90.

FENN, W. O. (1963). A comparison of respiratory and skeletal muscles. In: *Perspectives in Biology*, Eds. C. F. Cori, V. G. Foglia, L. F. Leloir, and S. Ochoa. Amsterdam: Elsevier.

FENN, W. O., and MARSH, B. S. (1935). Muscular force at different speeds of shortening. *J. Physiol.* (*Lond.*), **85**, 277–297.

FERRIS, B. G., JR., MEAD, J., and OPIE, L. H. (1964). Partitioning of respiratory flow resistance in man. *J. appl. Physiol.*, **19**, 653–658.

FISHER, A. B., DUBOIS, A. B., and HYDE, R. W. (1968). Evaluation of the forced oscillation technique for the determination of resistance to breathing. *J. clin. Invest.*, **47**, 2045–2057.

FREEDMAN, S. (1966). Prolonged maximum voluntary ventilation. *J. Physiol.* (*Lond.*), **184**, 42P–44P.

FRY, D. L. (1958). Theoretical considerations of the bronchial pressure-flow-volume relationships with particular reference to the maximum expiratory flow-volume curves. *Phys. in Med. Biol.*, **3**, 174–194.

FRY, D. L., and HYATT, R. E. (1960). Pulmonary mechanics. A unified analysis of the relationship between pressure, volume, and gasflow in the lungs of normal and diseased human subjects. *Amer. J. Med.*, **29**, 672–689.

GOLDMAN, M., KNUDSON, R. J., MEAD, J., PETERSON, N., SCHWABER, J. R., and WOHL, M. E. (1970). A simplified measurement of respiratory resistance by forced oscillation. *J. appl. Physiol.*, **28**, 113–116.

GREEN, J. H., and HOWELL, J. B. L. (1959). Correlation of intercostal muscle activity with respiratory airflow in conscious human subjects. *J. Physiol.* (*Lond.*), **149**, 471–476.

GRIMBY, G., TAKISHIMA, T., GRAHAM, W., MACKLEM, P., and MEAD, J. (1968). Frequency dependence of flow resistance in patients with obstructive lung disease. *J. clin. Invest.*, **47**, 1455–1465.

HALES, J. R. S., and FINDLAY, J. D. (1968). Respiration of the ox: normal values and the effects of exposure to hot environments. *Resp. Physiol.*, **4**, 333–352.

HALES, J. R. S., and WEBSTER, M. E. D. (1967). Respiratory function during thermal tachypnoea in sheep. *J. Physiol.* (*Lond.*), **190**, 241–260.

HEAF, P. J. D., and PRIME, F. J. (1954). The mechanical aspects of artificial pneumothorax. *Lancet*, **2**, 468–470.

HEMINGWAY, A. (1938). The panting response of normal anaesthetized dogs to measured dosages of diathermy heat. *Amer. J. Physiol.*, **121**, 747–754.

HEY, E. N., LLOYD, B. B., CUNNINGHAM, D. J. C., JUKES, M. G. M., and BOLTON, D. P. G. (1966). Effects of various respiratory stimuli on the depth and frequency of breathing in man. *Resp. Physiol.*, **1**, 193–205.

HULL, W. E., and LONG, E. C. (1961). Respiratory impedance and volume flow at high frequency in dogs. *J. appl. Physiol.*, **16**, 439–443.

HYATT, R. E., and FLATH, R. E. (1966). Relationship of air flow to pressure during maximal respiratory effort in man. *J. appl. Physiol.*, **21**, 477–482.

HYATT, R. E., SCHILDER, D. P., and FRY, D. L. (1958). Relationship between maximum expiratory flow and degree of lung inflation. *J. appl. Physiol.*, **13**, 331–336.

HYATT, R. E., and WILCOX, R. E. (1961). Extrathoracic airway resistance in man. *J. appl. Physiol.*, **16**, 326–330.

JAEGER, M. J., and MATTHYS, H. (1968). The pattern of flow in the upper human airways. *Resp. Physiol.*, **6**, 113–127.

JAEGER, M. J., and OTIS, A. B. (1964). Effects of compressibility of alveolar gas on dynamics and work of breathing. *J. appl. Physiol.*, **19**, 83–91.

KONNO, K., and MEAD, J. (1967). Measurement of the separate volume changes of rib cage and abdomen during breathing. *J. appl. Physiol.*, **22**, 407–422.

LOURENÇO, R. V., and MUELLER, E. P. (1967). Quantification of electrical activity in the human diaphragm. *J. appl. Physiol.*, **22**, 598–600.

MACKLEM, P. T., and MEAD, J. (1968). Factors determining maximum expiratory flow in dogs. *J. appl. Physiol.*, **25**, 159–169.

MACKLEM, P. T., and WILSON, N. J. (1965). Measurement of intrabronchial pressure in man. *J. appl. Physiol.*, **20**, 653–663.

MAIO, D. A., and FARHI, L. E. (1967). Effect of gas density on mechanics of breathing. *J. appl. Physiol.*, **23**, 687–693.

MARSHALL, R., LANPHIER, E. H., and DUBOIS, A. B. (1956). Resistance to breathing in normal subjects during simulated dives. *J. appl. Physiol.*, **9**, 5–10.

MCILROY, M. B., TIERNEY, D. F., and NADEL, J. A. (1963). A new method for measurement of compliance and resistance of lungs and thorax. *J. appl. Physiol.*, **18**, 424–427.

MEAD, J. (1956). Measurement of inertia of the lungs at increased ambient pressure. *J. appl. Physiol.*, **9**, 208–212.

MEAD, J. (1960). Control of respiratory frequency. *J. appl. Physiol.*, **15**, 325–336.

MEAD, J., and AGOSTONI, E. (1964). Dynamics of breathing. In: *Handbook of Physiology*. Section 3, Respiration. Vol. 1, pp. 411–427. Eds. W. O. Fenn and H. Rahn. Washington, D.C.: American Physiological Soc.

MEAD, J., and MILIC-EMILI, J. (1964). Theory and methodology in respiratory mechanics with glossary of symbols. In: *Handbook of Physiology*. Section 3, Respiration. Vol. 1, pp. 363–376. Eds. W. O. Fenn and H. Rahn. Washington D.C.: American Physiological Soc.

MEAD, J., TURNER, J. M., MACKLEM, P. T., and LITTLE, J. B. (1967). Significance of the relationship between lung recoil and maximum expiratory flow. *J. appl. Physiol.*, **22**, 95–108.

MERTON, P. A. (1954). Voluntary strength and fatigue. *J. Physiol.* (*Lond.*), **123**, 553–564.

MILES, S. (1966). *Underwater Medicine*, 2nd edit., pp. 90–95. London: Staples.

MILIC-EMILI, J., and CAJANI, F. (1957). La frequenza dei respiri in funzione della ventilazione polmonare durante il ristoro. *Boll. Soc. ital. Biol. sper.*, **33**, 821–825.

MILIC-EMILI, J., ORZALESI, M. M., COOK, C. D., and TURNER, J. M. (1964). Respiratory thoraco-abdominal mechanics in man. *J. appl. Physiol.*, **19**, 217–223.

MOGNONI, P., SAIBENE, F., SANT'AMBROGIO, G., and AGOSTONI, E. (1968). Dynamics of the maximal contraction of the respiratory muscles. *Resp. Physiol.*, **4**, 193–202.

OLAFSSON, S., and HYATT, R. E. (1969). Ventilatory mechanics and expiratory flow limitation during exercise in normal subjects. *J. clin. Invest.*, **48**, 564–573.

OLSON, H. F. (1958). *Dynamical Analogies.* New York: Van Nostrand.

OTIS, A. B. (1954). The work of breathing. *Physiol. Rev.*, **34**, 449–458.

OTIS, A. B., and BEMBOWER, W. C. (1949). Effect of gas density on resistance to respiratory gas flow in man. *J. appl. Physiol.*, **2**, 300–306.

OTIS, A. B., and CLARK, R. G. (1968). Ventilatory implications of phonation and phonatory implications of ventilation. *Ann. N.Y. Acad. Sci.*, **155,** 122–128.

OTIS, A. B., FENN, W. O., and RAHN, H. (1950). Mechanics of breathing in man. *J. appl. Physiol.*, **2,** 592–607.

OTIS, A. B., and GUYATT, A. R. (1968). The maximal frequency of breathing of man at various tidal volumes. *Resp. Physiol.*, **5**, 118–129.

PEDLEY, T. J., SCHROTER, R. C., and SUDLOW, M. F. (1970). The prediction of pressure drop and variation of resistance within the human bronchial airways. *Resp. Physiol.* (in press).

PETIT, J. M., MILIC-EMILI, J., and DELHEZ, L. (1960). Role of the diaphragm in breathing in conscious normal man: an electromyographic study. *J. appl. Physiol.*, **15**, 1101–1106.

PRIDE, N. B., PERMUTT, S., RILEY, R. L., and BROMBERGER-BARNEA, B. (1967). Determinants of maximal expiratory flow from the lungs. *J. appl. Physiol.*, **23**, 646–662.

ROHRER, F. (1915). Der Strömungswiderstand in den menschlichen Atemwegen und der Einfluss der unregelmässigen Verzweigung des Bronchialsystems auf den Atmungsverlauf in verschiedenen Lungenbezirken. *Pflügers Arch. ges. Physiol.*, **162**, 225–299.

SCHILDER, D. P., ROBERTS, A., and FRY, D. L. (1963). Effect of gas density and viscosity on the maximal expiratory flow-volume relationship. *J. clin. Invest.*, **42**, 1705–1713.

SHARP, J. T., HENRY, J. P., SWEANY, S. K., MEADOWS, W. R., and PIETRAS, R. J. (1964*a*). The total work of breathing in normal and obese men. *J. clin. Invest.*, **43,** 728–739.

SHARP, J. T., HENRY, J. P., SWEANY, S. K., MEADOWS, W. R., and PIETRAS, R. J. (1964*b*). Total respiratory inertance and its gas and tissue components in normal and obese men. *J. clin. Invest.*, **43,** 503–509.

SHEPHARD, R. J. (1967). The maximum sustained voluntary ventilation in exercise. *J. clin. Invest.*, **32**, 167–176.

TAYLOR, A. (1960). The contribution of the intercostal muscles to the effort of respiration in man. *J. Physiol.* (*Lond.*), **151**, 390–402.

TENNEY, S. M., and REESE, R. E. (1968). The ability to sustain great breathing efforts. *Resp. Physiol.*, **5**, 187–201.

WILKIE, D. R. (1950). The relation between force and velocity in human muscle. *J. Physiol.* (*Lond.*), **110**, 249–280.

WOOD, W. B. (1963). Ventilatory dynamics under hyperbaric states. In: *Second Symposium on Underwater Physiology*, pp. 108–123. Eds. C. J. Lambertsen and L. J. Greenbaum, Jr. Washington, D.C.: Nat. Acad. Sci.-Nat. Res. Counc.

ZECHMAN, F. W., JR., PECK, D., and LUCE, E. (1965). Effect of vertical vibration on respiratory airflow and transpulmonary pressure. *J. appl. Physiol.*, **20**, 849–854.

# Chapter V

# ENERGETICS

E. AGOSTONI, E. J. M. CAMPBELL AND S. FREEDMAN*

## THE MECHANICAL WORK OF BREATHING

### Introduction

MECHANICAL work is performed when a force ($F$) moves its point of application through a distance ($L$): the work ($W$) is given by $W = FL$, or more generally, if the direction of the force and that of the displacement differ by an angle $\alpha$, $W = FL \cos \alpha$. In case of a fluid system the force is measured in terms of pressure difference and the displacement in terms of volume change. Hence mechanical work is done when a pressure ($P$) changes a volume ($V$), and the work is volume times pressure. In fact, since pressure is force per unit area, force may be replaced by pressure $\times$ area ($A$), and therefore: $W = PAL$, and, since area $\times$ length $=$ volume, $W = PV$.

When a muscle contracts mechanical work is performed only if displacement takes place: the work is positive if the muscle shortens (miometric contraction), i.e. if the muscle does work on something, whereas the work is negative if the muscle lengthens (pliometric contraction), i.e. if something does work on the muscle, the direction of the displacement therefore being opposite to the direction of the force exerted by the muscle. For an equal force exerted by the active muscle the metabolic cost is considerably less for the negative work than for the positive work (see Chapter I). When the muscle contracts without changing its length (isometric contraction) no mechanical work is performed, but there is of course a metabolic cost for exerting the force, depending upon the force-time integral (see Chapter I). In many instances muscle activity involves these different types of contraction and therefore the energy required is the sum of that related to the mechanical work plus that related to the maintenance of the isometric force.

When the respiratory muscles contract they may: (*a*) change the volume of the respiratory system, expanding or compressing the gas and displacing it in and out of the lung, (*b*) change the shape of the respiratory system relative to the shape at the same lung volume when the muscles are relaxed, (*c*) sustain a pressure in the lung against closed airways, (*d*) antagonise the forces exerted by the agonist muscles (i.e. the muscles that shorten) while the volume and/or the shape of the respiratory system is changing.

* Department of Medicine, Royal Postgraduate Medical School, London.

In case (*a*) the respiratory muscles perform mechanical work that may be represented as an area on the volume-pressure diagram of the respiratory system. In case (*b*) they perform mechanical work that is not accounted for by the volume-pressure diagram (part of it could theoretically be represented in a volume-pressure diagram in which the volume contributed by the parts of the respiratory system placed in parallel is partitioned, as shown in Chapter III). In case (*c*) no mechanical work is performed, and the energy required to sustain the isometric force is all dissipated as heat. In case (*d*) the agonist muscles are performing work not only on the passive structure as in cases (*a*) and (*b*) but also on the antagonist muscles, that are lengthened and perform negative work. As previously shown it is not possible to determine the pressure contributed by the antagonist muscles, except for the diaphragm and the muscles of the abdominal wall; furthermore, it is difficult to determine the volume displacement contributed by these muscles.

The main mechanical effect of respiratory muscle activity is that of expanding or compressing the gas in the lung and of producing an air flow if the airways are open; hence most of the work done by them can be represented on the volume-pressure diagram.

The forces exerted by the respiratory muscles are opposed by (1) elastic, gravitational and surface forces (see Chapter III), that for brevity may be classified as elastic forces; (2) flow-resistive forces (see Chapter IV); (3) inertial forces, which are negligible except at high values of ventilation (see Chapter IV).

## Graphical Analysis of the Mechanical Work

As described in Chapters III and IV the mechanical work previously defined under (*a*), which is most of the work of breathing, may be analysed by means of volume-pressure diagrams. The area encompassed by a volume-pressure loop has the dimension of mechanical work. For instance, the areas encompassed by the loops of the volume-pressure diagrams of Figs. 38, 39 and 42 (Chapter IV) represent the mechanical positive work done by the respiratory muscles during the corresponding manoeuvres. The area encompassed by the continuous loop of Fig. 38 (or the outer loops of Figs. 39 and 42) corresponds to the maximum mechanical positive work that can be done during a breath under normal conditions and that can be represented in a volume-pressure diagram. If flow-resistance is increased the maximum work that can be done during a breath increases and tends to approach the area encompassed by the outer lines of Fig. 38 (see Chapter IV), which represents the potential maximum work.

In order to analyse in detail the work necessary to overcome the resistance of the different parts of the respiratory system and the exchange of energy between them it is convenient to use the volume-pressure diagram expressed in terms of pleural pressure. Though this diagram has already been introduced and used in the analysis of spontaneous breathing (see Chapter IV, Figs. 43–45) it is probably convenient to recapitulate its basis using the curves of Fig. 49. This diagram is similar to Fig. 43, but is drawn so as to display the various components of mechanical work. These curves indicate: (*a*) the pressure exerted by the relaxed chest wall

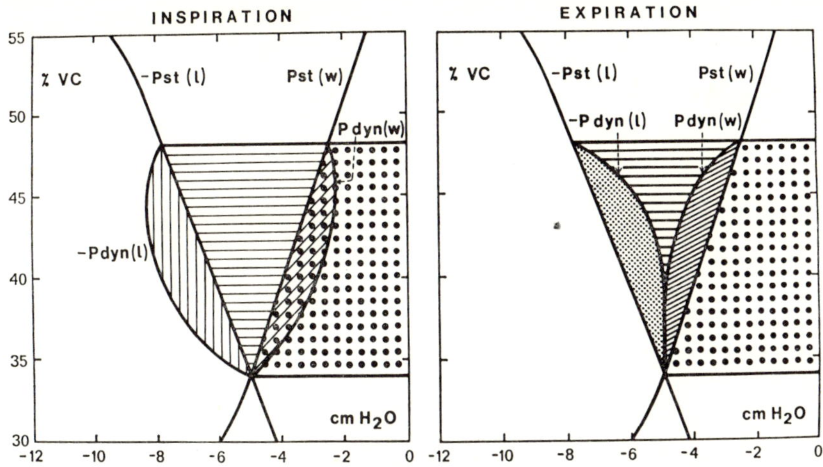


FIG. 49.—GRAPHICAL ANALYSIS OF THE WORK DONE DURING A QUIET BREATHING CYCLE THROUGH THE MOUTH (schematic drawing)

*Inspiration*—The vertically-hatched area represents the work done by the inspiratory muscles to overcome the flow-resistance of the lung (gas and tissue), this energy is lost as heat. The horizontally-hatched area represents the work done by the inspiratory muscles to overcome the elastic resistance of the lung, this energy is stored as elastic energy in the lung. The obliquely-hatched area represents the work done by the inspiratory muscles to overcome the flow-resistance of the chest wall, this energy is lost as heat. The coarsely-stippled area represents the elastic energy transferred from the chest wall to the lung.

*Expiration*—The finely-stippled area represents the work done by the elastic energy stored in the lung to overcome the flow-resistance of the lung (gas and tissue), this energy is lost as heat. The heavy horizontally-hatched area is the work done by the elastic energy stored in the lung to overcome the persistent activity of the inspiratory muscles, this energy is lost as heat. The obliquely-hatched area represents the work done by the elastic energy stored in the lung to overcome the flow-resistance of the chest wall, this energy is lost as heat. The coarsely-stippled area represents the elastic energy transferred from the lung to the chest wall.

under static and dynamic conditions, *P st* (*w*) and *P dyn* (*w*), respectively, (*b*) the pressure exerted by the lung under static and dynamic conditions, represented with negative sign since the lung recoil is "sensed" from the pleural surface and not from the alveolus, —*P st* (*l*) and —*P dyn* (*l*). The lung recoils inwards over all the VC, while the chest wall recoils inwards only above about 55 per cent of the VC. Hence below this volume the lung and the chest wall act as two opposing springs; when the volume increases the recoil of the lung increases and that of the chest wall decreases, and vice versa (see Chapter III).

The horizontal distance between —*P st* (*l*) and *P st* (*w*) indicates the pressure that the respiratory muscles must contribute to keep the respiratory system at a given lung volume with open airways. The pressure is contributed by the inspiratory muscles when —*P st* (*l*) is to the left of *P st* (*w*), and by the expiratory muscles when it is to the right (below FRC). The horizontal distance between *P dyn* and *P st* of either the lung or the chest wall indicates the pressure necessary to overcome the flow-resistance

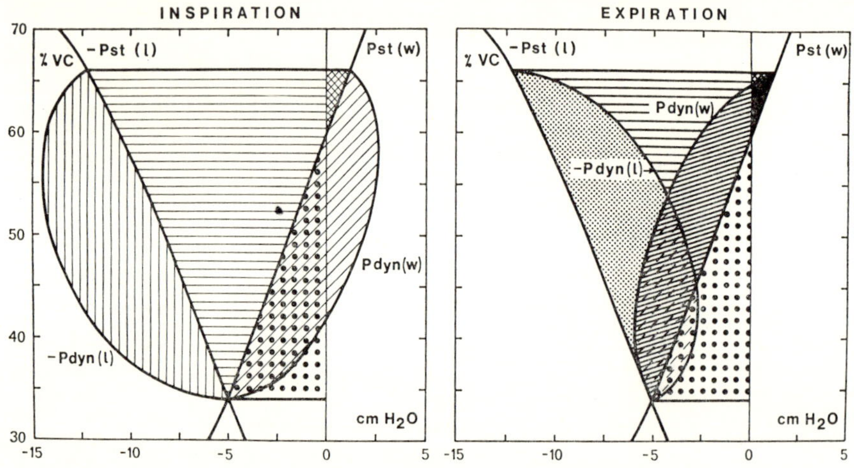


FIG. 50.—GRAPHICAL ANALYSIS OF THE WORK DONE DURING A BREATHING CYCLE AT INCREASED VENTILATION: THE CYCLE STARTS AT THE RESTING VOLUME OF THE RESPIRATORY SYSTEM (schematic drawing)

*Inspiration*—Vertically-hatched area: work done by the inspiratory muscles to overcome the flow-resistance of the lung. Horizontally-hatched area: work done by the inspiratory muscles to overcome the elastic resistance of the lung. Cross-hatched area: work done by the inspiratory muscles to overcome the elastic resistance of the chest wall. Obliquely-hatched area: work done by the inspiratory muscles to overcome the flow-resistance of the chest wall. Coarsely-stippled area: elastic energy transferred from the chest wall to the lung.

*Expiration*—Heavy horizontally and cross-hatched areas: work done by the elastic energy of the lung and, respectively, of the chest wall against the persistent activity of the inspiratory muscles. Finely-stippled area: work done by the elastic energy of the lung to overcome part of the flow-resistance of the lung. Broken-hatched area: work done by the expiratory muscles to overcome the rest of the flow-resistance of the lung (this according to the representation used, but of course this energy, here and in other cases, could be utilised to overcome part of the flow-resistance of the chest wall). Obliquely-hatched and black areas: work done by the elastic energy of the lung and, respectively, of the chest wall to overcome the flow-resistance of the chest wall. Coarsely-stippled area: elastic energy transferred from the lung to the chest wall.

of the lung or, respectively, of the chest wall. The horizontal distance between *P dyn* (*l*) and *P dyn* (*w*) gives the net pressure exerted by the muscles in dynamic conditions. The pressure is contributed by the inspiratory muscles when *P dyn* (*l*) is to the left of *P dyn* (*w*), and by the expiratory muscles when it is to the right. The loop described by —*P dyn* (*l*) in one breathing cycle corresponds to the volume-pressure loop obtained by measuring oesophageal pressure and lung volume change.

Once these concepts are clear one may consider the various components of the mechanical work and the exchange of energy between different parts of the respiratory system during the breathing cycle. Figure 49 illustrates a breathing cycle without intervention of the expiratory muscles; Figs. 50 and 51 illustrate breathing cycles at increased ventilation both of which involve the intervention of the expiratory muscles, but one starting

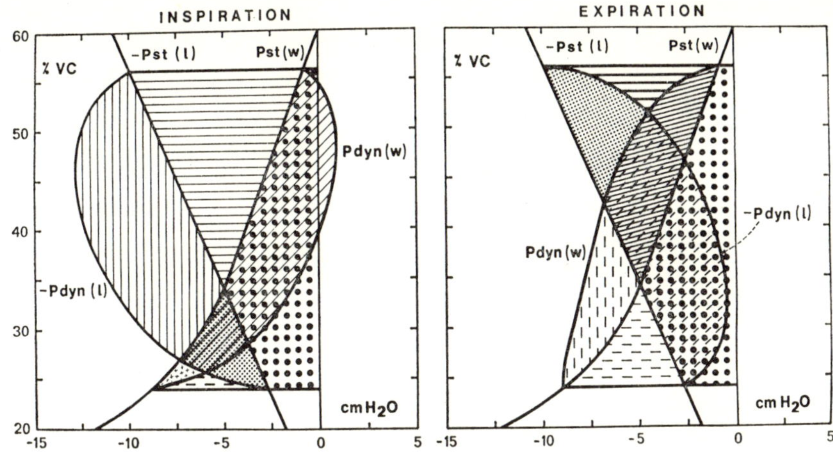


Fig. 51.—Graphical Analysis of the Work Done During a Breathing Cycle at Increased Ventilation: the Cycle Starts Below the Resting Volume of the Respiratory System (schematic drawing)

*Inspiration*—Heavy broken-hatched area: work done by the elastic energy of the chest wall to overcome the persistent activity of the expiratory muscles. Finely-stippled area: work done by the elastic energy of the chest wall to overcome part of the flow-resistance of the lung (gas and tissues). Area with small crosses: work done by the elastic energy of the chest wall to overcome part of the flow-resistance of the chest wall. Obliquely-hatched area: work done by the inspiratory muscles to overcome the rest of the flow-resistance of the chest wall. Vertically-hatched area: work done by the inspiratory muscles to overcome part of the flow-resistance of the lung. Horizontally-hatched area: work done by the inspiratory muscles to overcome the elastic resistance of the lung. Coarsely-stippled area: elastic energy transferred from the chest wall to the lung.

*Expiration*—Heavy horizontally-hatched area: work done by the elastic energy of the lung against the persistent activity of the inspiratory muscles. Finely-stippled area: work done by the elastic energy of the lung to overcome part of the flow-resistance of the lung. Obliquely-hatched area: work done by the elastic energy of the lung to overcome part of the flow-resistance of the chest wall. Vertically-broken-hatched area: work done by the expiratory muscles to overcome part of the flow-resistance of the chest wall. Obliquely-broken-hatched area: work done by the expiratory muscles to overcome part of the flow-resistance of the lung. Horizontally-broken-hatched area: work done by the expiratory muscles to overcome the elastic resistance of the chest wall. Coarsely-stippled area: elastic energy transferred from the lung to the chest wall.

at the resting volume of the respiratory system, i.e. where $-P\ st\ (l)$ and $P\ st\ (w)$ cross, and the other one below the resting volume. Detailed explanations of the meaning of the various areas of the volume-pressure diagrams illustrated by these figures are given in the legends.

**Theoretical Estimation of the Mechanical Work**

The mechanical work of breathing may be estimated from the equation of motion of the respiratory system (see Chapter IV) if the values of the constants are known and if breathing is assumed to be a sine-wave function of time (Otis *et al.*, 1950).

Since the compliance ($C$) and the constants related to the flow-resistance ($K_1$ and $K_2$) may be considered constant only within narrow limits this approach only gives a rough estimate. Furthermore it neglects the negative work.

The mechanical work necessary to overcome the accelerative resistance may be neglected, hence the work of a single inspiration (Otis *et al.*, 1950; Otis, 1964) is given by:

$$Winsp = \frac{1}{2C} V_T^2 + \frac{K_1}{4} \pi^2 f V_T^2 + \frac{2K_2}{3} \pi^2 f^2 V_T^3 \qquad (1)$$

where $C$ is the compliance of the respiratory system, $V_T$ the tidal volume and $f$ the breathing frequency.

The mechanical work of inspiration per unit time, i.e. the mechanical power, is then:

$$\dot{W}insp = \frac{1}{2Cf} \dot{V}^2 + \frac{K_1}{4} \pi^2 \dot{V}^2 + \frac{2K_2}{3} \pi^2 \dot{V}^3 \qquad (2)$$

If expiration, as during quiet breathing, does not require the intervention of the expiratory muscles this is also the total positive work of breathing per unit time. If, instead, the expiratory muscles intervene and all the elastic energy stored in the lung is utilised during the next expiration to overcome the flow resistance (a condition that is almost attained only at high values of ventilation), then the positive work per unit time is given by:

$$\dot{W} = 2\left(\frac{K_1}{4} \pi^2 \dot{V}^2 + \frac{2K_2}{3} \pi^2 \dot{V}^3\right) \qquad (3)$$

These formulas indicate that the mechanical work of breathing per litre of ventilation increases progressively with increasing ventilation, as can be expected from an analysis of the static and dynamic properties of the respiratory system. The kind of relationship between the rate of respiratory work and the ventilation described by equation (3) is similar to that obtained by determining the mechanical work of breathing during muscular exercise of various degrees (Margaria *et al.*, 1960).

## Measurements of the Mechanical Work

The whole mechanical work necessary to ventilate the respiratory system may be determined by ventilating a relaxed subject with an artificial respirator and measuring the differential pressure between the mouth and the body surface (Otis *et al.*, 1950). Since it is very difficult for a subject to relax his respiratory muscles completely when his respiratory system is driven by an artificial respirator, this procedure generally measures the total work of artificial breathing only if the muscles of the subject are paralysed. Sharp *et al.* (1964), however, did not find any electrical activity in the intercostal and abdominal muscles during this procedure. Furthermore since the pattern of spontaneous breathing is different from that produced by an artificial respirator the mechanical work of spontaneous breathing may be somewhat different.

The method more commonly used is that of measuring simultaneously changes in lung volume and in oesophageal pressure (with reference to mouth pressure). Changes in oesophageal pressure approximate closely changes in pleural surface pressure (see Chapter III). Work is calculated from the area enclosed by the pressure-volume loop so obtained. As can be seen from Figs. 49–51, and from the similar diagrams of Chapter IV this procedure does not measure the following parts of the mechanical work done by the respiratory muscles. (1) The work done to overcome the flow-resistance of the chest wall. Because the flow-resistance of the lung increases progressively as the ventilation increases, whereas that of the chest wall does not (see Fig. 35, Chapter IV), this work becomes a small percentage of the total work at high values of ventilation. (2) Part of the work done by the inspiratory muscles to overcome the elastic resistance of the lung, unless the static volume-pressure curve of the relaxed chest wall is determined. This amount of work is relatively large during quiet breathing: it is equivalent to the hatched areas of the expiration diagram of Fig. 49. (3) The negative work done by the inspiratory muscles during the first part of the expiration This work also becomes a small fraction as the ventilation increases. It appears from these considerations that the error involved in this procedure is greatest at low values of ventilation and becomes small at high values.

Because the gas in the lung is compressed and expanded during every breath the volume displacement measured at the airway openings is smaller than the volume displacement of the lung, as measured with a body plethysmograph. This difference is negligible in normal subjects at sea level, but in subjects with high flow-resistance and large functional residual capacity it may become appreciable particularly if the ventilation is high. Under these conditions the mechanical work of breathing is underestimated if the volume change is measured at the airway opening (Jaeger and Otis, 1964). Furthermore under these conditions, because of the appreciable phase shift between the movement of the chest wall and the displacement of air at the airway opening not all the elastic work done in expanding or compressing the gas may be utilised in displacing air in or out of the lung. The amount of elastic work not utilised to displace the gas is dissipated as heat balancing an equal amount of negative work done by the antagonist muscles (Jaeger and Otis, 1964).

The mechanical work per unit time (i.e. the power) during quiet breathing through the nose as in the diagrams of Figs. 40 and 43 (Chapter IV), is about 1 cal/min (or 0·4 kpm/min): of this 0·8 cal/min is the positive work done during inspiration and 0·2 cal/min the negative work during expiration.

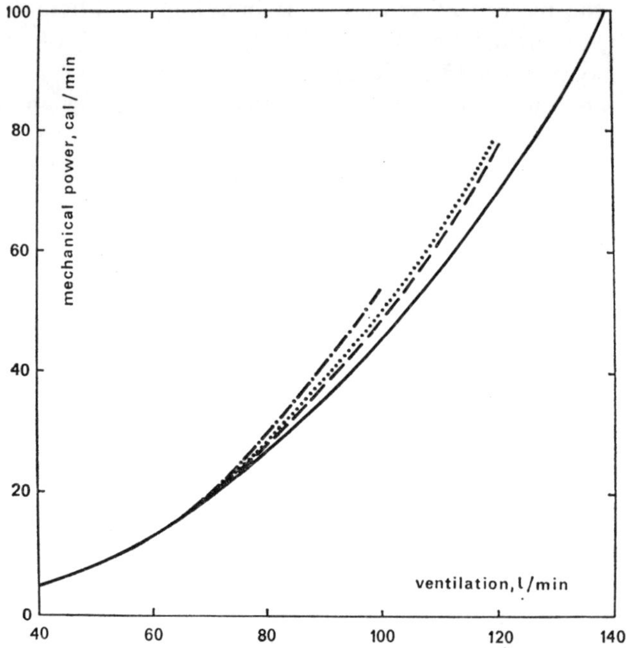


FIG. 52.—RATE OF MECHANICAL WORK OF BREATHING DURING MUSCULAR EXERCISE OF VARIOUS INTENSITIES

The work of breathing has been calculated by measuring the volume-pressure area enclosed by the loop obtained by recording the oesophageal pressure and the lung volume changes during the breathing cycle, and multiplying it by the breathing frequency. This work gives only part of all the work done by the respiratory muscles (see text). ·—·—· Rossier and Bülhmann (1959), 4 subjects, bicycle ergometer; ..... Margaria *et al.* (1960), 3 subjects, treadmill; - - - Cooper (1961), 8 subjects, treadmill; ——— Milic-Emili *et al.* (1962), 7 subjects, bicycle.

The mechanical power at medium and high values of ventilation during muscular exercise is shown in Fig. 52. These data refer only to the work that may be measured by determining the oesophageal pressure change during the breathing cycle, hence they do not account for all the mechanical work that may be included in the volume-pressure diagram. Nevertheless, as mentioned above, at high values of ventilation the part neglected should be small.

From the above data it appears that while the ventilation increases about 15 times from rest to maximum exercise, the mechanical power of breathing increases more than 150 times.

When the ventilation is increased by inhalation of $CO_2$, the mechanical power developed by the inspiratory muscles is linearly related to the alveolar $P_{CO_2}$; this relationship is not influenced by increasing the flow-resistance within a relatively wide range (Milic-Emili and Tyler, 1963). It is not clear whether the increase of muscle force occurring when the flow-resistance is increased is due only to the intrinsic properties of the muscle (force–velocity relationship) or also to reflexes (recruitments of more units, higher frequency of discharge). No consistent relationship has been found between power developed by the expiratory muscles and alveolar $P_{CO_2}$ (Milic-Emili and Tyler, 1963). The implications of these findings for the control of respiratory muscle activity are discussed in Chapters XI and XIII.

### Work of Deformation

The changes of shape of the chest wall produced by the action of the respiratory muscles have been illustrated in the Kinematics chapter. The study of the forces involved by these changes of shape is difficult and therefore it is difficult to determine even roughly the amount of mechanical work done in deformation.

From experiments in which the relaxed rib cage was squeezed along its lateral diameter to reproduce the deformation occurring during expiratory efforts, it has been estimated that during expiratory efforts at mid-lung volume against closed airways the work of deforming the rib cage is about one third of the compressive work done by the respiratory muscles as given by the volume-pressure diagram. On the other hand during the expiration against a mechanical load, the work of deforming the rib cage is about 1 per cent of the work done by the respiratory muscles as given by the volume-pressure diagram. During normal breathing the work of deformation is probably negligible, even at the values of ventilation achieved during heavy exercise (Agostoni *et al.*, 1966).

Mechanical work is also done in altering the distribution of volume changes between rib cage and abdomen. In order to measure this work, it is necessary to know the degree of interdependence of these two parallel volume-elastic elements. As rib cage and abdomen share a common structural boundary, the degree of interdependence is probably significant, and therefore the work done in this kind of deformation cannot at present be estimated (Konno and Mead, 1968). These authors have, however, made a theoretical analysis of the work of deformation for independent volume-elastic elements.

### Maximum Mechanical Power

In the Dynamics chapter the factors limiting maximum flow have been discussed and the maximum flow-pressure curves under various loads have been illustrated (Fig. 37). Since pressure times flow is equal to power, the maximum instantaneous power of the respiratory muscles under various loads can be determined from the data of Fig. 37. For the expiratory muscles the pattern is complicated by the compression of the airways, which introduces a marked change of the internal load: the analysis has been therefore limited to the inspiratory muscles (Fig. 53). At mid-lung volume the maximal inspiratory power reaches a peak under normal conditions (point B); if the flow-resistance is increased by adding an external resistance (point I), or decreased by giving 20 per cent $O_2$ in He to breathe (point E), the maximum inspiratory power decreases. As the peak is rather blunt, small variations in flow-resistance do not appreciably change the maximal power. The peak power occurs at a pressure that is almost half the isometric one, similarly the peak power of the flexors of the forearm occurs when the force is about 0·4 the isometric one (Wilkie, 1950). It is interesting to note that the intrinsic load of the respiratory system is optimally matched with the respiratory muscles to produce the peak power. On the other hand the limb muscles need an external load to reach peak power. In other words, the intrinsic load of the respiratory system is such that the speed of shortening of the muscles is optimal in terms of maximal power output, whereas in the limbs an external load is required to reach the optimal speed of shortening. One could speculate that the limbs are designed to perform external work, whereas the respiratory system is not.

Pengelly and his co-workers (1970) have shown that when diaphragm alone is active the maximum power output occurs at a flow rate lower than that during a maximum inspiratory effort. Furthermore, during contraction of one hemidiaphragm the maximum power output occurs at a flow rate that is about half that occurring during contraction of both hemidiaphragms. They suggested that the chest wall and the respiratory muscles

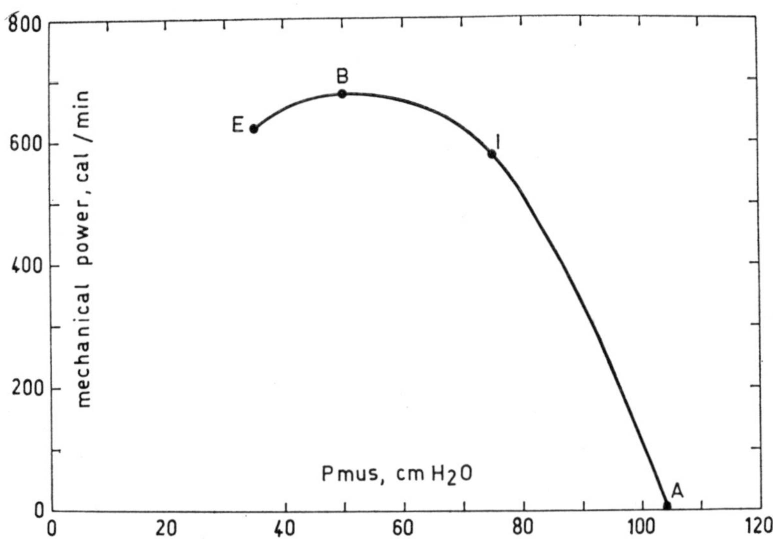


FIG. 53.—MAXIMAL INSTANTANEOUS POWER OF THE INSPIRATORY MUSCLES UNDER DIFFERENT LOADS

This relationship is obtained from the maximal flow-pressure relationship illustrated by Fig. 37. B: normal conditions, E: lowered flow-resistance (breathing 20 per cent $O_2$ in He), I: increased flow-resistance (breathing through a resistance), A: isometric conditions.

are arranged in such a way that the flow corresponding to the peak power increases as additional muscles are recruited. Thus the system is designed to work efficiently (in terms of power output) over a wide range of flow rates. Their data also show that when only the diaphragm or one hemi-diaphragm is active at mid-lung volume, peak power is still produced without additional load (see Chapter XIII).

## THE ENERGY COST OF BREATHING

The usual method of measuring the energy consumption of an organ or region is to measure its oxygen consumption by application of the Fick principle: oxygen consumption = blood flow × arteriovenous oxygen content difference. This approach cannot be applied to the respiratory muscles because there are so many blood vessels supplying them. The principle employed in all studies of the energy cost of breathing is that introduced by Liljestrand (1918). The total $O_2$ consumption of the body is measured at several different levels of ventilation under conditions which should not affect other sources of oxygen consumption. The oxygen consumption of the respiratory muscles can then be estimated by extrapolation. Thus, the ventilation at a constant level of physical activity is increased for example, voluntarily or by breathing $CO_2$ and the difference in total oxygen consumption is attributed to the respiratory muscles. This principle is difficult to apply because the oxygen consumption of the respiratory muscles is a small fraction of the total oxygen consumption. This means that great analytical accuracy is required and that it is difficult to be sure that any change in oxygen consumption is due only to the respiratory muscles, and not to other muscles, to the heart or to a change in basal metabolic rate (Karetsky and Cain, 1970).

Also, it is important to ensure that the work of breathing is not affected by the experimental procedure. Thus, hypocapnia produced by voluntary hyperventilation increases airway resistance (Newhouse *et al.*, 1964) and this will obviously increase the metabolic cost. Hyperventilation, and, in particular, voluntary hyperventilation, is accompanied by activity in many muscles which do not contribute directly to movements of the rib cage or abdomen. While the activity of some of these muscles (especially those concerned with maintenance of posture) might be considered an inevitable concomitant of increased ventilation, the oxygen they consume will be measured as part of the total, and this will lead to errors in extrapolation from one level of ventilation to another, and to a falsely low estimate of the efficiency of breathing (see below).

The oxygen cost of breathing as measured by the techniques outlined above increases progressively with increasing pulmonary ventilation (Fig. 54). There is a wide variation between the estimates of different authors and between results on different subjects in the same series. Some of these results are summarised in Table II. These differences may reflect differences in method, voluntary hyperventilation (Fritts *et al.*, 1959) being accompanied by more extraneous muscular activity than breathing through added dead space. The low values of Milic-Emili and Petit (1960) might be attributed to the fact that their subjects were supine. As the factors outlined above tend to create a bias towards overestimation, preference must perhaps be given to those studies reporting the lowest values and it appears that some subjects certainly require only 0·25 ml per litre ventilation at rest, and less than 1 ml per litre at a ventilation of 100 l/min. McKerrow

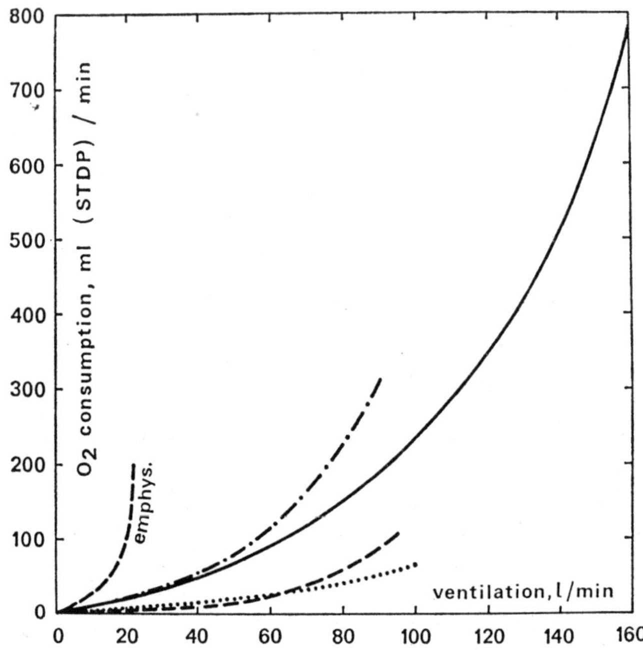


FIG. 54.—OXYGEN COST OF BREATHING AT VARIOUS LEVELS OF VENTILATION

The ventilation is increased either by breathing through a dead space or voluntarily (giving $CO_2$ to avoid hypocapnia). ·—·—· Cournand *et al.* (1954), dead space; – – – – Campbell *et al.* (1957), dead space; ——— Bartlett *et al.* (1958), voluntary; . . . . . . Milic-Emili and Petit (1960), dead space, supine.

and Otis (1956) measured the oxygen cost of high levels of voluntary ventilation over periods of 20–30 sec, and found values ranging from 2·4–7·9 ml per litre at ventilations of 200–270 l/min. Shephard (1966) found a cost of 4·3–4·4 ml per litre over the range 90–130 l/min during exercise.

Although there may be individual variations, as a fraction of the total oxygen consumption, the oxygen cost of breathing at rest is small even in subjects with severe chest disease. Much more striking is the difference between subjects in the rate at which the $O_2$ cost of breathing increases as ventilation is increased.

TABLE II

OXYGEN COST OF BREATHING, ML (STPD) PER LITRE (BTPS) VENTILATION

| Author | Means by which ventilation was increased | Pulmonary ventilation, 1/min | | |
|---|---|---|---|---|
| | | <20 | 20–50 | 50–100 |
| Liljestrand (1918) | Added dead space | 1 | 2–3 | |
| Nielsen (1936) | $CO_2$ | 0·33–0·80 | | |
| | Exercise | | 1·5 | |
| Cournand *et al.* (1954) | Dead space | 1 | 2 | 3·2 |
| Campbell *et al.* (1959) | Dead space | 0·25–0·5 | 0·5–1·5 | 0·5–1·0 |
| Cherniack (1959) | Dead space | 0·45–1·87 | | |
| Fritts *et al.* (1959) | Voluntary, with added $CO_2$ | 0·4–1·1 | 0·6–2·37 | 1·54–3·47 |
| Milic-Emili and Petit (1960) | Dead space | 0·25 | | 0·57 |

## THE MECHANICAL EFFICIENCY OF BREATHING

The difficulties of measuring not only the total mechanical work of breathing but also the oxygen cost imply that the determination of the efficiency of breathing is difficult and that published data are open to doubt.

In order to avoid the difficulty of measuring the total work of breathing Cain and Otis (1949) measured the external rate of work done by the respiratory system to overcome an external resistance. This may be easily measured, but if the resistance is small the extra $O_2$ consumption is so small that its measurement becomes critical. If the resistance is very great the work done within the respiratory system increases slightly (deformation work, unrecovered work of compression and expansion). By this approach Cain and Otis found an efficiency of about 3 per cent. With a similar approach Campbell *et al.* (1957 and 1959) found an efficiency of about 9 per cent.

Other estimates of the efficiency of breathing have been made in which the mechanical work was obtained from oesophageal pressure measurements. As illustrated above this procedure does not measure all the mechanical work, though at high values of ventilation the error should be relatively small. Fritts *et al.* (1959), during voluntary hyperventilation with added $CO_2$, found an efficiency ranging from 1 to 8 per cent, whereas Milic-Emili and Petit (1960) in supine subjects with added dead space, found an efficiency ranging from 19 to 25 per cent.

The mechanical efficiency of limb muscles during cycling or uphill walking is about 25 per cent: it seems unlikely that the respiratory muscles are intrinsically less efficient. Hence the very low values of efficiency found in some studies are probably due to inadequacies of the techniques resulting in great underestimates of the work done and/or overestimation of the

$O_2$ cost. On the other hand, the value of 25 per cent, obtained when the work to overcome the flow-resistance of the chest wall and the work of deformation are not taken into account, suggests that a small over-estimation of the work done and/or underestimation of the $O_2$ cost may also occur.

Mechanical efficiency seems to be approximately constant up to a ventilation of 100–120 l/min (Campbell *et al.*, 1957 and 1959; Milic-Emili and Petit, 1960); it decreases progressively at higher values of ventilation (Tenney and Reese, 1968).

In order to calculate the efficiency, the mechanical work done must be compared to the energy cost. The mechanical work is not necessarily the most revealing parameter with which to compare energy consumption. For instance when a subject supports a heavy object no mechanical work is done, though his energy consumption increases, probably as some function of the weight of the object (i.e. of the force exerted) and of the period that it was supported. Similarly, when a subject breathes through a resistance or when his flow-resistances are increased by disease there may be an extra increase of energy consumption with no corresponding mechanical work, as when the action of the antagonist muscles balances that of the agonist ones. McGregor and Becklake (1961) measured the $O_2$ cost of breathing during unobstructed hyperventilation and during breathing through a high flow-resistance, and found that the $O_2$ cost per unit mechanical work was greater for resistance breathing than for unobstructed hyperventilation, whereas the $O_2$ cost related to an index of the average respiratory force (the time integral of oesophageal pressure change over the breathing cycle), was about the same in the two cases.

The negative work of the respiratory muscles is small and generally is not measured: nevertheless it seems worthwhile here to recall that the $O_2$ consumption during negative work is much smaller than during positive work (see Chapter I).

### Metabolic Limits

During quiet breathing at rest the $O_2$ cost of breathing is probably 2 ml/min : this provides about 300 ml/min of $O_2$ to the rest of the body, i.e. a "gain" of 150 times. During muscular exercise, when the ventilation is about 100 l/min and the $O_2$ consumption about 3 l/min, the $O_2$ cost of breathing is about 100 ml/min, i.e. a "gain" of 30 times. When the ventilation increases still further the $O_2$ cost increases progressively and a stage is eventually reached at which the energy cost of breathing is such that a further increase of ventilation does not increase but rather decreases the $O_2$ supply to the rest of the body (Otis 1954, 1964). This critical level has been calculated by Otis to correspond to about 140 l/min. Margaria *et al.* (1960) obtained values of 130–170 l/min and Shephard (1966) a value of 120 l/min. These critical values are 30–40 per cent above the values of ventilation attained during maximum aerobic exercise and are about equal to the values of maximum exercise ventilation and to the maximum ventilation which normal subjects can, on average, sustain for a period of 3–4 minutes (Tenney and Reese, 1968; Freedman, 1970). But in patients with severe airways obstruction this "ceiling" may be met at a very small ventilation (Campbell *et al.* 1957; Levison & Cherniack, 1968).

Similarly, there is a critical value of ventilation above which more $CO_2$ is produced by the metabolism of the respiratory muscles than is removed by the increased alveolar ventilation (Otis, 1954). Since during maximum exercise ventilation the alveolar $P_{CO_2}$ is lower than normal, this critical value must be higher than was originally estimated.

It is generally agreed that maximum breathing capacity (MBC) is best estimated by measuring the ventilation which can be achieved by maximum voluntary ventilation (MVV) for a period of 15 or 30 seconds. As first suggested by Dripps and Comroe (1947), it has been found that when subjects are asked to ventilate maximally for longer periods, the fraction of the MBC which they can sustain falls off rapidly with time over the first 1–2 min and more slowly thereafter (Fig. 55) (Zocche *et al.*, 1960; Tenney and Reese, 1968; Freedman, 1970). These data show that the maximum voluntary ventilation that can be sustained indefinitely is about 50 per cent of the MBC: i.e. it corresponds to the ventilation reached

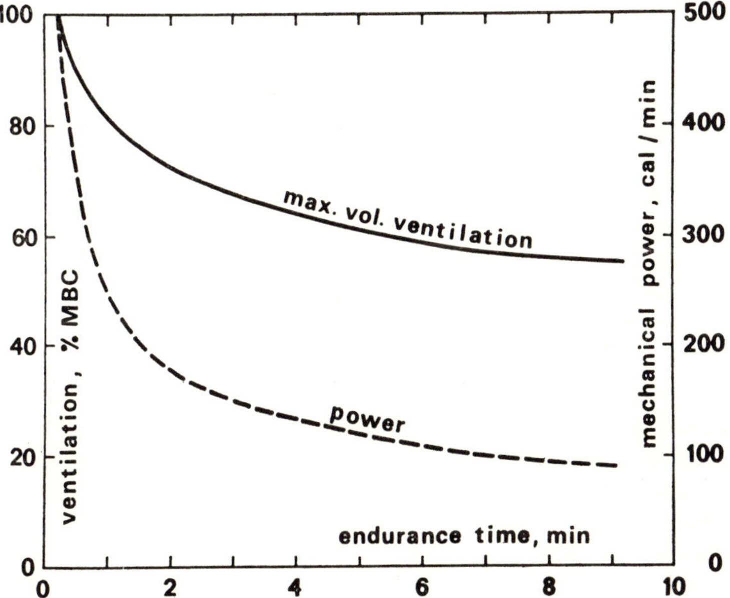


FIG. 55.—MAXIMAL VOLUNTARY VENTILATION AND MECHANICAL POWER AGAINST THE ENDURANCE TIME

From data of Tenney and Reese (1968).

during maximum aerobic exercise. On the other hand Shephard (1967) found that 70–80 per cent of the MBC can be sustained for 15 min when the subject is exercising at 80 per cent of his aerobic working capacity.

The long term limit to maximum voluntary ventilation could be set by fatigue or by the energy available to the respiratory muscles. Tenney and Reese (1968) by analysing the power requirements and the endurance times of maximum voluntary ventilation provided a metabolic interpretation of the relationship between maximum voluntary ventilation and endurance time. The total amount of energy available for respiratory work ($E_o$) is in part derived from a stored source ($S$) and in part supplied at a given rate ($\dot{E}_i$), hence:

$$E_o = S + \dot{E}_i t \qquad (4)$$

$\dot{E}_i$, which is primarily concerned with the delivery of $O_2$, replenishes $S$ as it is drawn upon, but if the rate at which energy is being used ($\dot{E}_o$) is great, $\dot{E}_i$ could be insufficient to keep the balance and $S$ will decrease and eventually disappear, producing exhaustion. This appears by dividing equation 4 by time:

$$\dot{E}_o = \frac{S}{t} + \dot{E}_i \qquad (5)$$

i.e. the energy store is being depleted at the rate ($\dot{E}_o - \dot{E}_i$) and will be exhausted at time $t = S/(\dot{E}_o - \dot{E}_i)$. When $\dot{E}_o = \dot{E}_i$, the ventilation can be prolonged indefinitely. The experimental data suggest that $\dot{E}_o$ equals $\dot{E}_i$ at about 50 per cent MBC. At this value of ventilation during exercise the blood flow to the respiratory muscles has been estimated to be about 13 per cent of the cardiac output (Tenney and Reese, 1968).

Freedman (1969) found an increase in systemic lactic acid concentration of 1–1·5 mmole/l during sustained hyperventilation without hypocapnia at 120–140 l/min for 4 minutes. Stoichiometrically, this is equivalent to an oxygen debt of approximately 90–120 ml/min, or a blood flow debt of 600–800 ml/min. For the same ventilation during exercise, this is equivalent to about 3 per cent of the cardiac output.

## Optimal Frequency of Breathing

The optimal breathing frequency is that involving the least energy cost. Equation 2 shows that elastic work is inversely related to frequency, whereas nonelastic work is independent of frequency; hence, within moderate values of ventilation, mechanical work for a given ventilation should be lower the higher the frequency, as shown by Liljestrand's (1918) measurements. At high values of ventilation, as shown by equation 3, the mechanical work is independent of frequency. There are, of course, physiological limits to this relationship: the lower limit to frequency being set by the maximal tidal volume, the upper one by the rate at which the neuromuscular apparatus can generate alternating movements (Otis,

1964). Furthermore if the tidal volume involves a large portion of the expiratory reserve the increase of airway flow-resistance appreciably increases the work of breathing (Milic-Emili *et al.*, 1960).

Hence, within limits, the work of breathing for a given ventilation would be minimised by breathing as shallowly and rapidly as possible. However, because of the anatomical dead space, the smaller the tidal volume the larger must be the total ventilation to maintain a suitable alveolar ventilation. Hence there must be an optimal frequency for each level of alveolar ventilation at which the mechanical work is minimal (Rohrer, 1925; Otis *et al.*, 1950). This is shown by substituting $(\dot{V}_A + fV_D)$ for $\dot{V}$ in equation 2:

$$\dot{W}\,insp = \frac{1}{2Cf}(\dot{V}_A + fV_D)^2 + \frac{K_1}{4}\pi^2(\dot{V}_A + fV_D)^2 + \frac{2K_2}{3}\pi^2(\dot{V}_A + fV_D)^3 \quad (6)$$

This equation permits calculation of the mechanical power, within moderate values of ventilation, for any given value of alveolar ventilation, frequency and dead space. For flow-resistive work the lowest frequency is the most economical. On the other hand frequency has two opposite effects on the elastic work: the increase of frequency increases the elastic work because the dead space ventilation increases, but it decreases the elastic work because a smaller tidal volume is required. Because of this there is an optimal frequency for the elastic work alone; it occurs when the tidal volume is equal to twice the dead space volume (Otis, 1964). For both elastic and resistive work the optimal frequency is lower than the optimal one for the elastic work alone. For any set of values of dead space, compliance and flow-resistance, the optimal frequency increases with increasing alveolar ventilation (Otis, 1964). The relationship between mechanical power and breathing frequency for an alveolar ventilation of 6 l/min and a dead space of 0·2 litres is shown in Fig. 56 (Otis *et al.*, 1950). One can see that there is a frequency, or better a band of frequencies, for which the mechanical power is minimal. When the elastic resistance increases it is more convenient to increase the frequency, whereas when the flow-resistance increases it is more convenient to decrease the frequency.

During spontaneous breathing subjects choose the optimal frequency (McIlroy *et al.*, 1954; Milic-Emili and Petit, 1959), and the breathing frequency of several species of different sizes corresponds to the optimal frequency (Agostoni *et al.*, 1959; Crosfill and Widdicombe, 1961). It appears therefore that the breathing frequency for a given alveolar ventilation is regulated according to the principle of minimum effort, as postulated by Rohrer (1925).

It seems, however, unlikely that there is a physiological mechanism setting the breathing frequency on the basis of the minimum mechanical work involved, both because this should involve a complicated integration of information and because under some conditions this is not the best index of the energy requirement (see above). Hence the correspondence

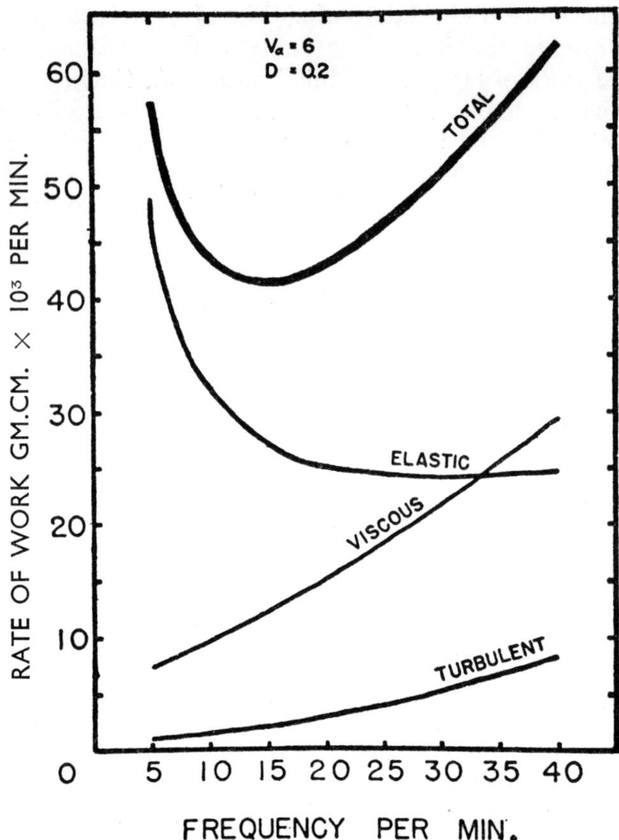


FIG. 56.—RELATIONSHIP OF ELASTIC, VISCOUS, TURBULENT, AND TOTAL WORK OF BREATHING/MIN TO FREQUENCY OF BREATHING

Alveolar ventilation is 6 1/min, and dead space is 200 ml. Curves calculated according to equation 6. (From Otis *et al.* 1950.)

between breathing frequency and minimum work may be somewhat coincidental (Mead, 1960). In fact both in man and guinea-pig the frequency during spontaneous breathing corresponds more closely to that associated with the minimum average force, rather than with that associated with the minimum work (Mead, 1960). However, because the optima are broad it makes little difference to the energy cost whether frequency is set for minimum work or for minimum muscle force (Otis, 1964). Since the average muscle force is more strictly related to $O_2$ consumption and involves a simple mechanism of detection, it seems likely that the organism regulates breathing frequency according to the minimum average muscle force.

Variations in airway calibre change the volume of the dead space and the airway resistance: there is a combination of values of these two variables that minimises the mechanical work of breathing or the average inspiratory force of breathing (Widdicombe and Nadel, 1963). The optimal values of dead space lie within the normal range: the nervously mediated airway tone in normal subjects could then represent an optimum adjustment of airway calibre (Widdicombe and Nadel, 1963).

## ENERGETICS OF PANTING

Many animals (dogs, cats, rabbits, sheep, cattle) seem to rely to some extent on heat loss by evaporation from the airways and the lungs to regulate body temperature. The amount of energy required by high ventilation might suggest that panting as a thermoregulatory mechanism is of limited value because of the heat produced by the respiratory muscles. On the other hand, since panting involves a frequency close to the resonant one (see Chapter IV) it may be that the energy required by this kind of hyperventilation is less than that of the hyperventilation involving great tidal volumes and relatively low frequencies.

The $O_2$ consumption during panting has been determined in dogs and cattle with the aim of determining the energy requirement of the respiratory muscles during this kind of hyperventilation. Considering that the technique of determining the $O_2$ consumption of the respiratory muscles during panting is even more difficult than under the conditions previously discussed, it is not surprising that the results are somewhat controversial.

In the dog, according to Hammel *et al.* (1958), the increase in $O_2$ consumption during panting is largely accounted for by the increase in the metabolic rate produced by the increase of the body temperature of the animal exposed to a hot environment. On the other hand, according to Albers (1961) and Thiele and Albers (1963), there is an increase of $O_2$ consumption of the respiratory muscles great enough as to limit the efficiency of panting as a mechanism for heat loss. Though Spaich and his co-workers (1968) have found that the $O_2$ cost of panting in dogs is smaller than that previously measured, according to them it could still limit the efficiency of panting as a mechanism for heat loss.

In the calf Hales and Findlay (1968) found that the $O_2$ consumption of the respiratory muscles during rapid shallow panting was not appreciably increased and was much less than for a similar value of ventilation induced by $CO_2$. In other words, dead space ventilation may be greatly increased with little energy. During deep panting, involving an increase of alveolar ventilation, they found an increase of the $O_2$ consumption of the respiratory muscles, but this was less than 1·2 ml per litre ventilation, whereas during a similar ventilation induced by $CO_2$ it was 3–4 ml $O_2$ per litre ventilation. Hence Hales and Findlay conclude that the efficiency of panting as a mechanism for heat loss is not limited by the heat produced by the respiratory muscles. In oxen, Whittow and Findlay (1968) induced panting by heating only the hypothalamus and found that the $O_2$ consumption was not significantly increased.

## Summary

Most of the mechanical work of breathing can be measured. The method commonly used of measuring the change of lung volume and of oesophageal pressure to calculate the work neglects a considerable part of the work, particularly at low values of ventilation. The mechanical work per unit time during quiet breathing through the nose is about 1 cal/min (or 0·4 kpm/min). The mechanical work per litre ventilation increases progressively with increasing ventilation. While the ventilation increases about 15 times from rest to maximum exercise, the mechanical power of breathing increases more than 150 times. The intrinsic load of the respiratory system at mid-lung volume is optimally matched with the respiratory muscles to produce the maximum power output. The respiratory muscles are probably arranged with the chest wall in such a way that the flow corresponding to the peak power increases as additional muscles are recruited.

The $O_2$ cost of breathing is difficult to measure particularly at low values of ventilation. During quiet breathing at rest the $O_2$ cost of breathing is probably 2 ml/min: this provides about 300 ml/min of $O_2$ to the rest of the body, i.e. a "gain" of 150 times. During maximum aerobic exercise the $O_2$ cost of breathing is about 100 ml/min and the body $O_2$ consumption 3 litres/min, i.e. a "gain" of 30 times. When the ventilation is increased still further a stage is eventually reached at which the energy cost of breathing is such that a further increase of ventilation does not increase but rather decreases the $O_2$ supply to the rest of the body. Under some circumstances the energy cost of breathing may be a factor limiting exercise. The mechanical efficiency of breathing at medium values of ventilation is 10–20 per cent. During spontaneous breathing the frequency chosen for a given value of alveolar ventilation involves the least energy cost.

## REFERENCES

Agostoni, E., Mognoni, P., Torri, G., and Miserocchi, G. (1966). Forces deforming the rib cage. *Resp. Physiol.*, **2**, 105–117.

Agostoni, E., Thimm, F. F., and Fenn, W. O. (1959). Comparative features of the mechanics of breathing. *J. appl. Physiol.*, **14**, 679–683.

Albers, C. (1961). Der Mechanismus des Wärmechechelns beim Hund. I. Die Ventilation und die arteriellen Blutgase während des Wärmehechelns. *Pflügers Arch. ges. Physiol.*, **274**, 125–147.

Bartlett, R. G., Brubach, H. F., and Specht, H. (1958). Oxygen cost of breathing. *J. appl. Physiol.*, **12**, 413–424.

Cain, C. C., and Otis, A. B. (1949). Some physiological effects resulting from added resistance to respiration. *J. Aviat. Med.*, **20**, 149–160.

CAMPBELL, E. J. M. (1957). The effects to increased resistance to expiration on the respiratory behaviour of the abdominal muscles and intra-abdominal pressure. *J. Physiol.* (*Lond.*), **136**, 556–562.

CAMPBELL, E. J. M., WESTLAKE, E. K., and CHERNIACK, R. M. (1957). Simple methods of estimating oxygen consumption and efficiency of the muscles of breathing. *J. appl. Physiol.*, **11**, 303–308.

CAMPBELL, E. J. M., WESTLAKE, E. K., and CHERNIACK, R. M. (1959). The oxygen consumption and efficiency of the respiratory muscles of young male subjects. *Clin. Sci.*, **18**, 55–65.

CHERNIACK, R. M. (1959). The oxygen consumption and efficiency of respiratory muscles in health and emphysema. *J. clin. Invest.*, **38**, 494–499.

COOPER, E. A. (1960). A comparison of the respiratory work done against an external resistance by man and by a sine wave pump. *Quart. J. exp. Physiol.*, **45**, 179–191.

COURNAND, A., RICHARDS, D. W., BADER, R. A., BADER, M. E., and FISHMAN, A. P. (1954). The oxygen cost of breathing. *Trans. Ass. Amer. Phycns*, **67**, 162–173.

CROSFILL, M. L., and WIDDICOMBE, J. G. (1961). Physical characteristics of the chest and lungs and the work of breathing in different mammalian species. *J. Physiol.* (*Lond.*), **158**, 1–14.

DRIPPS, R. D., and COMROE, J. H., JR. (1947). The respiratory and circulatory response of normal men to inhalation of 7·6 and 10·4 per cent $CO_2$ with a comparison of the maximal ventilation produced by severe muscular exercise, inhalation of $CO_2$ and maximal voluntary hyperventilation. *Amer. J. Physiol.*, **149**, 43–51.

FREEDMAN, S. (1969). The functional capacity of the respiratory muscles in man. Ph.D. Thesis. London Univ.

FREEDMAN, S. (1970). Sustained maximum voluntary ventilation. *Resp. Physiol.* **8**, 230–244.

FRITTS, H. W., JR., FILLER, J., FISHMAN, A. P., and COURNAND, A. (1959). The efficiency of ventilation during voluntary hyperpnea. *J. clin. Invest.*, **38**, 1339–1348.

HALES, J. R. S., and FINDLAY, J. D. (1968). The oxygen cost of thermally-induced and $CO_2$-induced hyperventilation in the ox. *Resp. Physiol.*, **4**, 353–362.

HAMMEL, H. T., WYNDHAM, C. H., and HARDY, J. D. (1958). Heat production and heat loss in the dog at 8°–36°C environmental temperature. *Amer. J. Physiol.*, **194**, 99–108.

JAEGER, M. J., and OTIS, A. B. (1964). Effects of compressibility of alveolar gas on dynamics and work of breathing. *J. appl. Physiol.*, **19**, 83–91.

KARETSKY, M. S., and CAIN, S. M. (1970). Effect of carbon dioxide on oxygen uptake during hyperventilation in normal man. *J. appl. Physiol.*, **28**, 8–12.

KONNO, K., and MEAD, J. (1968). Static volume-pressure characteristics of the rib cage and abdomen. *J. appl. Physiol.*, **24**, 544–548.

LEVISON, H., and CHERNIACK, R. M. (1968). Ventilatory cost of exercise in chronic obstructive pulmonary disease. *J. appl. Physiol.*, **25**, 21–27.

LILJESTRAND, G. (1918). Untersuchungen über die Atmungsarbeit. *Skand. Arch. Physiol.*, **35**, 199–293.

MARGARIA, R., MILIC-EMILI, G., PETIT, J. M., and CAVAGNA, G. (1960). Mechanical work of breathing during muscular exercise. *J. appl. Physiol.*, **15**, 354–358.

MCGREGOR, M., and BECKLAKE, M. R. (1961). The relationship of oxygen cost of breathing to respiratory mechanical work and respiratory force. *J. clin. Invest.*, **40**, 971–980.

MCILROY, M. B., MARSHALL, R., and CHRISTIE, R. V. (1954). The work of breathing in normal subjects. *Clin. Sci.*, **13**, 127–136.

MCKERROW, C. B., and OTIS, A. B. (1956). Oxygen cost of hyperventilation. *J. appl. Physiol.*, **9**, 375–379.

MEAD, J. (1960). Control of respiratory frequency. *J. appl. Physiol.*, **15**, 325–336.

MILIC-EMILI, J., and PETIT, J. M. (1959). Il lavoro meccanico della respirazione a varia frequenza respiratoria. *Arch. Sci. biol.* (*Bologna*), **43**, 326–330.

MILIC-EMILI, J., and PETIT, J. M. (1960). Mechanical efficiency of breathing. *J. appl. Physiol.*, **15**, 359–362.

MILIC-EMILI, J., PETIT, J. M., and DEROANNE, R. (1960). The effects of respiratory rate on the mechanical work of breathing during muscular exercise. *Int. Z. agnew. Physiol.*, **18**, 330–340.

MILIC-EMILI, J., PETIT, J. M., and DEROANNE, R. (1962). Mechanical work of breathing during exercise in trained and untrained subjects. *J. appl. Physiol.*, **17**, 43–46.

MILIC-EMILI, J., and TYLER, J. M. (1963). Relation between work output of respiratory muscles and end-tidal $CO_2$ tension. *J. appl. Physiol.*, **18**, 497–504.

NEWHOUSE, M. T., BECKLAKE, M. R., MACKLEM, P. T., and MCGREGOR, M. (1964). Effect of alterations in end-tidal $CO_2$ tension on flow-resistance. *J. appl. Physiol.*, **19**, 745–749.

NIELSEN, M. (1936). Die Respirationsarbeit bei Körperruhe und bei Muskel Arbeit. *Skand. Arch. Physiol.*, **74**, 299–316.

OTIS, A. B. (1954). The work of breathing. *Physiol. Rev.*, **34**, 449–458.

OTIS, A. B. (1964). The work of breathing. In: *Handbook of Physiology*. Section 3, Respiration. Vol. 1, pp. 463–476. Eds. W. O. Fenn and H. Rahn. Washington D.C.: American Physiological Soc.

OTIS, A. B., FENN, W. O., and RAHN, H. (1950). Mechanics of breathing in man. *J. appl. Physiol.*, **2**, 592–607.

PENGELLY, L. D., ALDERSON, A. M., and MILIC-EMILI, J. (1970). Mechanics of the diaphragm. *J. appl. Physiol.* (in press).

ROHRER, F. (1925). Physiologie der Atembewegung. In: *Handbuch der normalen und path. Physiologie*, Vol. 2, pp. 70–127. Eds. A. T. J. Bethe *et al.* Berlin: Springer.

SHARP, J. T., HENRY, J. P., SWEANY, S. K., MEADOWS, W. R., and PIETRAS, R. J. (1964). The total work of breathing in normal and obese men. *J. clin. Invest.*, **43**, 728–739.

SHEPHARD, R. J. (1966). The oxygen cost of breathing during vigorous exercise. *Quart. J. exp. Physiol.*, **51**, 336–350.

SHEPHARD, R. J. (1967). The maximum sustained voluntary ventilation in exercise. *Clin. Sci.*, **32**, 167–176.

SPAICH. P., USINGER, W., and ALBERS, C. (1968). Oxygen cost of panting in anaesthetized dogs. *Resp. Physiol.*, **5**, 302–314.

TENNEY, S. M., and REESE, R. E. (1968). The ability to sustain great breathing efforts. *Resp. Physiol.*, **5**, 187–201.

THIELE, P., and ALBERS, C. (1963). Die Wasserdampfabgabe durch die Atemwege und der Wirkungsgrad des Wärmehechelns beim wachen Hund. *Pflügers Arch. ges. Physiol.*, **278,** 31–6324.

WHITTOW, G. C., and FINDLAY, J. D. (1968). Oxygen cost of thermal panting. *Amer. J. Physiol.*, **214,** (94–99).

WIDDICOMBE, J. G., and NADEL, J. A. (1968). Airway volume, airway resistance, and work and force of breathing: theory. *J. appl. Physiol.*, **18,** 863–868.

WILKIE, D. R. (1950). The relation between force and velocity in human muscle. *J. Physiol.* (*Lond.*), **110,** 249–280.

ZOCCHE, G. P., FRITTS, H. W. JR., and COURNAND, A. (1960). Fraction of maximum breathing capacity available for prolonged hyperventilation. *J. appl. Physiol.*, **15**, 1073–1074.

# APPENDIX A

## SUBDIVISIONS OF GAS VOLUME CONTAINED IN THE LUNGS

(From Pappenheimer *et al.*, 1950; Agostoni and Mead, 1964).

*Total lung capacity* (TLC): volume of gas in the lungs and airways at the end of maximal inspiration.

*Residual volume* (RV): volume of gas in the lungs and airways at the end of maximal expiration.

*Vital capacity* (VC): maximal volume that can be expired after maximal inspiration.

*Functional residual capacity* (FRC): volume of gas in the lungs and airways at the end of a spontaneous expiration.

*Inspiratory capacity* (IC): volume that can be inspired from the end of a spontaneous expiration.

*Expiratory reserve volume* (ERV): volume that can be expired from the end of a spontaneous expiration.

*Tidal volume* ($V_T$): volume of a particular breath or the average volume of a series of breaths.

*Inspiratory reserve volume* (IRV): volume that can be inspired from end-tidal inspiration.

TABLE III

RATIOS OF THE VITAL CAPACITY, RESIDUAL VOLUME, AND TOTAL LUNG CAPACITY (BTPS) TO THE CUBE OF THE HEIGHT IN METRES*
(*From Jouasset*, 1960)

| Age | VC/ $(ht)^3$, litres/$m^3$ | RV/ $(ht)^3$, litres/$m^3$ | TLC/ $(ht)^3$, litres/$m^3$ |
|---|---|---|---|
| 18–19 | 0·990 | 0·240 | 1·230 |
| 20–29 | 1·025 | 0·275 | 1·300 |
| 30–34 | 1·020 | 0·300 | 1·300 |
| 35–39 | 1·010 | 0·310 | 1·320 |
| 40–44 | 1·000 | 0·320 | 1·320 |
| 45–49 | 0·990 | 0·330 | 1·320 |
| 50–54 | 0·970 | 0·350 | 1·320 |
| 55–59 | 0·950 | 0·370 | 1·320 |
| 60–64 | 0·930 | 0·390 | 1·320 |
| Normal limits | ± 17 per cent | ± 31 per cent | ± 22 per cent |

* These values refer to men. Comparatively few measurements have been made in women: from 20 to 30 years of age, values are on the average 10 per cent smaller than for men of the same age and size.

*Mid capacity*: Functional Residual Capacity plus one-half the Tidal Volume. It is therefore the average gas volume "around which" breathing takes place.

Volume changes affect the chest wall and the lungs equally, provided the thoracic blood volume is constant and provided there is no gas phase between the lung and the chest wall.

## GLOSSARY OF SYMBOLS USED IN RESPIRATORY MECHANICS

(From Mead and Milic-Emili, 1964)

*Major parts of the respiratory system:*

| | |
|---|---|
| *l* | lung inclusive of lung tissue, air spaces, airways (intrathoracic and extrathoracic, and gas contained) |
| *g* | gaseous part of *l* |
| *lt* | tissue part of *l* |
| *w* | chest wall including intrathoracic but extrapulmonary structures, rib cage, diaphragm, abdominal contents, and abdominal wall. As commonly used it refers to conditions in which the respiratory muscles are relaxed |
| *rs* | total respiratory system, i.e. $l + w$ |

*Boundaries of the major parts:*

| | |
|---|---|
| *ao* | airway opening, i.e. mouth, nose, tracheal cannula |
| *alv* | terminal air space in lungs |
| *pl* | pleural surface |
| *bs* | body surface |

*Other parts, places and conditions:*

| | |
|---|---|
| *aw* | airways |
| *uaw* | upper airways, larynx and above |
| *law* | lower airways, below larynx |
| *rc* | rib cage |
| *di* | diaphragm |
| *ab* | abdomen, including abdominal contents and abdominal wall |
| *abc* | abdominal contents |
| *abw* | abdominal wall |
| *es* | oesophagus |
| *ga* | stomach |
| *mus* | respiratory muscles |
| *i* | inspiration or inspiratory |
| *e* | expiration or expiratory |
| *st* | static: zero flow and zero volume acceleration |
| *dyn* | dynamic : flow and/or volume acceleration not zero |

*Measured quantities:*

| | |
|---|---|
| $V$ | volume in litres, l, or millilitres, ml |
| $\dot{V}$ | rate of volume change or flow : generally in litres per second |
| $\ddot{V}$ | volume acceleration: generally in litres per second per second |
| $P$ | pressure, in cm $H_2O$ relative to atmospheric unless otherwise specified. |

*Derived quantities:*

| | |
|---|---|
| *Pst* | static component of pressure |
| *Pres* | flow-resistive component of pressure |
| *Pin* | inertial component of pressure |
| *C* | *compliance*, the slope of a static volume-pressure curve at a point, or the linear approximation of a nearly straight portion of such a curve, expressed in l/cm $H_2O$ or ml/cm $H_2O$. |
| *E* | *elastance*, the inverse of *C* |
| *Cdyn* | *dynamic compliance*, the ratio of the tidal volume to change in pressure between the points of zero flow at the extremes of the tidal volume, in l/cm $H_2O$ or ml/cm $H_2O$. |
| *R* | *flow-resistance*, the ratio of the flow-resistive component of pressure to simultaneous flow, in cm $H_2O$/l/sec |
| *G* | *conductance*, the inverse of *R* |
| *I* | *inertance*, the ratio of the inertial component of pressure to simultaneous volume acceleration, in cm $H_2O/l/sec^2$ |

*Examples:*

| | |
|---|---|
| *Pao* | pressure at airway opening relative to atmospheric |
| *Pin* (*rs*) | inertial component of pressure of the respiratory system |
| *Cdyn* (*l*) | dynamic compliance of the lungs |
| *Pst* (*w*) | static component of chest wall pressure |
| *Rlt* | flow-resistance of lung tissue. |

For a treatment of the principles of mechanical analysis used in respiratory mechanics see Mead and Milic-Emili (1964) and Konno and Mead (1967).

## COMPARATIVE FEATURES WITHIN MAMMALS

The lung volume, at a transpulmonary pressure of 20 cm $H_2O$ (Tenney and Remners, 1963), the resting tidal volume (Tenney and Bartlett, 1967) and the compliance of the respiratory system within the resting tidal volume range (Drorbaugh, 1960; Spells, 1969) are proportional to body mass. The breathing frequency at rest is proportional to body mass to the $-0{\cdot}28$ power (Tenney and Bartlett, 1967). As pointed out by Spells (1969) the time constant, RC, is probably inversely related to the breathing frequency at rest; the dimensionless product RCf could then be a constant independent of body mass. Not enough information, however, is available on the relationship between flow-resistance and body mass in order to prove this hypothesis.

## TABLE IV

APPROXIMATE VALUES OF THE STATIC AND DYNAMIC PROPERTIES OF THE RELAXED RESPIRATORY SYSTEM AND ITS PARTS FOR AN ADULT MALE BREATHING AIR AT SEA LEVEL

| | | | |
|---|---|---|---|
| Compliance at | gas | Cg | 0·003 litres/cm $H_2O$ |
| mid-lung volume | lung tissue | Clt | 0·2 litres/cm $H_2O$ |
| (3 litres) | chest wall | Cw | 0·2 litres/cm $H_2O$ |
| | respiratory system | Crs | 0·1 litres/cm $H_2O$ |
| Flow-resistance at | gas by mouth (by nose) | Rg | 1·5 (4·5) cm $H_2O$/1/sec |
| mid-lung volume with | lung tissue | Rlt | 0·2 cm $H_2O$/1/sec |
| a flow of 1 litre/sec | lung (gas + tissue) by mouth | Rl | 1·7 cm $H_2O$/1/sec |
| | chest wall | Rw | 1·3 cm $H_2O$/1/sec |
| | respiratory system by mouth (by nose) | Rrs | 3·0 (6) cm $H_2O$/1/sec |
| Inertance | respiratory system | Irs | 0·01 cm $H_2O/1/sec^2$ |
| Time constant at mid-lung volume | respiratory system by mouth | Trs | 0·3 sec |
| Natural frequency | respiratory system | $f_n rs$ | 5 cycles/sec |
| Hysteresis* | respiratory system | H | 0·08 |

* Expressed as ratio of loop width to loop height over the vital capacity.

## REFERENCES

AGOSTONI, E., and MEAD, J. (1964). Statics of the respiratory system. In. *Handbook of Physiology*. Section 3, Respiration. Vol. 1, pp. 387–409. Eds: W. O. Fenn and H. Rahn. Washington, D.C.: American Physiological Soc.

DRORBAUGH, J. E. (1960). Pulmonary function in different animals. *J. appl. Physiol.*, **15**, 1069–1072.

JOUASSET, D. (1960). Normalisation des épreuves fonctionelles respiratoires dans les pays de la Communauté Européene du Charbon et de l'Acier. *Poumon*, **10**, 1145–1159.

KONNO, K., and MEAD, J. (1967). Measurement of the separate volume changes of rib cage and abdomen during breathing. *J. appl. Physiol.*, **22**, 407–422.

MEAD, J., and MILIC-EMILI, J. (1964). Theory and methodology in respiratory mechanics with glossary of symbols. In: *Handbook of Physiology*. Section 3, Respiration. Vol. 1, pp. 387–409. Eds. W. O. Fenn and H. Rahn. Washington D.C.: American Physiological Soc.

PAPPENHEIMER, J. R., COMROE, J. H., JR., COURNAND, A., FERGUSON, J. K. W., FILLEY, G. F., FOWLER, W. S., GRAY, J. S., HELMHOLZ, H. F., JR., OTIS, A. B., RAHN, H., and RILEY, R. L. (1950). Standardization of definitions and symbols in respiratory physiology. *Fed. Proc.*, **9**, 602–615.

SPELLS, K. E. (1969). Comparative studies in lungs mechanics based on a survey of literature data. *Resp. Physiol.*, **8**, 37–57.

TENNEY, S. M., and BARTLETT, D., JR. (1967). Comparative quantitative morphology of the mammalian lung: trachea. *Resp. Physiol.*, **3**, 130–135.

TENNEY, S. M., and REMNERS, J. E. (1963). Comparative quantitative morphology of the mammalian lung: diffusing area. *Nature* (*Lond.*), **197**, 54–56.

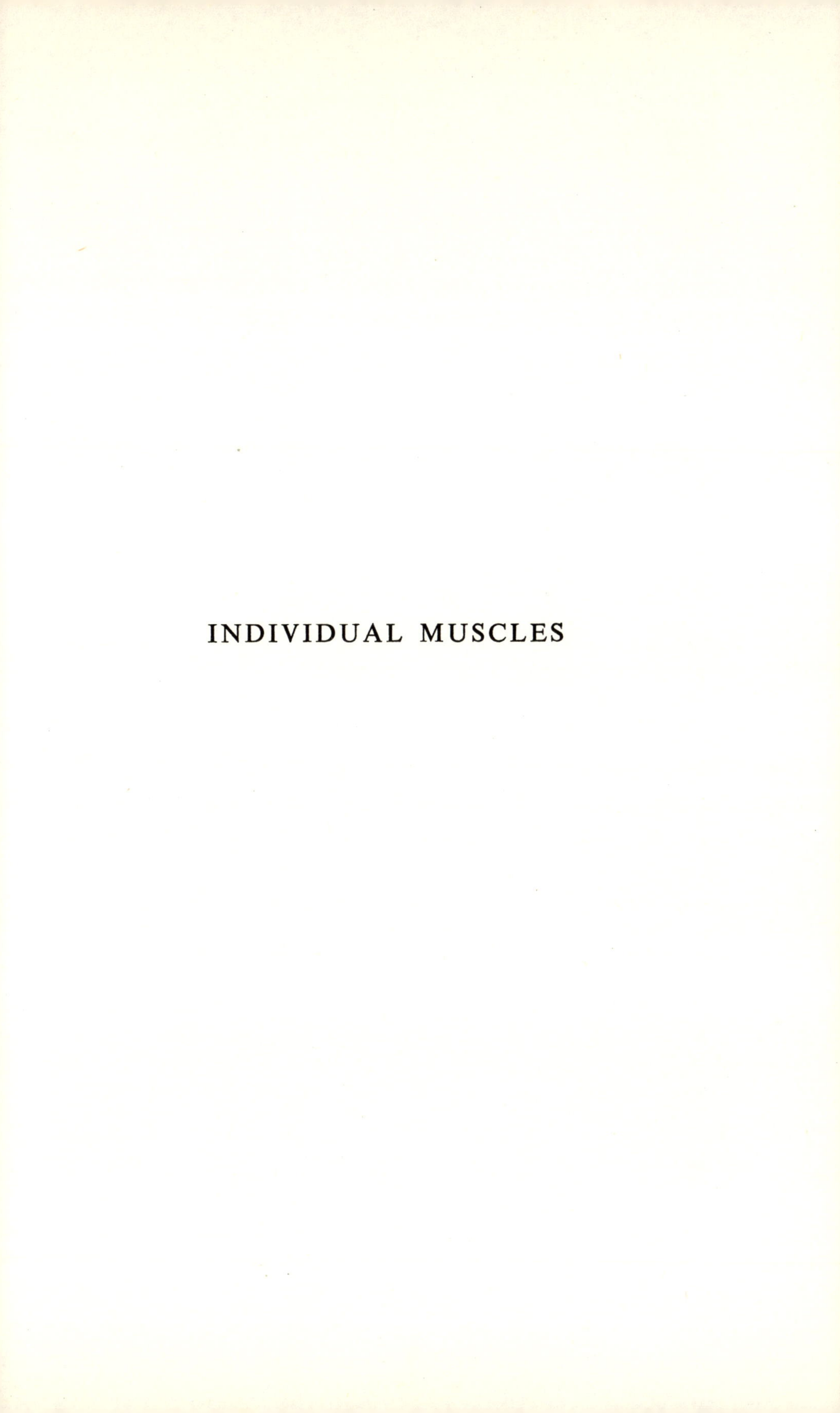

# INDIVIDUAL MUSCLES

# Chapter VI

# THE DIAPHRAGM

E. AGOSTONI AND G. SANT'AMBROGIO*

## Morphology

THE muscular fibres of the diaphragm are grouped into three parts: the vertebral, the costal and the sternal. The vertebral fibres arise from the second and third lumbar vertebrae, from the medial arcuate ligaments (psoas) and from the lateral arcuate ligaments (quadratus lumborum). The costal fibres arise from the side and the upper margin of the lower six ribs and interdigitate with those of the transversus abdominis. The sternal fibres arise from the back of the xiphoid process. All the fibres converge on the central tendon. In some subjects between the costal and the vertebral fibres there is a muscular hiatus and the parietal pleura is in contact with the abdominal cavity.

The diaphragm is a structural feature of the mammals; in fact only mammals and a few species of birds have a complete separation between the thoracic and the abdominal cavities. The embryological origin of the diaphragm is rather complex and controversial. It seems to originate from the septum transversum, which will eventually become the central tendon, from the pleuroperitoneal membranes and from the oesophageal mesentery. The muscular tissue appears in the pleuroperitoneal membranes at an early stage deriving from myoblasts migrating from cervical somites (Hamilton *et al.*, 1962, p. 262).

Histological evidence for the presence of two types of muscles fibres in the diaphragm of man and other mammals has been provided by Günther (1953). In very small mammals such as the shrew and very large ones such as the cow, the diaphragm is essentially homogeneous; in the shrew it is composed of small fibres rich in mitochondria, in the cow of large fibres with few mitochondria. The diaphragm of mammals of intermediate size (from rat to man) contains both kinds of fibres, and also fibres with intermediate characteristics (Gauthier and Padykula, 1966).

## Innervation

*Efferent.*—Whether or not the phrenic is the only motor nerve to the diaphragm has long been a matter of controversy. If during development the muscular part of the diaphragm migrates from the cervical region with

* Istituto di Fisiologia umana, Università di Milano, Milano.

the phrenic nerves (Goodrich, 1958; Hamilton *et al.*, 1962, p. 420), the only motor nerves should be the phrenic.

The old literature on the motor innervation of the diaphragm was reviewed by Felix (1922), who maintained that there is some extra-phrenic motor innervation, and by Kiss and Ballon (1929) who denied it. The muscle fibres of the diaphragm degenerate completely after cutting the phrenic nerves in dogs (Schlaepfer, 1926; Jefferson *et al.*, 1950), goats (Jansen, 1931), and man (Strauss, 1933). After bilateral thoracic phrenicotomy in the cat and the opossum, and after bilateral section of the ventral roots from $C_3$ to $C_7$ in the cat, respiratory electrical activity disappears in all parts of the diaphragm (Sant'Ambrogio *et al.*, 1963). In contrast to these findings, and to other less direct or complete evidence (Muller Botha, 1957; Ogawa, 1959; Wilson, 1963) showing that the only motor nerve to the diaphragm is the phrenic, Rosenblueth and his colleagues (1961) found some respiratory electrical activity in the diaphragm of cats, dogs, monkeys and rabbits after bilateral thoracic phrenicotomy. This activity, however, may be found under some conditions even after extensive crushing of the diaphragm around the recording site and therefore it is not of diaphragmatic origin (C. von Euler, personal communication).

The cervical neuromeres contributing the motor fibres to the phrenic nerves vary slightly among species: $C_3$, $C_4$ and $C_5$ in man (Hamilton *et al.*, 1962), ($C_4$), $C_5$, $C_6$ ($C_7$) in the cat (Sant'Ambrogio *et al.*, 1963), ($C_4$), $C_5$, ($C_6$) in the rabbit (Tosatti, 1938), $C_5$, $C_6$, $C_7$ in the dog (Cardin, 1939), $C_4$, $C_5$ in the rat (Rowett, 1957). The fibres coming from the upper segments innervate mainly the ventromedial part, those coming from the lower segments innervate the dorso-lateral part (Jansen, 1931; Cardin, 1939; Landau *et al.*, 1962; Sant'Ambrogio *et al.*, 1963). The diaphragm however contracts as a unit (Sant'Ambrogio *et al.*, 1963); its activity increases similarly in its various parts when an inspiratory load is added (Lourenço *et al.*, 1966).

An interchange of fibres between the phrenic nerves in the lower thorax has been shown through degeneration experiments by Warwick and Mitchell (1955) in monkeys. Furthermore Sant'Ambrogio and his coworkers (1963) gave functional evidence in the cat that a few fibres leave the left phrenic nerve in the lower thorax, reach the diaphragm with the right phrenic nerve and return to the left within the diaphragm. The branches of the phrenic nerves form a series of arcades within the diaphragm (Scott, 1965), but there is little cross innervation (Collis *et al.*, 1954).

In the dog Landau *et al.* (1962) found that the efferent fibres were 60 per cent of the phrenic myelinated fibres, a proportion similar to that known for muscle nerves in this and other species. On the other hand Hinsey *et al.* (1939) in the cat found that the efferent fibres were 90 per cent of the phrenic myelinated fibres. It must be remembered in this connection that part of the afferent innervation of the diaphragm is supplied by nerves from $T_6$ to $T_{12}$ (see below). The diameters of the efferent fibres in the phrenic nerve of the rabbit (Yasargil and Koller, 1964) and of the dog (Landau *et al.*, 1962) follow a unimodal distribution: the diameters of the myelinated fibres range between 2 and 16 $\mu$, with a peak between 9 and 11 $\mu$. The paucity of efferent fibres with diameters below 7 $\mu$, and hence the unimodal distribution, indicate that there are few gamma fibres in the phrenic nerve, in agreement with the paucity of muscle spindles in the diaphragm (see below) (Fernand and Young, 1951).

The number of muscle fibres per motor unit, or innervation ratio, is about 25 in the rat diaphragm (Krnjevic and Miledi, 1958), 110–120 in the sternal part of the rabbit diaphragm (Yasargil, 1967), and about 83 in the cat diaphragm (Oshima, 1939). A small innervation ratio, such as that found in the rat diaphragm, is typical of muscles performing fine movements, as, for instance, those of the eyes; such muscles usually contain many muscle spindles, but in contrast to this the diaphragm has only a few (see below).

The conduction velocity in specimens of human phrenic nerve was found to be 78 m/sec (Heinbecker *et al.*, 1936). The latency between the percutaneous stimulation of the phrenic nerves in the neck and the electrical activity of the diaphragm in man is 7·5–8 msec (Delhez, 1965; Newsom Davis, 1967).

During inspiration in the rabbit the first motoneurones to fire are those with the lowest axonal conduction velocity (30 m/sec), followed by those with progressively higher velocities, up to 90 m/sec (Yasargil and Koller, 1964).

Frequencies of discharge up to 50 and 100 per sec respectively have been recorded in the decerebrate cat (Gill, 1963) and in the rabbit (Yasargil, 1967). Evidence of recurrent inhibition in phrenic motoneurones was found by Baumgarten *et al.* (1963), but not by Gill and Kuno (1963). Phrenic motoneurones in the cat discharge at a frequency of 400 impulses/sec when the epipharynx is mechanically stimulated and when the glossopharyngeal nerve is stimulated electrically (Nail *et al.*, 1969).

*Afferent.*—The afferent innervation of the diaphragm is supplied by the phrenic nerves except for the marginal part which is supplied by nerves coming from $T_6$ to $T_{12}$ and most of the crura which are supplied by $T_{12}$ (Rasmussen, 1952; Hamilton *et al.*, 1962).

Both histological (Dogiel, 1902; Timofejew, 1902; Gregor, 1904; Masumoto, 1934; Winkler and Delaloye, 1957; Lanza, 1958; Barstad *et al.*, 1965) and physiological data (Glebovskii, 1962; Yasargil, 1962; Corda *et al.*, 1965) have shown in various species the paucity of proprioceptors in the diaphragm particularly in its costal and sternal part. Slowly adapting discharges have been recorded in the peripheral cut end of the phrenic nerve which presumably originate from diaphragmatic proprioceptors

(Cardin, 1944). Sensory free endings and Pacinian corpuscles have been shown to be present and widely scattered over the entire diaphragm of different species (Lanza, 1958): these receptors might account for the rapidly adapting discharges on probing the diaphragm recorded in the 5th and 6th cervical dorsal roots of cats by Glebovskii (1962), Yasargil (1962) and Corda *et al.* (1965). These fast adapting discharges may derive also from peritoneal sensory endings.

Corda and his colleagues (1965) found that the ratio between afferents from muscle spindles and those from tendon organs is low (0·8) as compared to that found in the intercostals (2·9) and the limb muscles (about 1·6) (Barker, 1962). Since the muscle spindles in the diaphragm are very few and the ratio between them and the tendon organs is not particularly low, the concentration of tendon organs in the diaphragm should be less than in other muscles.

The afferent fibres in the phrenic nerve of the cat are about 10 per cent of the myelinated fibres (Hinsey *et al.*, 1939), whereas in the dog they are 43 per cent (Landau *et al.*, 1962). In both these species, as well as in the rabbit (Fernand and Young, 1951), there are only a few fibres with a diameter above 14 $\mu$. This is in keeping with the paucity of muscle spindles in the diaphragm.

**Contractile Properties**

Data on the contractile properties of the diaphragm have been obtained by Goffart and Ritchie (1952) on the isolated rat diaphragm, and by Glebovskii (1961) on cat diaphragm *in situ* with one insertion cut to measure the force and the other fixed. Mognoni *et al.* (1968) and Sant'Ambrogio and Saibene (1970) determined the transdiaphragmatic pressure in dogs, cats, rabbits and rats. This technique leaves the diaphragm in its normal state, but the time course of the pressure change may be influenced by the deformation of the rib cage and may therefore differ somewhat from the time course of the force exerted by the diaphragm. The data obtained with these different methods all indicate that the overall behaviour of the diaphragm is intermediate between "fast" and "slow" muscles, being in fact the resultant of fast and slow fibres.

In the cat the rise time to peak force in a single twitch (contraction time) is 27 msec for "fast muscles" (average of several limb muscles), and 70 msec for "slow muscles" (crureus and soleus) (Buller *et al.*, 1960). Glebovskii (1961) reported that for the diaphragm the contraction time was 48·8 msec. Sant'Ambrogio and Saibene (1970) subsequently found a value of 38·9 $\pm$ 1·8 (SE) msec. The time interval between stimuli at which the half tetanic force is reached (stimulus interval) is 30 msec according to Glebovskii (1961) and 45·5 msec according to Sant'Ambrogio and Saibene (1970).

In the rat the contraction time of the diaphragm has been found to be 16 msec by Goffart and Ritchie (1952) and 18·3 ± 0·8 msec by Sant'Ambrogio and Saibene (1970), being 11 msec for the extensor digitorum longus and 35 msec for the soleus (Close, 1964).

The mean contraction time of the diaphragm in the rabbit is 32 ± 3·5 msec and in the dog 64·7 ± 4·6 msec, i.e. shorter and longer respectively than the cat (Sant'Ambrogio and Saibene, 1970). The frequency of stimulation-pressure relationship in these animals is accordingly shifted, with respect to the cat, to the right in the rabbit and to the left in the dog. The stimulus interval is 33·3 msec in the rabbit and 77 msec in the dog.

During tetanic stimulation of one or both phrenic nerves with closed airways the transdiaphragmatic pressure changes approximately in an exponential way. The half time of the pressure change found by Mognoni *et al.* (1968) was 27·5 ± 1·5 msec in cats and 61·9 ± 3·6 msec in dogs; Sant'Ambrogio and Saibene (1970) recorded values of 32·0 ± 5·4 msec in rabbits, 34·3 ± 2 msec in cats and 57·0 ± 5/6 msec in dogs. The slower time course in the dog could be due in part to a smaller deformation of its chest wall and in part to different features of the passive and/or active part of the diaphragm (Mognoni *et al.*, 1968; Sant'Ambrogio and Saibene, 1970).

When the cat diaphragm, tetanically stimulated, contracts against graded flow resistance the transdiaphragmatic pressure decreases as the flow increases, i.e. as the resistance decreases: with no added resistance the pressure is about half the isometric one (Mognoni *et al.*, 1968; Pengelly *et al.*, 1970), as in man during maximum efforts (see Chapter IV).

Glebovskii (1961) has pointed out that the time course of the single contraction suggests that this muscle contains fibres with different rates of contraction, some behaving as "fast" and others as "slow" fibres. Evidence for this composite structure of the diaphragm was also given by Decandia *et al.* (1966), who recorded action potentials from diaphragmatic single fibres during static inspiratory efforts in the rabbit, and found two maximal rates of discharge: 20 and 70 per sec. This finding has been confirmed also for the cat by Nail *et al.* (1969) who found two peak frequencies of 290 and 400 per sec. Histological evidence for the presence of two types of muscle fibres in the human and in the mammalian diaphragm have also been found (see above).

### Mechanical Action

The contraction of the diaphragm decreases the intrathoracic pressure, increases the lung volume, and increases the abdominal pressure. It is easy to understand that these effects are produced by the lowering of the diaphragmatic dome, which may be demonstrated fluoroscopically in man. On the other hand it is difficult to assess whether these effects are favoured or hindered by the direct action of the diaphragm on the ribs.

This problem is a classical source of dispute: Galen maintained that the diaphragm expands the rib cage, but this view was challenged by Borelli (1680), who reasoned that the diaphragm constricts it. The historical controversies were reviewed by Duchenne (1867), who showed that the contraction of the diaphragm expands the rib cage provided that the normal relationships of the diaphragm are preserved and the abdomen is closed. This view is generally accepted; in fact, since the costal fibres of the diaphragm run rather vertically, their main action should be that of raising the ribs and hence also of moving them outwards (see Chapter II). On the other hand the complexity of the diaphragm geometry suggests that such reasoning may be an oversimplification. Furthermore the effect on the rib cage of man and standing animals could be different from that on the rib cage of quadrupeds.

In the supine rabbit the rib cage circumference during spontaneous breathing does not change appreciably; it decreases when the phrenic nerves are stimulated, and increases when the phrenic nerves are blocked (Provini *et al.*, 1965; Mognoni *et al.*, 1969*a*). During tetanic stimulation of the diaphragm with open airway the lateral diameter of the rib cage at the xiphoid level decreases markedly, while the dorso-ventral diameter increases slightly: the cross sectional area of the rib cage at this level decreases (Mognoni *et al.*, 1969*a*). The lateral and dorso-ventral diameters of the rib cage have been measured in two rhesus and one cinocephalus monkeys in the supine position (Sant'Ambrogio, Saibene and Mognoni, unpublished observations). When the phrenic nerves were stimulated to produce an increase of lung volume similar to the normal tidal volume the cross sectional area of the rib cage at the xiphoid level decreased slightly in the rhesus monkeys, whereas it increased in the cinocephalus. Since the intrathoracic pressure falls during inspiration and therefore the transmural pressure tends to collapse the passive rib cage, the finding that in one case the cross sectional area increased suggests that the direct action of the diaphragm on the ribs is an expanding one, but that this action is small in comparison with the effects produced by lowering the diaphragmatic dome. In fact the cross sectional area of the rib cage decreased markedly in all cases when the transmural pressure was increased by stimulating the diaphragm with closed airways. In these monkeys the lateral diameter is smaller than the dorso-ventral, as in quadrupeds, hence the extrapolation of these results to man may be misleading.

In subjects with low cervical cord transection or with spinal anesthesia producing motor block up to about $T_1$ it has been observed that the lower half of the rib cage moves out with inspiration, while the upper half moves in (Eisele *et al.*, 1968). Zechman (personal communication) showed that in man with low cervical transection the rib cage circumference during inspiration decreases in the cranial part, but increases at the xiphoid level. This could be consistent with the previous interpretation on the direct expanding action of the diaphragm on the ribs, but unfortunately the changes of circumference are not a reliable index of the change of the cross section, because if this latter becomes more circular an increase of cross section could be accompanied by a decrease of circumference, and vice versa.

As the lung volume increases and the costal fibres of the diaphragm run more horizontally the direct action of the diaphragm on the rib cage should gradually become constrictive. The flattened diaphragm such as is seen in patients with chronic airway obstruction is not only a cause of disability because of its diminished vertical excursion, but also because it opposes the other muscles of inspiration which attempt to move the ribs outwards. In this way the contraction of the diaphragm contributes to the paradoxical inward movement of the base of the thorax during inspiration which is seen in many of these patients (Godfrey *et al.*, 1969).

The transdiaphragmatic pressures exerted during maximum voluntary efforts at different lung volumes are dealt with in Chapter III, while those exerted during the breathing cycle are dealt with in Chapter IV. It is worth while considering here the transdiaphragmatic pressure exerted during electrical stimulation of one or both phrenic nerves while the other respiratory muscles are inactive. Experiments on supine dogs and cats (Marshall, 1962; Pengelly *et al.*, 1970) showed that the transdiaphragmatic pressure decreases almost linearly with increasing lung volume above FRC. Sant'Ambrogio and Saibene (1970) made similar experiments on supine rabbits on a wider volume range. They found that the decrease of pressure with lung volume becomes smaller as the lung volume increases. The transdiaphragmatic pressure was —45 cm $H_2O$, at a lung volume corresponding to an alveolar relaxation pressure of —15 cm $H_2O$, —30 cm $H_2O$ at FRC, and zero at a lung volume corresponding to an alveolar relaxation pressure of 30 cm $H_2O$; at larger volumes the transdiaphragmatic pressure reversed, i.e. the action of the diaphragm become expiratory. Pengelly and his co-workers (1970) found a behaviour similar to the cat in the sitting man in whom one phrenic nerve was stimulated submaximally. The slope of the relationship between transdiaphragmatic pressure and lung volume was substantially independent of the degree of stimulation. The decrease of the pressure exerted by the diaphragm with increasing lung volume depends both upon the force-length relation of the muscle, since the fibres of the diaphragm become shorter as the lung volume increases, and upon its geometry, because as the lung volume increases the radius of curvature of the diaphragm increases and therefore its ability to exert pressure for a given force of its fibres diminishes according to the Laplace law (Marshall, 1962). The decrease of the pressure exerted by the diaphragm with increasing lung volume implies that for a given output of the phrenic motoneurones the tidal volume should be smaller during positive pressure breathing and in patients with diffuse airways obstruction and increased

FRC. This could explain in part the reduced ventilatory response to $CO_2$ in such patients (Pengelly *et al.*, 1970).

**Circumstances of Contraction**

The electrical activity of the diaphragm has been studied in human subjects, both in the supine and sitting postures, by means of needle electrodes inserted through the body wall in the costal part (Rossier *et al.*, 1956; Koepke *et al.*, 1958; Murphy *et al.*, 1959; Taylor, 1960; Fink *et al.*, 1960), or by means of surface electrodes through a bipolar oesophageal lead picking up the action potentials of the vertebral part (Draper *et al.*, 1959; Agostoni *et al.*, 1960*a*; Petit *et al.*, 1960; Grønbaek and Skouby, 1960; Lourenço and Mueller, 1967; Sears *et al.*, 1969).

During quiet breathing the activity increases progressively throughout inspiration and decreases in the early part of expiration to become nil at about half the expiration time (Fig. 57). During the last part of expiration the diaphragm is relaxed even in the horizontal position despite its marked distension (see Statics chapter): this suggests that the stretch reflex in the diaphragm is weak or absent, as could be expected owing to the paucity of muscle spindles (see above). Furthermore, diaphragm activity is not altered by postural reflexes, which affect other respiratory muscles (Massion *et al.*, 1960). At high values of ventilation the activity during the expiratory phase decreases more quickly, to begin again before the expiration is over (Agostoni *et al.*, 1960*a*; Petit *et al.*, 1960).

The diaphragm is active during expulsive efforts (Agostoni *et al.*, 1960*a*, see Statics chapter), i.e. when it acts synergistically with the abdominal muscles to raise the abdominal pressure as probably occurs in parturition, defaecation and vomiting. Electrical activity of the diaphragm during vomiting has been recorded in dog (Jiménez-Vargas *et al.*, 1967). Some activity is also found during strong expiratory efforts particularly at low lung volume (see Statics chapter), and therefore at the end of a maximum expiration (Fig. 58) (Agostoni and Torri, 1962; Delhez *et al.*, 1964). Furthermore diaphragm activity has been shown in coughing, and sneezing (Agostoni *et al.*, 1960*b*). During phonation the activity of the diaphragm diminishes and rapidly ceases during the first two or three seconds of an

---

FIG. 57 (SEE OPPOSITE).—ACTIVITY OF THE DIAPHRAGM DURING QUIET BREATHING

From bottom to top: (1) electrical activity of the vertebral part of the human diaphragm, recorded with a bipolar oesophageal lead during quiet breathing in the upright posture; the heart electrical activity is also recorded; (2) respiratory flow; (3) tidal volume; (4) transdiaphragmatic pressure. (From Agostoni, 1964.)

FIG. 58 (SEE OPPOSITE).—ACTIVITY OF THE DIAPHRAGM DURING VITAL CAPACITY

From bottom to top: (1) spirometric record; (2) pneumotachogram; (3) electrical activity of the diaphragm; (4) electrical activity of the external oblique, during the vital capacity. (From Delhez *et al.*, 1964.)

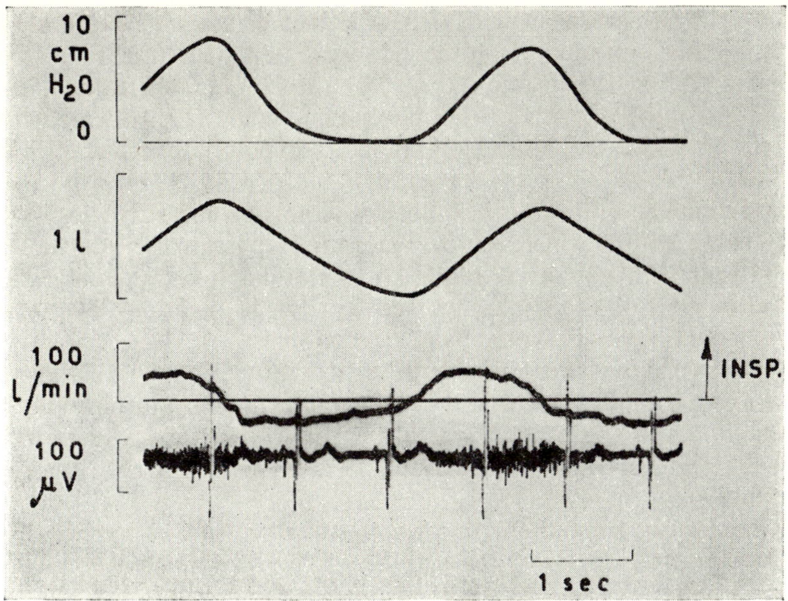
10
cm
$H_2O$
0
1 l
100
l/min
100
μV
INSP.
1 sec

FIG. 57.

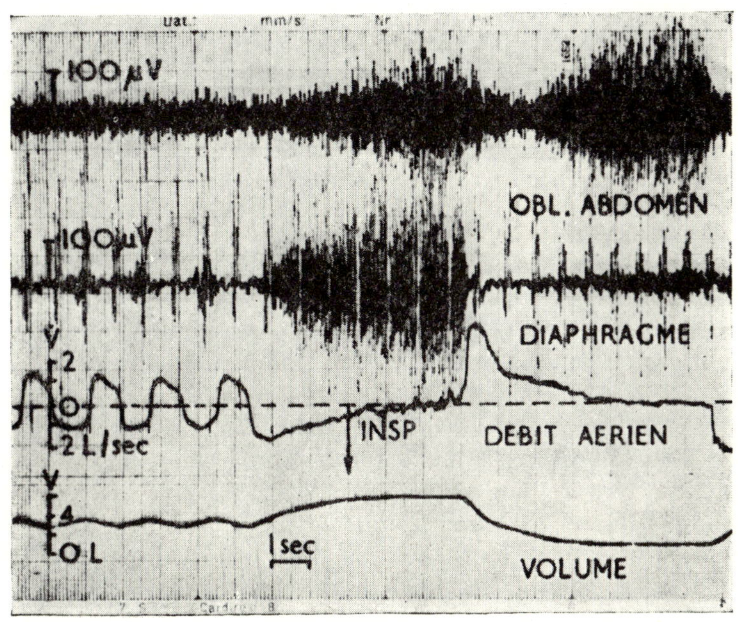
100 μV
OBL. ABDOMEN
100 μV
DIAPHRAGME
2
0
2 L/sec
INSP
DEBIT AERIEN
4
0 L
1 sec
VOLUME

FIG. 58.

utterance after a maximal inspiration (Draper *et al.*, 1959). All these findings are in agreement with the behaviour of the transdiaphragmatic pressure after subtraction of the passive component (see Fig. 57 and Chapters III and IV). In subjects claiming to be able to breathe with the diaphragm alone or without the diaphragm, Wade (1954) demonstrated that the movements of the diaphragmatic dome relative to the rib cage were the same. On the other hand Stigol and Cuello (1966) were able to record in one out of three trained physiotherapists some active inspirations from FRC without increase of the transdiaphragmatic pressure, i.e. inspirations with voluntary inhibition of the diaphragm activity. Hence voluntary control of diaphragm activity during breathing is probably possible within certain limits, but is exceptional.

During positive pressure breathing in conscious man the electrical activity of the diaphragm persists even at values at which inspiration could be completely passive according to the volume-pressure diagram of the relaxed respiratory system (Agostoni, 1962). This is not the case during anaesthesia.

When in the anaesthetised cat a positive pressure is applied to the airways during expiration, the activity of the diaphragm increases notwithstanding the concurrent increase of lung volume; this increase begins 5–6 breathing cycles after the load is applied and does not disappear immediately when the load is removed (Bishop, 1967). On the other hand if a threshold expiratory load (i.e. one which requires that a given pressure is reached before air will flow) is applied to the anaesthetised dog the activity of the diaphragm increases inconsistently (Lourenço *et al.*, 1966).

During breath holding, in most subjects, the activity of the diaphragm is at first absent; discharges then take place at progressively higher rates up to the breaking point. The onset of diaphragm activity during breath holding is strictly related to the values of $P_{CO_2}$ and $P_{O_2}$; it does not seem to be affected by neurogenic factors related to lung volume or respiratory movements (Agostoni, 1963).

**Respiratory Function**

The diaphragm is the principal muscle of inspiration, but is not essential for breathing. The movements of the diaphragmatic dome and the approximate volume contribution of the diaphragm in man are dealt with in Chapter II. During quiet breathing the diaphragm contribution to the tidal volume has been estimated to be about two-thirds in the sitting and standing posture, and three-quarters in the supine posture. About two-thirds of the VC has been estimated to be contributed by the diaphragm in all postures; 60 per cent of this depends, however, upon the passive displacement of the diaphragm over the ERV caused by the activity of the muscles of the abdominal wall.

The contribution of the diaphragm to ventilation of various intensities has been determined in anaesthetised vagotomised rabbits in the supine position by measuring the decrease of tidal volume occurring immediately after blocking the conduction in the phrenic nerves with an electrotonic current (Mognoni *et al.*, 1969*a*). The relative contribution of diaphragm to the tidal volume was about 90 per cent during quiet breathing and about 75 per cent for the maximum tidal volume attained during rebreathing (Fig. 59). For an inspiratory drive smaller than that during quiet breathing only the diaphragm seems to contribute a volume change in the anaesthetised rabbit.

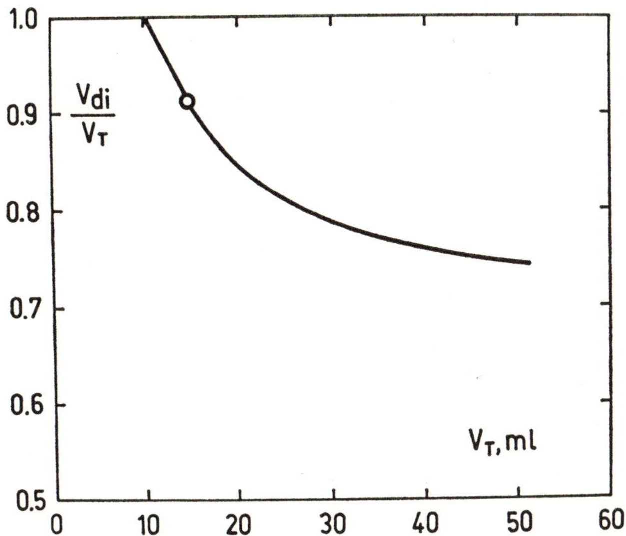


FIG. 59.—RELATIVE CONTRIBUTION OF THE DIAPHRAGM ($V_{di}$) TO THE TIDAL VOLUME ($V_T$)

Spontaneous breathing in the anaesthetised vagotomised rabbits in the supine position. The tidal volume is changed by increasing or decreasing the dead space; the circle refers to the normal dead space. The curve has been calculated from data of Mognoni, Saibene and Sant'-Ambrogio (1969*a*) on the immediate fall of tidal volume after blocking the conduction on the phrenic nerves.

Anaesthesia may vary the relative contribution of the diaphragm; during deep anaesthesia there may be paralysis of the intercostals without any marked effect on the diaphragm (Gillsepie, 1943; Mognoni *et al.*, 1969*b*).

Normal ventilation and even hyperventilation may be maintained in animals and man by periodic stimulation of one phrenic nerve through an implanted or surface electrode (Sarnoff *et al.*, 1948; Whittenberger *et al.*, 1949).

There has been no study on the mechanics of breathing before and after complete diaphragmatic paralysis in subjects with normal respiratory apparatus. McCredie *et al.* (1962) in three subjects with diaphragm paralysis found a marked decrease of the vital capacity as compared to the predicted values. Fackler and his colleagues (1967) studied 14 subjects before and after unilateral phrenicectomy; the vital capacity was significantly reduced during the first six months and then returned to normal.

After bilateral phrenicotomy in the supine anaesthetised rabbit ventilation falls so markedly that the arterial $P_{CO_2}$ increases above 60 mm Hg. Hence under these conditions the extradiaphragmatic muscles do not provide an adequate ventilation and the ventilatory response to $CO_2$ is essentially nil. One to two months after bilateral phrenicectomy the ventilatory response to $CO_2$ reappears, but the degree of recovery varies greatly (Sant'Ambrogio *et al.*, 1968). On the other hand in cats the decrease of ventilation after bilateral phrenicotomy is not so marked as in rabbits and a ventilatory response to $CO_2$ is still present (unpublished observations).

## REFERENCES

AGOSTONI, E. (1962). Diaphragm activity and thoracoabdominal mechanics during positive pressure breathing. *J. appl. Physiol.*, **17**, 215–220.

AGOSTONI, E. (1963). Diaphragm activity during breath holding: factors related to its onset. *J. appl. Physiol.*, **18**, 30–36.

AGOSTONI, E. (1964). Action of the respiratory muscles. In: *Handbook of Physiology*. Section 3, Respiration. Vol. 1, pp. 377–386. Eds. W. O. Fenn and H. Rahn. Washington, D.C.: American Physiological Soc.

AGOSTONI, E., SANT'AMBROGIO, G., and DEL PORTILLO CARRASCO, H. (1960*a*). Electromyography of the diaphragm in man and transdiaphragmatic pressure. *J. appl. Physiol.*, **15**, 1093–1097.

AGOSTONI, E., SANT'AMBROGIO, G. and DEL PORTILLO CARRASCO, H. (1960*b*). Ellettromiografia del diaframma e pressione transdiaframatica durante la tosse, lo sternuto ed il riso. *Atti Accad. Naz. Lincei, Rend.*, **28**, 493–496.

AGOSTONI, E., and TORRI, G. (1962). Diaphragm contraction as a limiting factor to maximum expiration. *J. appl. Physiol.*, **17**, 427–428.

BARKER, D. (1962). The structure and distribution of muscle receptors. In: *Symposium on Muscle Receptors*, pp. 227–240. Ed. Barker, D. Hong Kong Univ. Press.

BARSTAD, J. A. B., KRISTOFFERSEN, A., LILLEHEIL, G., and STAALAND, H. (1965). Muscle spindles in the rat diaphragm. *Experientia* (*Basel*), **21**, 533–534.

BAUMGARTEN, R. v., SCHMIEDT, H., DODICH, N. (1963). Microelectrode studies of phrenic motoneurons. *Ann. N.Y. Acad. Sci.*, **109**, 536–546.

BISHOP, B. (1967). Diaphragm and abdominal muscle responses to elevated airway pressures in the cat. *J. appl. Physiol.*, **22**, 959–965.

BORELLI, G. A. (1680). *De Motu Animalium.* Roma: Bernabò.

BULLER, A. J., ECCLES, J. C., and ECCLES, R. M. (1960). Differentiation of fast and slow muscles in the cat hind limb. *J. Physiol.* (*Lond.*), **150**, 399–416.

CARDIN, A. (1939). Il muscolo diaframma e le sue componenti cinestesiche. *Arch. Sci. biol.* (*Bologna*), **25**, 51–88.

CARDIN, A. (1944). Recettori di tensione nel diaframma. *Arch. Sci. biol.* (*Bologna*), **30**, 9–22.

Close, R. (1964). Dynamic properties of fast and slow skeletal muscles of the rat during development. *J. Physiol.* (*Lond.*), **173**, 74–95.

Collis, J. L., Satchwell, L. M., and Abrams, L. D. (1954). Nerve supply to the crura of the diaphragm. *Thorax*, **9**, 22–25.

Corda, M., von Euler, C., and Lennerstrand, G. (1965). Proprioceptive innervation of the diaphragm. *J. Physiol.* (*Lond.*), **178**, 161–177.

Decandia, M., Gantchev, G. N., and Sant'Ambrogio, G. (1966). Correlation between intrapulmonary pressure and rate of firing of diaphragmatic motor units during inspiratory efforts. *J. Physiol.* (*Lond.*), **186**, 3–4P.

Delhez, L. (1965). Modalités, chez l'homme normal, de la réponse électrique des piliers du diaphragme à la stimulation électrique des nerfs phréniques par des chocs uniques. *Arch. int. Physiol., Biochim.*, **73**, 832–839.

Delhez, L., Troquet, J., Damoiseau, J., Pirnay, F., Deroanne, R., and Petit, J. M. (1964). Influence des modalités d'exécution des manoeuvres d'éxpiration forcée et d'hyperpression thoraco-abdominale sur l'activité électrique du diaphragme. *Arch. int. Physiol., Biochim.*, **72**, 76–94.

Dogiel, A. S. (1902). Nervenendigungen in Bauchfell, in den Sehnen, den Muskelspindeln und dem centrum tendineum des Diaphragms beim Menschen und bei Saugethieren. *Arch. mikr. Anat.*, **59**, 1–31.

Draper, M. H., Ladefoget, P., and Whitteridge, D. (1959). Respiratory muscles in speech. *J. Speech Res.*, **2**, 16–27.

Duchenne, G. B. A. (1867). *Physiologie des mouvements démontrée à l'aide de l'expérimentation électrique et de l'observation clinique, et applicable à l'étude des paralysies et des déformations.* Paris: Baillière.

Eisele, J., Trenchard, D., Burki, N., and Guz, A. (1968). The effect of chest wall block on respiratory sensation and control in man. *Clin. Sci.*, **35**, 23–33.

Fackler, C. D., Perret, G. E., and Bedell, G. N. (1967). Effect of unilateral phrenic nerve section on lung function. *J. appl. Physiol.*, **23**, 923–926.

Felix, W. (1922). Anatomische, experimentelle und Klinische Untersuchungen über den Phrenicus und über die Zwerchfellinnervation. *Dtsch. Z. Chir.*, **171**, 283–397.

Fernand, V. S. V., and Young, J. Z. (1951). The sizes of the nerve fibres of muscle nerves. *Proc. roy. Soc.*, **139B**, 38–58.

Fink, R., Hanks, E. C., Holaday, D. A., and Ngai, S. H. (1960). Monitoring of ventilation by integrated diaphragmatic electromyogram. *J. Amer. med. Ass.*, **172**, 1367–1371.

Gauthier, G. F., and Padykula, H. A. (1966). Cytological studies of fiber types in skeletal muscle. *J. cell. Biol.*, **28**, 333–354.

Gill, P. K. (1963). The effects of end-tidal $CO_2$ on the discharge of individual phrenic motoneurones. *J. Physiol.* (*Lond.*), **168**, 239–257.

Gill, P. K., and Kuno, M. (1963). Excitatory and inhibitory actions on phrenic motoneurones. *J. Physiol.* (*Lond.*), **168**, 274–289.

Gillespie, N. A. (1943). The signs of anesthesia. *Curr. Res. Anesth.*, **22**, 275–282.

Glebovskii, V. D. (1961). Contractile properties of respiratory muscles in fully grown and neonate animals. *Sechenov. physiol. J. U.S.S.R.*, **47**, 470–480.

Glebovskii, V. D. (1962). Stretch receptors of the diaphragm. *Fiziol. Zh.* (*Mosk.*), **48**, 545. (Trans. in *Fed. Proc.* (1963), **22**, II T405–T410).

GODFREY, S., EDWARDS, R. H. T., CAMPBELL, E. J. M., ARMITAGE, P., and OPPENHEIMER, E. A. (1969). Repeatability of physical signs in airways obstruction. *Thorax*, **24**, 4–9.

GOFFART, M., and RITCHIE, J. M. (1952). The effect of adrenaline on the contraction of mammalian skeletal muscle. *J. Physiol.* (*Lond.*), **116**, 357–371.

GOODRICH, E. S. (1958). *Studies on the Structure and Development of Vertebrates*, pp. 643–656. New York: Dover.

GREGOR, A. (1904). Ueber die Vertheilung der Muskelspindeln in den Musculatur des menschlichen Fötus. *Arch. Anat. Physiol., Anat. Abt.*, 112–191.

GRØNBAEK, P., and SKOUBY, A. P. (1960). The activity pattern of the diaphragm and some muscles of the neck and trunk in chronic asthmatics and normal controls. *Acta med. scand.*, **168**, 413–425.

GÜNTHER, P. G. (1953). Das muskuläre substrat der bewegungs und halteleistung des menschlichen zwerchfells. *Acta anat.* (*Basel*), **17**, 348–352.

HAMILTON, W. S., BOYD, J. D., and MOSSMAN, H. W. (1962). *Human Embryology*, 3rd edit. Cambridge: Heffer.

HEINBECKER, P., BISHOP, G. H., and O'LEARY, J. L. (1936). Functional and histologic studies of somatic and autonomic nerves of man. *Arch. Neurol. Psychiat.* (*Chic.*), **35**, 1233–1255.

HINSEY, J. C., HARE, K., and PHILLIPS, R. A. (1939). Sensory components of the phrenic nerve of the cat. *Proc. Soc. exp. Biol.* (*N.Y.*), **41**, 411–414.

JANSEN, J. Z. (1931). Beitrag zur Kenntnis der Zwerchfellinnervation. *Z. Anat. Entwickl.-Gesch.*, **96**, 624–657.

JEFFERSON, N. C., PHILLIPS, C. W., and NECHELES, N. (1950). Left phrenophrenic anastomosis. *J. appl. Physiol.*, **3**, 161–163.

JIMÉNEZ-VARGAS, H., ASIRÓN, M., VOLTAS, J. and ONAINDIA, J. (1967). Electromiografia de músculos respiratorios en la tos, en los reflejos de la glotis y en el vómito. *Rev. esp. Fisiol.*, **23**, 65–74.

KISS, F., and BALLON, H. C. (1929). Contribution to the nerve supply of the diaphragm. *Anat. Rec.*, **41**, 285–298.

KOEPKE, G. H., SMITH, E. M., MURPHY, A. J., and DICKINSON, D. G. (1958). Sequence of action of the diaphragm and intercostal muscles during respiration. I-Inspiration. *Arch. phys. Med.*, **39**, 426–430.

KRNJEVIC, K., and MILEDI, R. (1958). Motor units in the rat diaphragm. *J. Physiol.* (*Lond.*), **140**, 427–439.

LANDAU, B. R., AKERT, K., and ROBERTS, T. S. (1962). Studies on the innervation of the diaphragm. *J. comp. Neurol.*, **119**, 1–10.

LANZA, G. G. (1958), Ricerche istologiche comparative sulla innervazione sensitiva e su quella vegetativa del diaframma di alcuni mammiferi. *Boll. Soc. ital. Biol. sper.*, **34**, 501–504.

LOURENÇO, R. V., CHERNIACK, N. S., MALM, J. R., and FISHMAN, A. P. (1966). Nervous output from the respiratory center during obstructed breathing. *J. appl. Physiol.*, **21**, 527–533.

LOURENÇO, R. V., and MUELLER, E. P. (1967). Quantification of electrical activity in the human diaphragm. *J. appl. Physiol.*, **22**, 598–600.

MARSHALL, R. (1962). Relationships between stimulus and work of breathing at different lung volumes. *J. appl. Physiol.*, **17**, 917–921.

MASSION, J., MEULDERS, M., and COLLE, J. (1960). Fonction posturale des muscles respiratoires. *Arch. int. Physiol., Biochem.*, **68**, 314–326.

MASUMOTO, K. (1934). Histologische Studien über die peripheren Nerven des Zwerchfells. Mitteilung 1. *Mitt. med. Akad. Kioto.*, **10**, 1015–1018.

MCCREDIE, M., LOVEJOY, F. W., JR., and KALTREIDER, N. L. (1962). Pulmonary function in diaphragmatic paralysis. *Thorax*, **17**, 213–217.

MOGNONI, P., SAIBENE, F., SANT'AMBROGIO, G., and AGOSTONI, E. (1968). Dynamics of the maximal contraction of the respiratory muscles. *Resp. Physiol.*, **4**, 193–202.

MOGNONI, P., SAIBENE, F., and SANT'AMBROGIO, G. (1969*a*). Contribution of the diaphragm and the other inspiratory muscles to different levels of tidal volume and static inspiratory efforts in the rabbit. *J. Physiol.* (*Lond.*), **202**, 517–534.

MOGNONI, P., SAIBENE, F., and SANT'AMBROGIO, G. (1969*b*). Effect of anaesthesia on the contribution of the diaphragm to ventilation in the rabbit. *Atti Accad. Naz. Lincei, Rend., Classe Sci.: Fis. Mat. Nat.*, **46**, 99–100.

MULLER BOTHA, G. S. (1957). The anatomy of phrenic nerve termination and the motor innervation of the diaphragm. *Thorax*, **12**, 50–56.

MURPHY, A. J., KOEPKE, G. H., SMITH, E. M. and DICKINSON, D. G. (1959). Sequence of action of the diaphragm and intercostal muscles during respiration. II: Expiration. *Arch. phys. Med.*, **40**, 337–342.

NAIL, B. S., STERLING, G. M., and WIDDICOMBE, J. G. (1969). Some properties of single phrenic motoneurones. *J. Physiol.* (*Lond.*), **200**, 137P–138P.

NEWSOM DAVIS, J. (1967). Phrenic nerve conduction in man. *J. Neurol. Neurosurg. Psychiat*, **30**, 420–426.

OGAWA, T. (1959). Studies on the phrenic nerves and the diaphragm of the dog. *Amer. J. Surg.*, **97**, 744–748.

OSHIMA, T. (1939). Zahlenverhältnis zwischen den Muskel und Nervenfasern bei der Innervation des Muskelindividuums. *Jap. J. med. Sci.*, **7**, 57–76.

PENGELLY, L. D., ALDERSON, A. M., and MILIC-EMILI, J. (1970). Mechanics of the diaphragm. *J. appl. Physiol.* (in press).

PETIT, J. M., MILIC-EMILI, G., and DELHEZ, L. (1960). Role of the diaphragm in breathing in conscious normal man: an electromyographic study. *J. appl. Physiol.*, **15**, 1101–1106.

PROVINI, L., DECANDIA, M., and SANT'AMBROGIO, G. (1965). Variazioni del perimetro toracico del coniglio durante la respirazione normale e a diaframma paralizzato. *Boll. Soc. ital. Biol. sper.*, **41**, 1547–1548.

RASMUSSEN, A. T. (1952). *The Principal Nervous Pathways*, p. 43. New York: Macmillan.

ROSENBLUETH, A., ALANIS, J., and PILAR, G. (1961). The accessory motor innervation of the diaphragm. *Arch. int. Physiol., Biochem.*, **69**, 19–25.

ROSSIER, P. M., NIEPORENT, H. J., PIPBERGER, H., and KÄLIN, Z. (1956). Electromyographic studies on respiratory muscle function in normal volunteers. *Z. ges. exp. Med.*, **127**, 39–52.

ROWETT, H. G. R. (1957). *The Rat as a Small Animal*, p. 66. London: Murray.

SANT'AMBROGIO, G., BONANNI, M. V., and CAMPORESI, E. (1968). Risposta ventilatoria all'ipercapnia nel coniglio frenicotomizzato. *Boll. Soc. ital. Biol. sper.*, **44**, 1055–1056.

SANT'AMBROGIO, G., FRAZIER, D. T., WILSON, M. F., and AGOSTONI, E. (1963). Motor innervation and pattern of activity of cat diaphragm. *J. appl. Physiol.*, **18**, 43–46.

SANT'AMBROGIO, G., and SAIBENE, F. (1970). Contractile properties of the diaphragm in some mammals. *Resp. Physiol.* (in press).

SARNOFF, S. J., HARDENBERGH, E., and WHITTENBERGER, J. L. (1948). Electrophrenic respiration. *Amer. J. Physiol.*, **155**, 1–9.

SCHLAEPFER, K. (1926). A further note on the motor innervation of the diaphragm. *Anat. Rec.*, **32**, 143–150.

SCOTT, R. (1965). Innervation of the diaphragm and its practical aspects in surgery. *Thorax*, **20**, 357–361.

SEARS, T. A., MEAD, J., LEITH, D., KELLOGG, R., KNUDSON, R., and GOLDMAN, M. D. (1968). The role of the diaphragm in stabilizing ventilation. *Proc. int. Un. Physiol. Sci.*, **7**, 394.

STIGOL, L. C., and CUELLO, A. C. (1966). Voluntary control of the diaphragm in one subject. *J. appl. Physiol.*, **21**, 1911–1912.

STRAUSS, L. H. Z. (1933). Beitrag zur motorischen Innervation des Zwerchfeller bei Menschen und bei Tieren. *Z. ges. exp. Med.*, **86**, 244–257.

TAYLOR, A. (1960). The contribution of the intercostal muscles to the effort of respiration in man. *J. Physiol.* (*Lond.*), **151**, 390–402.

TIMOFEJEW, D. A. (1902). Über die Nervenendigungen im Bauchfelle und in dem Diaphragma der Säugetiere. *Arch. mikr. Anat.*, **59**, 629–646.

TOSATTI, E. (1938). L'incrociamento del respiro diaframmatico dopo frenicotomia ed emisezione del midollo cervicale. *Arch. Fisiol.*, **38**, 533–564.

WADE, O. L. (1954). Movements of the thoracic cage and diaphragm in respiration. *J. Physiol.* (*Lond.*), **124**, 193–212.

WARWICK, R., and MITCHELL, G. A. G. (1955). The phrenic nucleus of the rhesus monkey. *J. Anat.* (*Lond.*), **89**, 562–563.

WHITTENBERGER, J. L., SARNOFF, S. J., and HARDENBERGH, E. (1949). Electrophrenic respiration II. Its use in man. *J. clin. Invest.*, **28**, 124–128.

WILSON, A. S. (1963). Studies on the innervation of the diaphragm in cats and rodents. *J. Anat.* (*Lond.*), **97**, 482–483.

WINKLER, G., and DELALOYE, B. (1957). A propos de la présence de fuseaux neuro-musculaires dans le diaphragme humain. *Acta Anat.* (*Basel*), **29**, 114–116.

YASARGIL, G. M. (1962). Proprioceptive Afferenzen im N. phrenicus der Katze. *Helv. physiol. pharmacol. Acta*, **20**, 39–58.

YASARGIL, G. M. (1967). Systematische Untersuchung der motorischen Innervation des Zwerchfells beim Kaninchen. *Helv. physiol. pharmacol. Acta*, Suppl. XVIII, 24.

YASARGIL, G. M., and KOLLER, E. A. (1964). Über die motorische Innervation des Zwerchfells beim Kaninchen. *Helv. physiol. pharmacol. Acta*, **22**, 137–147.

## Chapter VII

# THE INTERCOSTAL MUSCLES AND OTHER MUSCLES OF THE RIB CAGE

E. J. M. Campbell and J. Newsom Davis

## INTERCOSTAL MUSCLES

### Morphology

*The external intercostals* extend from the tubercles of the ribs to the costo-chondral junction where they become continuous with the anterior intercostal membrane. They are thicker posteriorly than anteriorly (Bryce, 1923) and thicker than the internal intercostals (Johnston and Whillis, 1949). The fibres slope obliquely downwards and forwards from the upper rib to the one below.

*The internal intercostals* extend from the anterior end of the intercostal space to the angles of the ribs posteriorly, where they become continuous with the posterior intercostal membrane. They are thicker anteriorly than posteriorly. The fibres slope obliquely downwards and backwards from the upper rib to the one below. The internal intercostals can be subdivided into a posterior or interosseous portion where the ribs slope downwards and forwards and an anterior or intercartilaginous portion (the "parasternals") where the costal cartilages slope upwards and forwards.

Three types of intercostal muscle fibre have been differentiated (see Chapter I). The small red fibres have a high oxidative enzymatic activity and low phosphorylase activity, while these properties are reversed in the large white fibres. Some fibres are intermediate in size and in their enzyme characteristics (Ogata *et al.*, 1963; Nishiyama, 1966). Fibre differentiation is not present at birth in the kitten, but by 25 days all three fibre types can be distinguished (Schwieler, 1968).

### Innervation

The intercostal nerves arise from the first to the twelfth thoracic segments, and are derived from the ventral primary ramus. In man, each main intercostal nerve, which supplies filaments to the intercostal muscles, lies deep to the internal intercostal muscle, giving off a collateral branch early in its course, a lateral cutaneous branch and a terminal anterior cutaneous branch. Individual nerves may communicate by fine branches with the intercostal nerves of the adjacent spaces. The lower intercostal nerves, which supply the abdominal muscles after penetrating the diaphragm (to which they give a few sensory branches; see Chapter VI), freely communicate with one another over the abdominal wall (Davies *et al.*, 1932). In the cat, each intercostal nerve branches into the external and internal intercostal nerves which separately supply the two layers of intercostal muscle by giving off fine filaments along their course (Sears, 1964*a*).

The intercostal alpha motoneurones in the cat and monkey are situated in the lateral and central nuclei of the anterior horn, inspiratory motoneurones being mainly central and expiratory motoneurones lateral (Sprague, 1951; G. L. Coffey, personal communication).

The fibre calibre spectrum of the intercostal nerves is known for the cat (Sears, 1964*a*). Both the internal intercostal nerve (a mixed nerve) and the external intercostal nerve (a pure muscle nerve) contain motor fibres whose diameters range from 2–20 $\mu$. The distribution of fibre diameters shows bimodal grouping, consistent with the presence of alpha fibres and fusimotor fibres, but the division between these groups is less prominent than in limb nerves. The bimodal distribution can also be demonstrated in the filaments of the nerves. No information is available about the innervation ratio of the intercostal muscles.

The intercostal muscles are well supplied with proprioceptors. Histological studies have demonstrated muscle spindles, tendon organs and Pacinian corpuscles (Huber, 1902; Barker, 1962). Huber (1902) counted the spindles in the intercostal spaces of one side in the cat. Each of the upper six contained 60–100, the seventh to the tenth spaces somewhat fewer and the lower two spaces 28 and 18 respectively.

Physiological methods have confirmed the presence of receptors with discharge characteristics consistent with muscle spindle primary and secondary endings, and tendon organs (Critchlow and von Euler, 1963; von Euler and Peretti, 1966). The significance of this rich proprioceptive supply is discussed in Chapter XI. The largest afferent fibres (16–20 $\mu$) in the internal intercostal nerve are almost entirely of muscular origin and must contain many fibres arising from muscle spindle primary endings (Sears, 1964*a*).

Recording the action potential in the intercostal nerves (Bronk and Ferguson, 1935) or the electrical activity of motor units in the muscle itself (Gesell *et al.*, 1940) has shown that intercostal muscle contraction is graded by variation in the discharge frequency and by the number of active units (see p. 9). The discharge frequency of units recruited early rose to higher values than of those recruited late. The gradation was particularly apparent in the progressive hyperpnoea induced by rebreathing; in addition to the progressive increase of the discharge frequency of active units and the recruitment of other units, there was an earlier initiation of activity in the inspiratory phase (Gesell *et al.*, 1940). The frequency of unit discharge remained somewhat lower in the intercostal muscles than in the diaphragm under conditions of chemical driving, and did not exceed

30/sec. The spinal and supraspinal control mechanisms which determine the gradation of muscle contraction are considered in Chapters XI and XII.

**Contractile Properties**

Values for the mean contraction times of intercostal muscles in the cat have been given as 48·9 msec (Glebovskii, 1961), 30–35 msec (Biscoe, 1962) and 26·3 msec (Andersen and Sears, 1964). The intercostal muscles contain both fast and slow units. The mean contraction time for fast units is 24·6 $\pm$ 4·0 SD msec and for slow units 47·0 $\pm$ 5·1 SD msec (Andersen and Sears, 1964; see also Glebovskii, 1961).

The apparent discrepancy between the values for mean contraction time in the cat given by different workers has been accounted for in a study using analogue computer modelling (Biscoe and Taylor, 1967), in which it was shown that the form of the muscle twitch was determined by the proportions of fast and slow units. From analysis of further data on single unit twitches provided by Andersen and Sears, the composition of cat intercostal muscle appeared to be 23·5 per cent fast units and 76·5 per cent slow units.

The contractile properties of the human external intercostal muscle have been studied in biopsy material. The mean contraction time was 71·5 msec, and the presence of fast and slow units was confirmed (Hofmann *et al.*, 1966).

**Mechanical Action**

The function of the intercostal muscles has been a source of controversy throughout medical history. The following summary of the various theories which have been held is compounded from the works of Beau and Maissiat (1843), Duchenne (1867), Luciani (1911) and Fleisch (1934).

1. Both the external and internal intercostals are inspiratory in function.
2. Both groups are expiratory in function.
3. The external intercostals are inspiratory in function, and the internal are expiratory, with the exception of the intercartilaginous portion of the internal intercostals, which is inspiratory.
4. The external intercostals are expiratory and the internal are inspiratory.
5. Both groups are inspiratory and expiratory at the same time.
6. Both groups act together, but their functions vary in different parts of the chest. They are inspiratory in function in some spaces and expiratory in others.
7. They are not concerned with either inspiration or expiration but regulate the tension in the intercostal spaces.

Up to the middle of the last century more than thirty authorities had divided their support fairly evenly between these various theories. Evidence was produced from many sources: dissection, geometry, mechanical models, experiments on animals, clinical observation and the effects of electrical stimulation.

A detailed examination of each theory would be a work of scholasticism rather than of physiology. One of them (the third), however, deserves special attention because of the widespread acceptance that it gained towards the end of the nineteenth century and because much subsequent experimental work supports it. This theory, which since the early eighteenth century has been associated with the name of Hamberger, is the most popular among standard modern texts of anatomy and physiology. Hamberger supported his theory by a mechanical model.

Hamberger's theory maintains that as the external intercostals and the intercartilaginous portion of the internal intercostals shorten they elevate the ribs, whereas shortening of the interosseous portion of the internal intercostals depresses the ribs. The most cogent critic of the theory was Hoover (1922), who argued that it neglects movement about the antero-posterior axis. However, movement about this axis is improbable (p. 23).

Although the detailed actions of the individual intercostal muscles are not fully understood, the groups which are active during normal inspiration probably raise the ribs. The evidence for this comes from two sources. There have been no objective studies of the effects of paralysis using methods for recording the movements of the rib cage; inspection reveals, however, that paralysis of the intercostal muscles in man decreases the amplitude of the rib movements in breaths of normal depth. This effect was shown by Alexander (1929), Cetrangolo (1930), and Joly and Vincent (1937), who destroyed the intercostal nerves as collapse therapy for patients with pulmonary tuberculosis. The second source of evidence comes from electromyographic studies which, as will be reported later, show that the intercostal muscles contract during inspiration, in breaths of normal depth.

This demonstration that some intercostal muscles raise the ribs does not exclude the possibility that the same groups may also lower the ribs when acting synergistically with other muscles such as those of the abdominal wall.

Apart from producing movements of the ribs, contraction of the intercostal muscles increases tension in the intercostal spaces and prevents them from bulging or receding under the influence of varying intrathoracic pressure. Furthermore, their contraction probably facilitates the action of other muscles on the rib cage.

### Circumstances of Contraction

Several workers (Martin and Hartwell, 1879; Hoover, 1922; Schafer and MacDonald, 1925; Briscoe, 1927; Primrose, 1952) have described animal experiments designed to demonstrate the phasic behaviour of the intercostal muscles during respiration. The usual procedure was to isolate two adjacent ribs and to record the movements when one or other intercostal muscle was removed. Results obtained by these workers were contradictory.

Electrical recordings from the intercostal nerves (Bronk and Ferguson, 1935) and muscles (Anderson and Lindsley, 1935) in anaesthetised cats showed that the phasic behaviour of these muscles was in accord with Hamberger's theory, and has since been amply confirmed (e.g. Eklund *et al.*, 1964; Sears, 1964*b*). Gesell (1936) examined dogs electromyographically and found that, although the usual pattern was in accord with the theory, there were numerous individual variations. Because of species variation, observations made in animals cannot be reliably applied to man.

In order to investigate intercostal EMG activity in man, it is necessary to overcome the technical problem of recording selectively from thin layers of muscle. Accounts of intercostal activity based on recording with surface electrodes (Jones *et al.*, 1953; Campbell, 1955; Jones and Pauly, 1957) will accordingly be of only limited value. More information can be obtained with the use of unipolar needle electrodes (Tokizane *et al.*, 1952; Rossier *et al.*, 1956; Koepke *et al.*, 1955 and 1958; Murphy *et al.*, 1959) but this type of recording does still not allow electrical activity to be readily distinguished in the individual layers (see p. 201). For this purpose, bipolar needle electrodes are required.

With this technique, Taylor (1960) demonstrated in healthy subjects, both when supine and when almost erect on a tilting table, that the external intercostal muscle is purely inspiratory, whereas the internal intercostal is expiratory except for its parasternal intercartilaginous part which is inspiratory (see also Tokizane *et al.*, 1952).

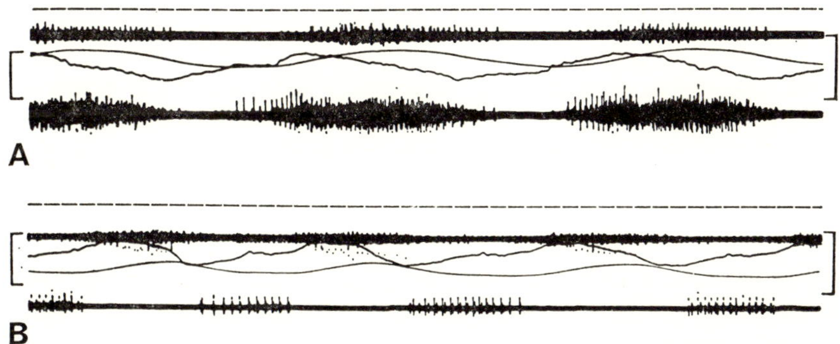


FIG. 60.—INTERCOSTAL EMG DURING QUIET BREATHING RECORDED WITH BIPOLAR NEEDLE ELECTRODES

(A) Inspiratory activity of parasternal internal intercostal (3 ICS, upper trace) and of diaphragm (lower trace) in supine subject.

(B) Expiratory activity of internal intercostals in 8 ICS in the mid-clavicular line (upper trace) and inspiratory activity in the diaphragm (lower trace).

*Calibration:* left hand, volume record (smooth trace) 1 litre; right hand, pneumotachogram 20 l/min. Time scale: uppermost trace interrupted three times a second. (From Taylor, 1960.)

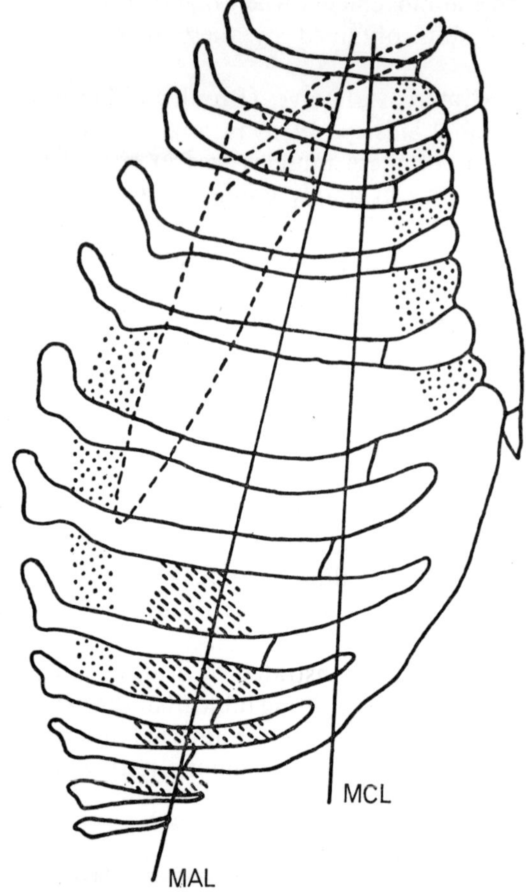


FIG. 61.—DIAGRAMMATIC PROJECTION OF THE RIGHT SIDE OF THE CHEST TO SHOW WHERE INTERCOSTAL ACTIVITY IS FOUND IN QUIET BREATHING (DOTTED AREA, INSPIRATORY; HATCHED AREA, EXPIRATORY). THE MID-CLAVICULAR AND MID-AXILLARY LINES ARE INDICATED (Modified from Taylor, 1960.)

Inspiratory activity during quiet breathing is limited to the parasternal intercostals (Fig. 60a; Taylor, 1960) and to the external intercostal muscle layer posteriorly in the upper spaces where it is accessible for EMG recording between the angle of the ribs and the medial border of the scapula (Newsom Davis and Sears, unpublished), as indicated in Fig. 61. In more vigorous breathing, either voluntarily or with rebreathing, inspiratory activity can be recorded throughout the external layer (Taylor, 1960), and there is evidence that this activity spreads from the upper spaces downwards with increasing size of the breath (Koepke *et al.*, 1958).

Expiratory activity during quiet breathing can be recorded in the lower four spaces (Figs. 60b and 61) between the angle of the ribs and the anterior axillary line (Taylor, 1960), indicating that, even under resting conditions, expiration is not an entirely passive event. With more vigorous breathing, forced expiration, coughing (Fig. 62) or phonation, expiratory activity occurs throughout the interosseous internal intercostals (Taylor,

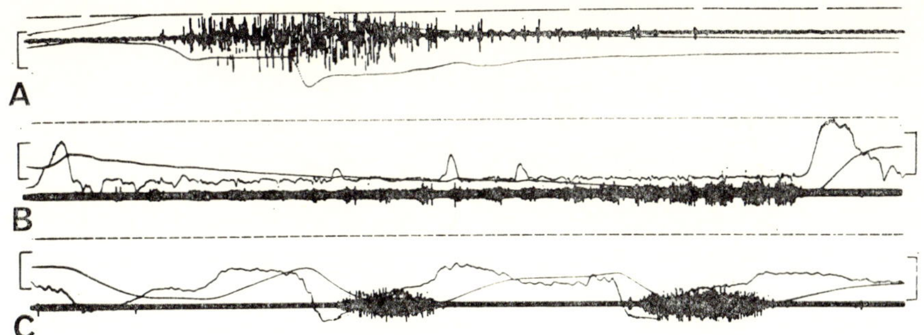


FIG. 62.—INTERCOSTAL EMG RECORDED WITH BIPOLAR NEEDLE ELECTRODES

(A) Activity in the internal intercostal (6 ICS, MCL) during a voluntary cough.
(B) Action of the internal intercostal during speech (counting from 1 to 20 in a single breath).
(C) Expiratory activity in the internal intercostals (7 ICS, anterior axillary line) during voluntary overbreathing. Flow and volume trace calibrations as Fig. 60. (From Taylor, 1960.)

1960), this activity spreading from the lower spaces upwards with increasing size of the breath in contrast to the direction of spread in the external layer during inspiration (Murphy *et al.*, 1959). Variation of the absolute lung volume threshold for the onset of activity between different recording sites is a notable feature during EMG sampling of intercostal muscles (Newsom Davis and Sears, to be published).

During phonation and hyperventilation, activity in the external intercostals operates to check the recoil forces of the lungs and chest wall, regressively from higher lung volumes to the relaxation point, in accordance with the volume pressure diagram (see Chapter IV); this is followed by progressively increasing activity in the internal intercostals as residual

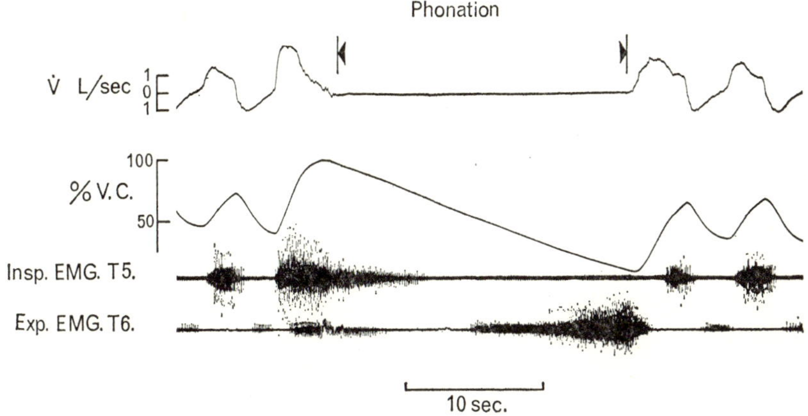


FIG. 63.—INTERCOSTAL EMG ACTIVITY DURING PHONATION

Recorded with wire electrodes in the external intercostal (5 ICS) just lateral to the angle of the rib, and in the internal intercostal (6 ICS) in the anterior axillary line. It is possible that the activity apparently recorded from the expiratory intercostal muscle close to TLC was due to contamination from the inspiratory layer. Note the 'checking' action of the inspiratory intercostals during expiration at high lung volumes, and of the expiratory intercostals during inspiration at low lung volumes. (From Sears and Newsom Davis, 1968.)

volume is approached (Draper *et al.*, 1960; Sears and Newsom Davis, 1968). This is illustrated in Fig. 63, where recording has been made from the 5th external intercostal posteriorly (inspiratory) and the 6th internal intercostal in the anterior axillary line (expiratory), during the singing of a note of constant pitch and intensity following a maximal inspiration. The figure also shows, at the onset of inspiration following the period of phonation, the checking action of the expiratory intercostal on the net recoil forces which act in an inspiratory direction at low lung volumes (see Chapters IV and V). Reciprocal activity can be seen in the two layers of muscle in the breaths that precede and follow the period of phonation. It is interesting to note the constant flow rate, and thus constant volume change, achieved during the sung note.

## Relation between Intercostal EMG and Mechanical Activity

In a recent study in man, the EMG of the intercostal muscles has been compared to their mechanical activity (Viljanen, 1967; see also Delhez and Petit, 1966). The impulse frequency of the external intercostal muscle, recorded with surface electrodes in the sixth intercostal space in the mid-clavicular line, was used as a measure of electrical activity (see p. 202). The impulse frequency recorded for a one-third of a second period at the end of inspiration had a linear relationship with the inspiratory muscular work, and the value four seconds after the end of inspiration with the subject holding his breath with glottis open was directly proportional to the relaxation pressure *Pst*(*rs*) (see also Sears and Newsom Davis, 1968).

The total number of impulses (*n*) in a single breath bears a linear relation both to the first time integral of the inspiratory muscular work ($\int W \times dt$), except for deep breaths, confirming earlier studies in the intercostal muscles of the rat (Bergstrom and Kerttula, 1961), and to the "pressure impulse" (*I*) of the inspiratory muscles ($\int Pmus \times dt$) (Fig. 64). The slope between (*n*) and (*I*) varies in different types of inspiration, but remains constant.

From these observations, it is deduced that throughout inspiration the amount of electrical activity is in linear relation with the total inspiratory force (*Pmus*), the slope of this relationship being determined at the onset of inspiration, and that the EMG impulse frequency at the end of inspiration, under stable conditions, is in logarithmic relation with the tidal volume. It thus appears as if the central nervous system decides in advance which inspiratory muscles are to take part in the inspiratory movement, and programmes each inspiration to reach a predetermined end-state (Viljanen, 1967).

## Respiratory Function

The intercostal muscles, as muscles of inspiration, are of secondary importance to the diaphragm (see Chapter VI). Their precise contribution to larger inspiratory volumes is not known, but in patients with lesions of the lower cervical cord, which cause paralysis of the intercostal and abdominal muscles and leave the diaphragm and accessory muscles intact, TLC may be reduced (Stone and Keltz, 1963), providing an indirect indication of intercostal respiratory function. The reduction in inspiratory volume is more apparent in acute transection in which an initially flaccid paralysis of the affected muscles develops, with sucking in of the intercostal spaces during inspiration as occurs in patients with a T1 block

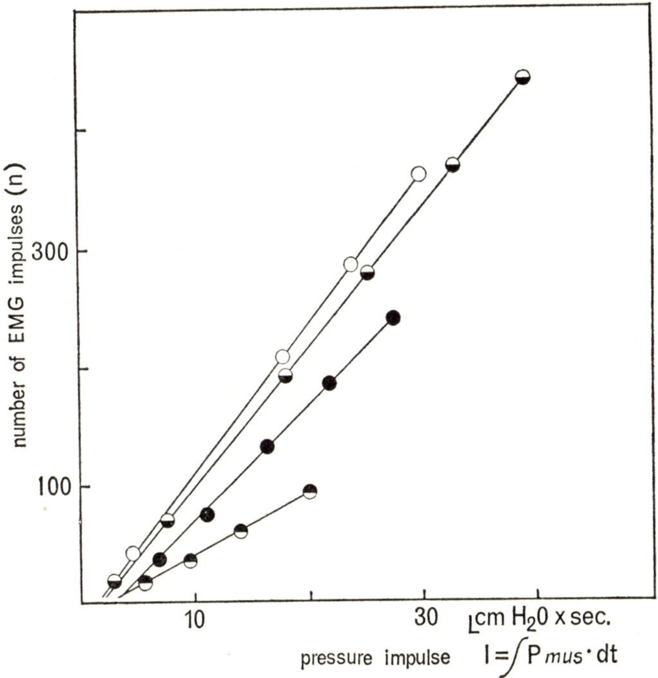


Fig. 64.—Number of Electromyogram Impulses (n) as a Function of the Pressure Impulse (I) Produced by the Inspiratory Muscles During one and the Same Inspiration

The figure relates to four different inspirations. (From Viljanen, 1967.)

induced by local anaesthesia (Eisele *et al.*, 1968). With the development of spasticity, hyperactive stretch reflexes render the rib cage a more rigid structure (Guttmann and Silver, 1965).

Restriction of the rib cage as an experimental procedure (Caro *et al.*, 1960) reduced TLC and VC by about 20 per cent in a group of normal subjects, but minute volume was unaffected. Only part of this reduction, however, can be attributed to the effects of restriction upon the mechanical actions of the intercostal muscles.

In a subject whose diaphragm was paralysed, Campbell (1954) observed contraction of the intercostal muscles and expansion of the thoracic cage in the absence of activity in other inspiratory muscles. More studies on such subjects are needed for further assessment of the respiratory function of intercostal muscles.

## TRANSVERSUS THORACIS

This muscle is the thoracic equivalent of the transversus abdominis. It is an incomplete layer, lying deep to the internal intercostal muscle, consisting of the sternocostalis anteriorly, the intercostales intimi laterally and the subcostales posteriorly.

Expiratory activity has been recorded in the sternocostalis and probably also in the intercostalis intimi; it therefore seems likely that the subcostalis is expiratory as well (Taylor, 1960).

## LEVATORES COSTARUM

These fan-shaped muscles arise from the transverse processes of the seventh cervical and upper eleven thoracic vertebrae, and are inserted between the tubercle and the angle of the rib immediately caudal to the vertebra from which they take origin. They are innervated by the lateral division of the posterior primary ramus of the thoracic nerves (Morrison, 1954). Afferent discharges from muscle spindles and tendon organs in levatores costarum have been identified in thoracic dorsal root filaments (Godwin-Austen, personal communication). The probable action of these muscles is to elevate and abduct the ribs and Primrose (1952) has described an animal experiment which supports this view. Polgar (1949) regarded the levatores costarum as the most important muscles producing inspiratory movements of the ribs. Both Primrose and Polgar were, however, inclined to overestimate the importance of these muscles because neither believed that the intercostals have any significant inspiratory function. There have been no adequate studies of their behaviour in man. On anatomical grounds it would appear unlikely that they can make a powerful contribution to the mechanics of breathing. The proximal position of these muscles in the intercostal space, however, make them well placed for exerting a proprioceptive control over intercostal muscle activity.

## Summary

1. The intercostal muscles contain muscle spindles, tendon organs and Pacinian corpuscles, and the fibre calibre spectrum of their nerve supply is consistent with proprioceptive function.

2. The mean contraction time of (biopsied) human intercostal muscle is 71·5 msec. The muscle is a mixture of fast and slow units.

3. The mechanical action of the individual intercostal muscles has not yet been definitely established. The most acceptable theory is that the

external intercostals and intercartilaginous internal intercostals raise the ribs, and the interosseous internal intercostals depress the ribs.

4. In quiet breathing, the intercartilaginous internal intercostals of the upper three to five spaces (the "parasternals") and the external intercostals posteriorly contract during inspiration, and the interosseous internal intercostals in the lower four spaces over the lateral chest wall contract during expiration.

5. With increased ventilation, activity is present throughout the external intercostals as well as in the parasternals during inspiration, and throughout the interosseous internal intercostals during expiration, i.e. a pattern of inspiratory activity in the outer layer and expiratory activity in the inner layers.

6. The transversus thoracis muscle is expiratory and the levatores costarum are inspiratory in function. Neither group makes a large contribution to the mechanics of breathing.

## REFERENCES

ALEXANDER, J. (1929). Multiple intercostal neurectomy for pulmonary tuberculosis. *Amer. Rev. Tuberc.*, **20**, 637–684.

ANDERSEN, P., and SEARS, T. A. (1964). The mechanical properties and innervation of fast and slow motor units in the intercostal muscles of the cat. *J. Physiol.* (*Lond.*), **173**, 114–129.

ANDERSON, F. M., and LINDSLEY, D. B. (1935). Action potentials from intercostal muscles before and after unilateral pneumectomy. *J. Lab. clin. Med.*, **20**, 623–628.

BARKER, D. (1962). The structure and distribution of muscle receptors. In: *Symposium on Muscle Receptors*, pp. 227-240. Ed. Barker, D. Hong Kong: Univ. Press.

BEAU, J. H. S., and MAISSIAT, J. H. (1842–43). Recherches sur le mecanisme des mouvements respiratoires. *Arch. gén. Méd.*, Ser. 3, **15**, 397–420; Ser. 4, **1**, 265–295; Ser. 4, **2**, 257–282; Ser. 4, **3**, 249–284.

BERGSTRÖM, R. M., and KERTTULA, Y. (1961). On the neural control of breathing as studied by electromyography of the intercostal muscles of the rat. *Ann. Acad. Sci. fenn.* A5, 79.

BISCOE, T. J. (1962). The isometric contraction characterisitics of cat intercostal muscle. *J. Physiol.* (*Lond.*), **164**, 189–199.

BISCOE, T. J., and TAYLOR, A. (1967). The effect of admixture of fast and slow muscle in determining the form of the muscle twitch. *Med. biol. Engin.*, **5**, 473–479.

BRISCOE, C. (1927). Lumleian lectures on the muscular mechanism of respiration and its disorders. *Lancet*, **1**, 637–643; 749–753; 857–862.

BRONK, D. W., and FERGUSON, L. K. (1935). The nervous control of intercostal respiration. *Amer. J. Physiol.*, **110**, 700–707.

BRYCE, T. H. (1923). In *Quain's Elements of Anatomy*, 11th edit., Vol. 4, pt. 2. London: Longmans.

CAMPBELL, E. J. M. (1954). The muscular control of breathing in man. (Ph.D. Thesis, Univ. of London.)

CAMPBELL, E. J. M. (1955). An electromyographic examination of the role of the intercostal muscles in breathing in man. *J. Physiol.* (*Lond.*), **129**, 12–26.

CARO, C. G., BUTLER, J., and DUBOIS, A. B. (1960). Some effects of restriction of chest cage expansion on pulmonary function in man: an experimental study. *J. clin. Invest.*, **39**, 573–591.

CETRANGOLO, A. A. (1930). La movilidad de la pared costal despues de la neurectomia intercostal multiple. *Rev. méd. lat.-amer.*, **15**, 1224–1234.

CRITCHLOW, V., and EULER, C. VON (1963). Intercostal muscle spindle activity and its gamma motor control. *J. Physiol.* (*Lond.*), **168**, 820–847.

DAVIES, F., GLADSTONE, R. J., and STIBBE, E. P. (1932). The anatomy of the intercostal nerves. *J. Anat.* (*Lond.*), **66**, 323–333.

DELHEZ, L., and PETIT, J. M. (1966). Données actuelles de l'ectromyographie respiratoire chez l'homme normal. *Electromyography*, **6**, 101–146.

DRAPER, M. H., LADEFOGED, P., and WHITTERIDGE, D. (1960). Expiratory pressures and air flow during speech. *Brit. med. J.*, **1**, 1837–1843.

DUCHENNE, G. B. A. (1867). *Physiologie des mouvements démontrée à l'aide de l'expérimentation électrique et de l'observation clinique, et applicable à l'étude des paralysies et des déformations.* Paris: Baillière. Also translation: *Physiology of Motion* by Kaplan, E. B. (1949), pp. 443–503. Philadelphia: Lippincott.

EISELE, J., TRENCHARD, D., BURKI, N., and GUZ, A. (1968). The effect of chest wall block on respiratory sensation and control in man. *Clin. Sci.*, **35**, 23–33.

EKLUND, G., EULER, C. VON, and RUTKOWSKI, S. (1964). Spontaneous and reflex activity of intercostal gamma motoneurones. *J. Physiol.* (*Lond.*), **171**, 139–163.

EULER, C. VON, and PERETTI, G. (1966). Dynamic and static contributions to the rhythmic $\gamma$ activation of primary and secondary spindle endings in external intercostal muscle. *J. Physiol.* (*Lond.*), **187**, 501–516.

FLEISCH, A. (1934). Neuere Ergebnisse über Mechanik und propriozeptive Steuerung der Atmungsbewegung. *Ergebn. Physiol.*, **36**, 252–256.

GESELL, R. (1936). Individuality of breathing. *Amer. J. Physiol.*, **115**, 168–180.

GESELL, R., ATKINSON, A. K. and BROWN, R. C. (1940). The gradation of the intensity of inspiratory contractions. *Amer. J. Physiol.*, **131**, 659–673.

GLEBOVSKII, V. D. (1961). Contractile properties of respiratory muscles in fully grown and neonate animals. *Sechenov physiol. J. U.S.S.R.*, **47**, 470–480.

GUTTMANN, L., and SILVER, J. R. (1965). Electromyographic studies on reflex activity of the intercostal and abdominal muscles in cervical cord lesions. *Paraplegia*, **3**, 1–22.

HOFMANN, W. W., ALSTON, W., and ROWE, G. (1966). A study of individual neuro-muscular junctions in myotonia. *Electroenceph. clin. Neurophysiol.*, **21**, 521–537.

HOOVER, C. F. (1922). The functions and integration of the intercostal muscles. *Arch. intern. Med.*, **30**, 1–33.

HUBER, G. C. (1901–2). Neuro-muscular spindles in the intercostal muscles of the cat. *Amer. J. Anat.*, **1**, 520–521.

JOHNSTON, T. B., and WHILLIS, J. (1949). *Gray's Anatomy*, 30th edit., pp. 557–584. London: Longmans.

JOLY, H., and VINCENT, PH.-A. (1937). Role respiratoire des muscles scalenes et inter-costaux etudie en fonction de la collapsotherapie pulmonaire. *Arch. méd.-chir. Appar. resp.*, **12**, 392–404.

JONES, D. S. BEARGIE, R. J., and PAULY, J. E. (1953). An electromyographic study of some muscles of costal respiration in man. *Anat. Rec.*, **117**, 17–24.

JONES, D. S., and PAULY, J. E. (1957). Further electromyographic studies on muscles of costal respiration in man. *Anat. Rec.*, **128**, 733–746.

KOEPKE, G. H., MURPHY, A. J., RAE, J. W., and DICKINSON, D. G. (1955). An electromyographic study of some of the muscles used in respiration. *Arch. phys. Med.*, **36**, 217–222.

KOEPKE, G. H., SMITH, E. M., MURPHY, A. J., and DICKINSON, D. G. (1958). Sequence of action of the diaphragm and intercostal muscles during respiration: I. Inspiration. *Arch. phys. Med.*, **39**, 426–430.

LUCIANI, L. (1911). *Human Physiology*, translated by Welby, F. A. Vol. 1, pp. 402–439. London: Macmillan.

MARTIN, H. N., and HARTWELL, E. M. (1879). On the respiratory function of the internal intercostal muscles. *J. Physiol.* (*Lond.*), **2**, 24–27.

MORRISON, A. B. (1954). The levatores costarum and their nerve supply. *J. Anat.* (*Lond.*), **88**, 19–24.

MURPHY, A. J., KOEPKE, G. H., SMITH, E. M., and DICKINSON, D. G. (1959). Sequence of action of the diaphragm and intercostal muscles during respiration: II. Expiration. *Arch. phys. Med.*, **40**, 337–342.

NISHIYAMA, A. (1966). Histochemical studies on the red, white and intermediate muscle fibres of some skeletal muscles. III. Histochemical demonstration of oxidative enzymes, phosphorylase and glycogen in respiratory muscle fibres. *Acta Med. Okayama*, **20**, 137–146.

OGATA, T., KAWASHIMA, T., and NISHIYAMA, A. (1963). Histochemical demonstration of three types of muscle fibres of the intercostal muscles. A study on oxidative enzymes. *Acta Med. Okayama*, **17**, 257–258.

POLGAR, F. (1949). Studies on respiratory mechanics. *Amer. J. Roentgenol.*, **61**, 637–657.

PRIMROSE, W. B. (1952). Chest movements and the intercostal muscles. *Brit. J. Anaesth.*, **24**, 3–24.

ROSSIER, P. H., NIEPORENT, H. J., PIPBERGER, H., and KÄLIN, R. (1956). Elektromyographische Untersuchungen der Atemmuskelfunktion an normalen Versuchspersonen, *Z. ges. exp. Med.*, **127**, 39–52.

SCHAFER, E. S., and MACDONALD, A. D. (1925). The action of the intercostal muscles. *J. Physiol.* (*Lond.*), **60**, 25–26.

SCHWIELER, G. H. (1968). Respiratory regulation during postnatal development in cats and rabbits and some of its morphological substrate. *Acta physiol. scand.*, Suppl., 304.

SEARS, T. A. (1964*a*). The fibre calibre spectra of sensory and motor fibres in the intercostal nerves of the cat. *J. Physiol.* (*Lond.*), **172**, 150–161.

SEARS, T. A. (1964*b*). Efferent discharges in alpha and fusimotor fibres of intercostal nerves of the cat. *J. Physiol.* (*Lond.*), **174**, 295–315.

SEARS, T. A., and NEWSOM DAVIS, J. (1968). The control of respiratory muscles during voluntary breathing. *Ann. N.Y. Acad. Sci.*, **155**, 183–190.

SPRAGUE, J. M. (1951). Motor and propriospinal cells in the thoracic and lumbar ventral horn of the rhesus monkey. *J. comp. Neurol.*, **95**, 103–123.

STONE, D. J., and KELTZ, H. (1963). The effect of respiratory muscle dysfunction on pulmonary function. *Amer. Rev. resp. Dis.*, **88**, 621–629.

TAYLOR, A. (1960). The contribution of the intercostal muscles to the effort of respiration in man. *J. Physiol. (Lond.)*, **151**, 390–402.

TOKIZANE, T., KAWAMATA, K., and TOKIZANE, H. (1951–2). Electromyographic studies on the human respiratory muscles. *Jap. J. Physiol.*, **2**, 232–247.

VILJANEN, A. A. (1967). The relation between the electrical and mechanical activity of human intercostal muscles during voluntary inspiration. *Acta physiol. scand.*, Suppl. 296.

# Chapter VIII

# THE ABDOMINAL MUSCLES

E. AGOSTONI AND E. J. M. CAMPBELL

## Morphology

**External oblique.**—This muscle arises from the outer surfaces of the lower eight ribs. It is therefore superficial to the intercostal muscles in the lower spaces. The dorsal fibres pass downwards to the iliac crest. The rest of the muscle slopes obliquely downwards and forwards, and passes into a fibrous aponeurosis which forms part of the rectus sheath and fuses with its fellow of the other side in the linea alba. The lower border of the aponeurosis forms the inguinal ligament.

**Internal oblique.**—This muscle arises from the lumbar fascia, the iliac crest and the lateral part of the inguinal ligament. The dorsal fibres pass almost vertically upwards to the last three ribs. The rest of the fibres from the lumbar fascia pass forwards as a fan into a fibrous aponeurosis similar in extent to that of the external oblique.

**Transversus abdominis.**—This muscle arises from the costal cartilages of the lower six ribs, the lumbar fascia, the iliac crest and the lateral part of the inguinal ligament. The main part of the muscle passes horizontally forwards into an aponeurosis similar in extent to that of the external oblique. The transversus abdominis is the most deep of these muscles.

**Rectus abdominis.**—The rectus abdominis muscle arises from the pubic symphysis and crest and passes vertically upwards in a sheath formed by the aponeurosis of the oblique and transverse muscles to the superficial sides of the fifth, sixth, and seventh costal cartilages.

## Innervation and Contractile Properties

The external oblique and the rectus abdominis are supplied by the lower 5 intercostal nerves (T7–11), the internal oblique and the transversus by the lower 5 intercostal, the subcostal (T12), the iliohypogastric (L1) and the ilioinguinal (L1) nerves.

The abdominal muscles contain the components necessary for the monosynaptic stretch reflex. In man, an abdominal tendon jerk can usually be elicited (see Kugelberg and Hagbarth, 1958). In the cat, a stretch reflex can only be elicited when the abdominal motoneurone pool is functionally released from respiratory control (Bishop, 1966).

Eberstein and Goodgold (1968) showed in biopsy specimens that human abdominal muscles are composed of fast and slow fibres like other skeletal muscles. The contraction time ranged from 50 to 140 msec.

## Mechanical Action

The external and internal obliques compress the abdominal contents, flex the trunk, and depress the lower ribs. The transversus abdominis compresses the abdominal content and depresses the lower ribs. The rectus abdominis draws the ventral part of the rib cage nearer the pubis and reduces the craniocaudal curvature of the ventral wall of the abdomen; it helps in compressing the abdominal content. The abdominal muscles have therefore an expiratory action. They are able to exert a pressure of 300–400 cm $H_2O$ on the abdominal contents (see Chapter III).

## Circumstances of Contraction

Some postural activity of the muscles of the abdominal wall is almost always present during quiet breathing in the sitting position, but this can be decreased or abolished by adjustment of the posture. If some activity remains and a respiratory rhythm is found, it decreases during inspiration and increases during expiration (Campbell, 1952; Campbell and Green, 1955). In the supine position no electrical activity is found during quiet breathing (Campbell, 1952; Freund *et al.*, 1964); activity appears toward the end of expiration when the ventilation reaches about 40 litres/min. The slight difference in the pattern of activity at low levels of ventilation sometimes found (Jones *et al.*, 1953; Tokizane *et al.*, 1952; Weddel *et al.*, 1964) is probably attributable to postural effects or to the presence of needle electrodes.

Great activity of the abdominal muscles appears, both in the sitting and supine position, only when the ventilation reaches 70–90 litres/min (Campbell and Green, 1953*b*; Campbell and Green, 1955). The activity of the abdominal muscles during quiet and increased breathing is illustrated diagrammatically in Fig. 65. At high values of ventilation the abdominal pressure toward the end of expiration becomes greater than during inspiration because of the strong contraction of the abdominal muscles (see Chapter IV). During maximum voluntary ventilation the abdominal muscles probably enter into activity before the end of inspiration (Agostoni and Torri, 1967, see Chapter IV). This does not seem to be the case in the supine position (Campbell and Green 1953*b*).

Few patients with chronic airways obstruction use the abdominal muscles during expiration even when the ventilation is so increased by $CO_2$ or exercise that the patients are very dyspnoeic (Campbell and Friend, 1955). An increase of pleural pressure in the other patients would possibly not be effective in increasing expiratory airflow because of collapse of the airways (see Chapter IV).

When expiration is opposed by a positive pressure, in most normal subjects contraction of the abdominal muscles at mid-lung volume occurs only when the pressure is higher than 10 cm $H_2O$ (Campbell, 1957).

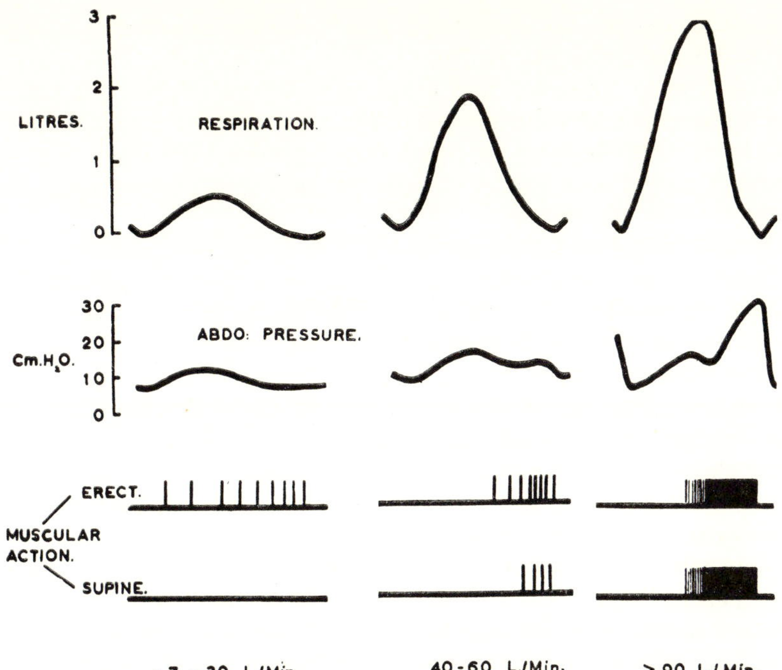


FIG. 65.—SCHEMATIC PATTERN OF THE ACTIVITY OF THE MUSCLES OF THE ABDOMINAL WALL, OF THE ABDOMINAL PRESSURE AND OF THE SPIROGRAM AT VARIOUS LEVELS OF VENTILATION

During graded expiratory or expulsive efforts at a given lung volume the electrical activity of the muscles of the abdominal wall is roughly proportional to the pressure exerted; as the lung volume increases part of the pressure is contributed by the passive structures, and the activity of the abdominal muscles, for a given alveolar pressure, decreases (Campbell and Green, 1953*a* and *b*).

The abdominal muscles are of course active during a maximum expiration and, in most subjects, also at the end of a full inspiration (Mills, 1950; Campbell, 1952; Campbell and Green, 1953*b*; Delhez *et al.*, 1959). In many subjects, however, the abdominal pressure does not increase markedly at full inspiration (Mead *et al.*, 1963), probably because, for geometrical reasons, the contraction of the abdominal muscles has little effect on the abdominal pressure at full inspiration (J. Mead, personal communication). In these subjects, according to Mead and his colleagues (1963), the antagonist contraction of the abdominal muscles does not contribute to limit the upper volume extreme. On the other hand, if the lack of increase of abdominal pressure at full inspiration indicates that the diaphragm is not able to pull further on the lung, it does not exclude the possibility that the contraction of the abdominal muscles antagonises the action of other inspiratory muscles besides the diaphragm and therefore contributes to set the upper volume extreme (see p. 70).

According to Campbell and Green (1953*a*), the electrical activity of the abdominal muscles during maximum expiratory or expulsive effort is markedly smaller than that recorded during maximal flexion of the trunk, whereas Agostoni and Sant'Ambrogio (unpublished observations) found that the activity was similar.

The abdominal muscles contract vigorously during coughing, parturition, vomiting and defaecation, i.e. acts requiring thoracic and abdominal pressures much higher than those provided by the passive structure even at large lung volume. Furthermore, they are active during phonation below mid-lung volume (Draper *et al.*, 1959, see Chapter IV).

In anaesthetised supine human subjects the abdominal muscles may be active throughout expiration (Campbell *et al.*, 1957; Freund *et al.*, 1964). In anaesthetised dogs the abdominal muscles contract in a steady manner which is interrupted during inspiration. This has been called the abdominal compression reaction and seems to be produced by a vagal reflex originating probably from atrial volume receptors (Youmans *et al.*, 1963). Bishop (1963, 1964) has found activity of the abdominal muscles during expiration in anaesthetised supine cats under positive pressure breathing. This activity which may persist during inspiration, is abolished by cervical vagotomy, but not by abdominal vagotomy. It can be distinguished from the inflation Hering-Breuer reflex (Bishop, 1964): it increases proportionally to the pressure applied during expiration, but it is not affected by the pressure applied during inspiration or by the lung volume (Bishop, 1967). Furthermore, abdominal activity ceases abruptly when pressure breathing is discontinued, whereas the diaphragm activity does not return to normal for several breaths (Bishop, 1967, see Chapter XIII).

**Respiratory Function**

The muscles of the abdominal wall are the most powerful expiratory and expulsive muscles. Studies of their control (see Chapters XI and XII) indicate that they are subject to a rhythmic respiratory activation in addition to their involvement in postural and other functions. During mouth breathing their contribution to expiration is probably negligible up to about 40 l/min, whereas during nose breathing their contribution becomes appreciable at ventilation slightly above quiet breathing (see Chapter IV). As the ventilation increases above these levels the expiratory contribution of the abdominal muscles increases progressively. On the other hand, as shown in the Dynamics chapter, a high pleural pressure cannot help to increase the ventilation because of the compression of the airways. Hence the great power of the abdominal muscles is not used for respiratory functions except in coughing where it increases the kinetic energy of the air stream. Moreover, during spontaneous breathing an increased load to expiration is not met primarily by an increase in abdominal muscle activity, but rather by utilising the recoil of the respiratory system by an increase in end-inspiratory volume through an increase of the inspiratory muscle activity (see Chapter XIII).

## REFERENCES


AGOSTONI, E., and TORRI, G. (1967). An analysis of the chest wall motions at high values of ventilation. *Resp. Physiol.*, **3**, 318–332.

BISHOP, B. (1963). Abdominal muscle and diaphragm activities and cavity pressures in pressure breathing. *J. appl. Physiol.*, **18**, 37–42.

BISHOP, B. (1964). Reflex control of abdominal muscles during positive-pressure breathing. *J. appl. Physiol.*, **19**, 224–232.

BISHOP, B. (1966). The stretch reflex of the abdominal wall. *Physiologist*, **9**, 139.

BISHOP, B. (1967). Diaphragm and abdominal muscle responses to elevated airway pressures in the cat. *J. appl. Physiol.*, **22**, 959–965.

CAMPBELL, E. J. M. (1952). An electromyographic study of the role of the abdominal muscles in breathing. *J. Physiol.* (*Lond.*), **117**, 222–233.

CAMPBELL, E. J. M. (1957). The effects of increased resistance to expiration on the respiratory behaviour of the abdominal muscles and intra-abdominal pressure. *J. Physiol.* (*Lond.*), **136**, 556–562.

CAMPBELL, E. J. M., and FRIEND, J. (1955). Action of breathing exercises in pulmonary emphysema. *Lancet*, **1**, 325–329.

CAMPBELL, E. J. M., and GREEN, J. H. (1953*a*). The expiratory function of the abdominal muscles in man. An electromyographic study. *J. Physiol.* (*Lond.*), **120**, 409–418.

CAMPBELL, E. J. M., and GREEN, J. H. (1953*b*). The variations in intra-abdominal pressure and the activity of the abdominal muscles during breathing; a study in man. *J. Physiol.* (*Lond.*), **122**, 282–290.

CAMPBELL, E. J. M., and GREEN, J. H. (1955). The behaviour of the abdominal muscles and the intra-abdominal pressure during quiet breathing and increased pulmonary ventilation. A study in man. *J. Physiol.* (*Lond.*), **127**, 423–426.

CAMPBELL, E. J. M., HOWELL, J. B. L., and PECKETT, B. W. (1957). The pressure-volume relationships of the thorax of anaesthetized human subjects; a comparison of the effects of expiratory resistance and positive pressure inflation. *J. Physiol.* (*Lond.*), **136**, 563–568.

DELHEZ, L., PETIT, J. M., and MILIC-EMILI, J. (1959). Influence des muscles expirateurs dans la limitation de l'inspiration. (Étude électromyographique chez l'homme). *Rev. franc. Étud. clin. biol.*, **4**, 815–818.

DRAPER, M. H., LADEFOGED, P., and WHITTERIDGE, D. (1959). Respiratory muscles in speech. *Hearing Res.*, **2**, 16–27.

EBERSTEIN, A., and GOODGOLD, J. (1968). Slow and fast twitch fibres in human skeletal muscle. *Amer. J. Physiol.*, **215**, 535–541.

FREUND, F., ROOS, A., and DODD, R. B. (1964). Expiratory activity of the abdominal muscles in man during general anesthesia. *J. appl. Physiol.*, **19**, 693–697.

JONES, D. S., BEARGIE, R. J., and PAULY, J. E. (1953). An electromyographic study of some muscles of costal respiration in man. *Anat. Rec.*, **117**, 17–24.

KUGELBERG, E., and HAGBARTH, K. E. (1958). Spinal mechanism of the abdominal and erector spinae skin reflexes. *Brain*, **81**, 290–304.

MEAD, J., MILIC-EMILI, J., and TURNER, J. M. (1963). Factors limiting depth of a maximal inspiration in human subjects. *J. appl. Physiol.*, **18**, 295–296.

MILLS, J. N. (1950). The nature of the limitation of maximal inspiratory and expiratory efforts. *J. Physiol.* (*Lond.*), **111**, 376–381.

TOKIZANE, T., KAWAMATA, K., and TOKIZANE, H. (1952). Electromyographic studies on the human respiratory muscles. *Jap. J. Physiol.*, **2**, 232–247.

WEDDEL, G., FEINSTEIN, B., and PATTLE, R. E. (1944). The electrical activity of voluntary muscle in man under normal and pathological conditions. *Brain*, **67**, 178–257.

YOUMANS, W. B., MURPHY, Q. R., TURNER, J. K., DAVIS, L. D., BRIGGS, D. I., and HOYE, A. S. (1963). Activity from abdominal muscles elicited from the circulatory system. *Amer. J. phys. Med.*, **42**, 1–70.

# Chapter IX

# ACCESSORY MUSCLES

E. J. M. Campbell

Of all the muscles which are generally thought to act as accessory muscles of inspiration, only the scaleni and the sternomastoids show significant respiratory activity in man (Campbell, 1954). These will, therefore, be described in detail.

## THE SCALENI

### Anatomy

The scaleni arise from the transverse process of the lower five cervical vertebrae and pass downwards to be inserted into the upper surface of the first rib (scalenus anterior and medius) and second rib (scalenus posterior). The scalenus medius is the largest. They are supplied by the lower five cervical nerves.

### Mechanical Action

The action of importance in respiration is elevation or fixation of the first two ribs; they have a good mechanical advantage.

### Circumstances of Contraction

In both the supine and erect postures the scaleni have a marked tendency to show continuous non-rhythmic activity, which can usually be abolished by attention to the posture of the shoulders and neck. In many subjects some inspiratory activity can be found during quiet breathing when supine or seated (Fig. 67) (Jones *et al.*, 1953; Campbell, 1955; Jones and Pauly, 1957; Raper *et al.*, 1966); in other subjects the scaleni are inactive at rest (Thompson *et al.*, 1964; Varene *et al.*, 1963) but activity appears when the tidal volume is 2 litres and ventilation over 40 l/min (Campbell, 1955). At greater ventilations they are active in all subjects and their activity may precede the onset of inspiration (Raper *et al.*, 1966).

Campbell (1955) observed a relationship between EMG activity and graded voluntary inspiratory efforts; Raper *et al.* (1966) also found a significant correlation but a closer one with lung volume. And any increase in end-expiratory lung volume whether produced acutely in normal subjects or chronically by disease increases their participation in inspiration.

During static expiratory efforts resulting in intra-alveolar pressures of +20 to +30 cm $H_2O$ the scaleni are inactive. Spontaneous activity if present is suppressed. However, during efforts of +40 cm $H_2O$ and above the scaleni contract quite vigorously (Fig. 66).

**Respiratory Function**

The importance of the scaleni as muscles of inspiration is variously assessed by different authorities. Some regard them as muscles of inspiration at least equal in importance to the intercostals. Others class them with the sternomastoids as accessory muscles. As described above, they are

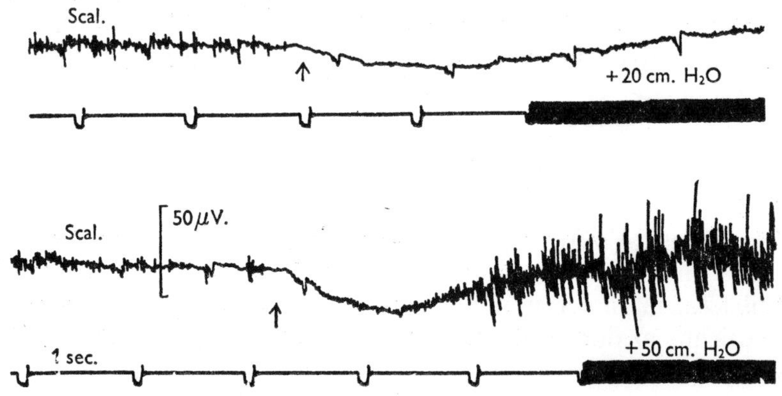


FIG. 66.—THE SCALENE MUSCLES: BEHAVIOUR DURING GRADED VOLUNTARY EXPIRATORY EFFORTS

A healthy young male subject, supine. *Scal.*—the electromyogram recorded from the scalene muscles. Time in sec. At the arrow he began to make an expiratory effort. The signal marks the period of maintenance of the expiratory pressure indicated. This subject showed continuous irregular activity in the scalene muscles, as seen at the beginning of each record. An expiratory effort of +20 cm $H_2O$ (intra-alveolar pressure 20 cm $H_2O$ above atmospheric) was associated with a decrease in background activity: an effort of +50 cm $H_2O$ was associated with a marked increase (see text). (Campbell, 1955.)

often active during quiet breathing whereas the sternomastoids are only employed at very high levels of ventilation (see below). In intensity of activity, therefore, they resemble the intercostals more than the sternomastoids, and should, perhaps, be classified with the ordinary rather than the accessory muscles of breathing (Thompson *et al.*, 1964).

These observations do not establish the importance of the scaleni in respiration because the presence of activity does not of itself give much information about the force contributed to the mechanics of respiration. Thus, Fick (1923) calculated that the scaleni are only potentially one-fifth as important as the intercostals. Duchenne (1867) observed that costo-superior breathing continued in a patient who had lost most of his scaleni. Joly and Vincent (1937) observed that scalenotomy performed in patients with pulmonary tuberculosis did not reduce the amplitude of rib movements in breaths of normal depth (whereas paralysis of the intercostals did). Giauni (1936) observed that although scalenotomy caused an immediate decrease in the vital capacity, considerable recovery occurred later and in some cases there was a complete return to normal.

We may reasonably conclude that, while the scaleni are readily employed as muscles of inspiration, their total contribution to the force of inspiration is probably not great compared with those of other muscles. In view of their anatomy it is not surprising that the scaleni are more active at larger lung volumes because it is then that the upper chest participates in thoracic expansion.

In quiet and increased natural breathing the scaleni play no part in expiration, but the finding of considerable activity during the maintenance of voluntary expiratory pressures suggests the possibility that the scaleni may be of importance during coughing and other expulsive efforts. Their action under these conditions may be to fix the upper ribs and prevent them from being pulled down by the traction of the abdominal muscles on the lower ribs, or it may be to provide support for the apex of the lung to prevent it from bulging upwards through the suprapleural membrane (Sibson's fascia). Herniation of the apex of the lung does, in fact, sometimes occur in patients with a chronic cough (Fenichel and Epstein, 1955).

## THE STERNOMASTOIDS (STERNOCLEIDOMASTOIDS)

### Anatomy

The sternomastoid arises by two heads from the manubrium sterni and the medial part of the clavicle. The fibres of these two heads fuse into a single belly which is inserted into the mastoid process and the occipital bone. The muscle is supplied by the spinal accessory and second cervical nerves.

### Mechanical Action

The mechanical action of the sternomastoids, of importance in respiration, is elevation of the sternum which increases the antero-posterior diameter of the thorax.

### Circumstances of Contraction

In both supine and erect postures the sternomastoids are usually inactive during quiet breathing, there being a notable lack of postural activity in contrast to that of the scaleni (Campbell, 1955) (Fig. 67). They always contract towards the end of a voluntary maximum inspiration at lung volumes corresponding to 70–80 per cent of the vital capacity (Raper *et al.*, 1966).

During graded inspiratory pressures the sternomastoids become progressively more active as the intra-alveolar pressure is lowered. Campbell (1955) observed contraction of the sternomastoids when the intra-alveolar pressure is reduced to $-10$ cm $H_2O$ at FRC.

There is no significant activity during static expiratory efforts or during maximum expiration.

During increased pulmonary ventilation most subjects when supine are able to attain tidal volumes greater than 2·5 litres and ventilation rates of over 60 l/min without using the sternomastoids (Fig. 68). In the erect posture the values are slightly lower and more variable (Campbell, 1955; Delhez and Petit, 1966).

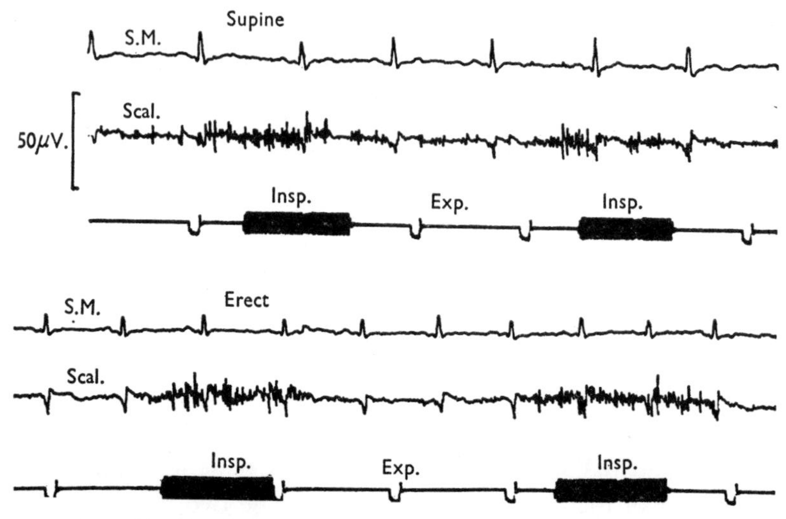


FIG. 67.—THE STERNOMASTOID AND SCALENE MUSCLES: BEHAVIOUR DURING QUIET BREATHING IN THE SUPINE AND ERECT POSTURES

A healthy young male subject. Electromyograms recorded from the sternomastoid (*S.M.*) and scalene muscles (*Scal.*). The phases of respiration were signalled by an observer. Time in sec. No activity in sternomastoids. Activity of scaleni present in both supine and erect posture; there is marked increase in activity with inspiration. (Campbell, 1955.)

## Respiratory Function

The sternomastoids are probably the most important accessory muscles of inspiration and their participation in breathing in states of dyspnoea is a commonplace clinical observation. In patients with chronic lung disease causing chronic airway obstruction the use of the sternomastoids is probably due to the expansion of the rib cage rather than to the increased resistance to breathing, because, as the rib cage becomes expanded, the efficiency and mechanical advantage of the intercostals and diaphragm become progressively less and they probably cannot develop inspiratory pressures equal to those they are capable of in a normally shaped chest (see p. 310).

The mechanical importance of the sternomastoids is difficult to assess. Anatomically they appear powerful, but their mechanical advantage as inspiratory muscles would not seem to be very favourable. Through their

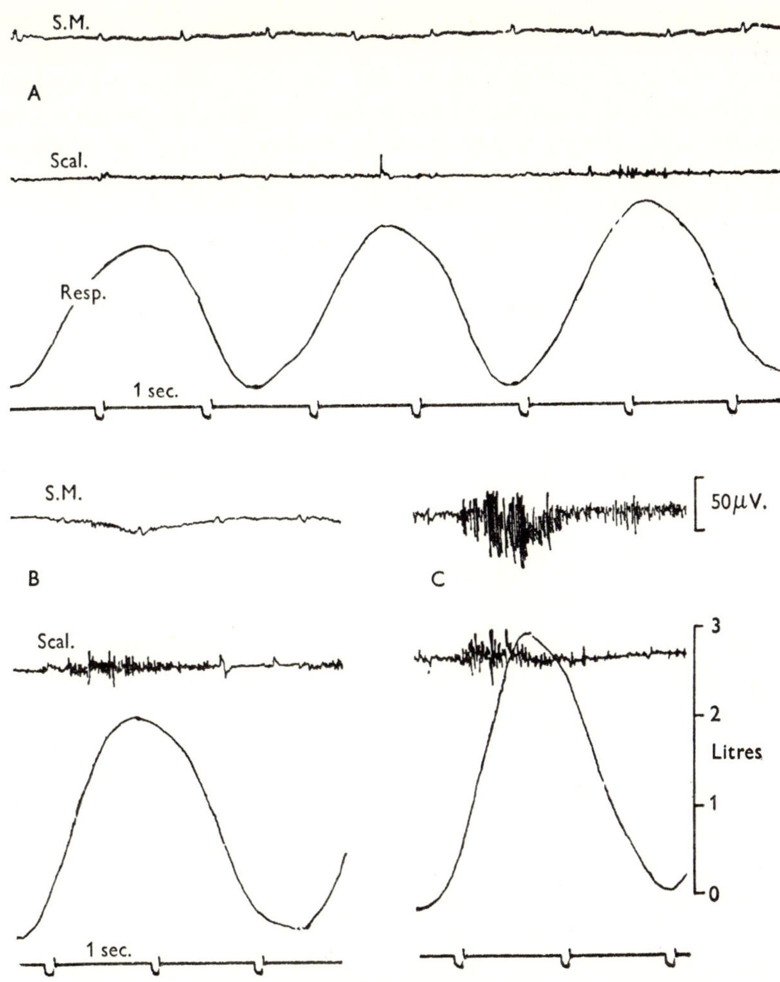


FIG. 68.—THE STERNOMASTOID AND SCALENE MUSCLES: BEHAVIOUR DURING INCREASED BREATHING

A healthy young male subject, supine. Electromyograms recorded from the sternomastoid (*S.M.*) and scalene (*Scal.*) muscles. *Resp.* the record of respiration (inspiration upwards). Time in sec. The subject rebreathed expired air from a spirometer with no $CO_2$ absorber. *A* shows the onset of inspiratory activity in the scaleni; *B* and *C* show the development of activity in the sternomastoid. At the beginning of *A* the ventilation rate was 42 l/min. and at the end it was 52 l/min. In *B* the ventilation rate was 60 l/min. and in *C*, 78 l/min.

The high level of breathing achieved without the use of the scalene muscles is in contrast with that shown by the subject from whom Fig. 67 was taken. (Campbell, 1955.)

action on the sternum they are able to move the ribs upwards, but not outwards, thus largely accounting for the "up and down" movement with little lateral expansion observed in patients with chronic airways obstruction. In extreme cases some of the antero-posterior expansion of the thorax is offset by an actual indrawing of the lower ribs.

Duchenne (1867) recounted the case of a young man with a high cervical transection of the spinal cord who breathed for some weeks apparently by means of his sternomastoids alone. He was very cyanose and when he was given artificial respiration contraction of the sternomastoids ceased. On stopping the artificial respiration the contraction only returned when he again became very cyanose. If it was true that the sternomastoids were the only muscles in action, then it appears that they can support some degree of pulmonary ventilation.

Lemon (1929) showed that a dog can tolerate paralysis of all the ordinary muscles of breathing and that respiration can then be maintained without great distress, during moderate activity, by the neck and shoulder muscles.

## OTHER MUSCLES

There are many muscles which may in some circumstances participate in respiratory acts. Most of the muscles which are discussed in this section are unlikely on anatomical grounds to be of importance in the mechanics of breathing. An electromyographic survey of the thorax and adjacent part of the trunk, which would have detected significant respiratory activity in most of them if any were present, failed to do so (Campbell, 1954). This conclusion is supported by subsequent work (see Delhez and Petit, 1966). Anatomical descriptions are not given unless they are important in the functional analysis.

### Trapezius

The trapezius is generally described as an auxillary muscle of inspiration by virtue of two actions: (i) the upper fibres extend the neck and facilitate the action of the sternomastoid; (ii) the rest of the muscle fixes the shoulder girdle and facilitates the action of the pectoral muscles.

In dyspnoeic patients the occipito-clavicular part of the muscle can be felt to contract during inspiration. Tokizane *et al.* (1952) reported that in normal subjects most of the muscle was active in inspiration during quiet breathing, but other workers have found that it contracts only at the end of maximal inspiration. Ventilations greater than 100 l/min can be attained without using it (Campbell, 1954; Delhez, 1963).

### Pectoralis Major

When the shoulder and upper arm are fixed, the pectoralis major may draw the upper ribs outwards and upwards, and it is usually therefore classified as an auxiliary muscle of inspiration.

Campbell (1954) found that the only respiratory activity in normal subjects is at the end of a maximum inspiration. Grombaek and Skouby (1960) found activity during inspiration in forced but not quiet breathing.

Contraction can sometimes be felt during inspiration in the pectoralis major of very dyspnoeic patients.

**Pectoralis Minor**

The actions and function of this muscle are probably similar to those of the pectoralis major. Duchenne (1867) and Beevor (1903) suggested that the pectoralis minor is more important than the major as an auxiliary muscle of inspiration. Tokizane *et al.* (1952) and Campbell (1954), however, found no evidence of greater respiratory activity in the minor than in the major.

**Subclavius**

Anatomically this would appear to be a muscle of inspiration. There are no data on its behaviour and it cannot be very powerful.

**Latissimus Dorsi**

The latissimus dorsi contains fibres which arise from the lower three or four ribs and which might be able to elevate these ribs, thus facilitating inspiration. Contraction of the muscle as a whole, however, compresses the lower thorax and therefore assists expiration.

The main mass of the muscle can be felt to contract quite vigorously during coughing (Beevor, 1903). Campbell (1954) found no evidence of activity during quiet or moderately increased breathing. Tokizane *et al.* (1952), however, reported inspiratory activity during deep breathing. Gronbaek and Skouby (1960) report various patterns in erect normal subjects during forced breathing.

**Serratus Anterior**

Duchenne (1867) suggested that this muscle can raise and evert the ribs if the scapula is fixed by the rhomboids. He claimed to have produced this action by stimulating the muscle electrically.

Catton and Gray (1951) examined this muscle electromyographically in normal subjects and failed to detect any activity during voluntary deep breathing, breathing through a narrow tube or coughing. Campbell (1954) also found no activity under similar conditions. Tokizane *et al.* (1952) classify the serratus anterior as expiratory during deep breathing. Gronbaek and Skouby (1960) report various patterns in erect normal subjects during forced breathing.

The absence of respiratory activity and the indirect mode of action make the serratus anterior unlikely to be important in the mechanics of breathing.

### Serratus Posterior Superior

Anatomically this muscle could elevate the ribs and should therefore be a muscle of inspiration. Its respiratory behaviour does not appear to have been examined.

### Serratus Posterior Inferior

Anatomically this muscle could depress the lower ribs. It is therefore generally regarded as a muscle of expiration. Electromyographic recordings taken with surface electrodes (Campbell, 1954) detected no marked respiratory activity. The activity that was recorded was no greater than that found in the sacrospinalis group which interfere with recordings taken at this site.

### Quadratus Lumborum

Anatomically the quadratus lumborum is potentially an active respiratory muscle. During inspiration it could fix the last ribs and facilitate the action of the diaphragm. During expiration it could co-operate with the other muscles of the abdominal wall in raising the intra-abdominal pressure. There are no studies of its behaviour in the literature. Its relatively small size and small attachment to the thorax make it unlikely to be of much importance in the mechanics of breathing.

### Sacrospinalis

This large mass of muscle and its upward extensions has a complex anatomy which, from the standpoint of respiratory function, can be subdivided into those components which are inserted into ribs and those which are inserted into vertebrae. Those such as the iliocostalis and part of the longissimus thoracis which are inserted into the ribs would be expected to have an expiratory function because their attachments enable them to depress the ribs. On the whole, these groups are lateral to those which are inserted only into the vertebrae. These latter groups, such as the spinalis thoracis, are extensors of the back. As extension of the vertebral column commonly occurs during deep inspiration, these muscles would be expected to show inspiratory activity.

From the foregoing description the lateral components of the sacrospinalis might be expected to show expiratory activity and the medial components might be expected to show inspiratory activity. Tokizane *et al.* (1952) reported rhythmic activity during quiet breathing in each of the components of the sacrospinalis of a pattern in agreement with these anatomical predictions. Neither Campbell (1954), Silver (1954, personal communication) nor Gronbaek and Skouby (1960) found any persistent respiratory rhythm in any part of the sacrospinalis group during quiet breathing.

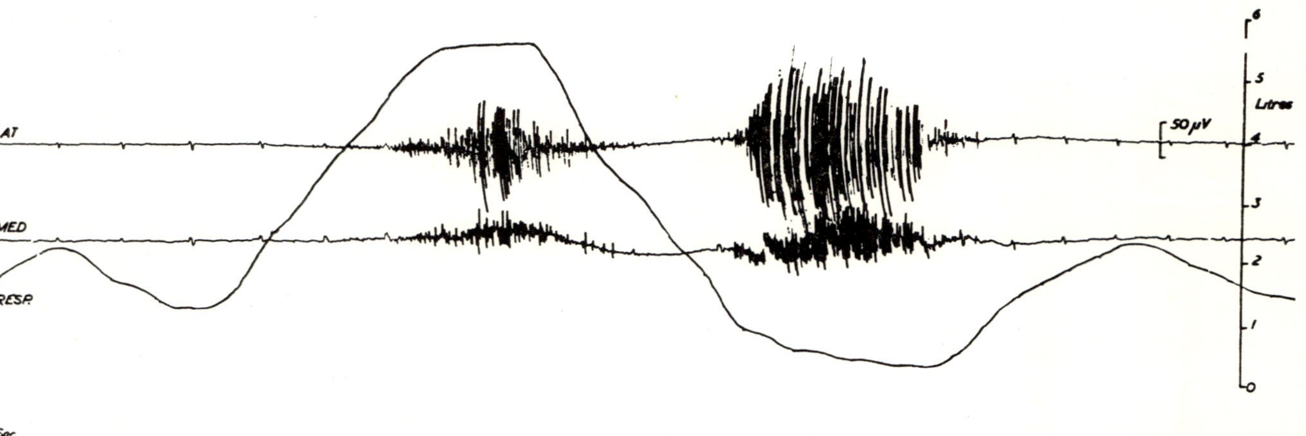


FIG. 69.—THE BACK MUSCLES: BEHAVIOUR DURING MAXIMUM INSPIRATION AND EXPIRATION

Record taken from a healthy young male subject in the supine posture. Two pairs of electrodes were placed vertically at the level of the 2nd and 4th lumbar vertebrae. *Lat.*—the axis of this pair was 9·5 cm from the midline. *Med.*—the axis of this pair was 4 cm from the midline. *Resp.*—Respiration, inspiration upwards.

There is no activity during quiet breaths at the beginning and end of the record. At both sites there is marked activity both at the end of maximum inspiration and during maximum expiration. This activity is greater during expiration than during inspiration. The bulk of this activity probably arose in the sacrospinalis group, the activity during maximum inspiration being associated with extension of the back, and the activity during expiration being associated with the contraction of the other muscles of the abdominal wall. For further discussion see text. (Campbell, 1954.)

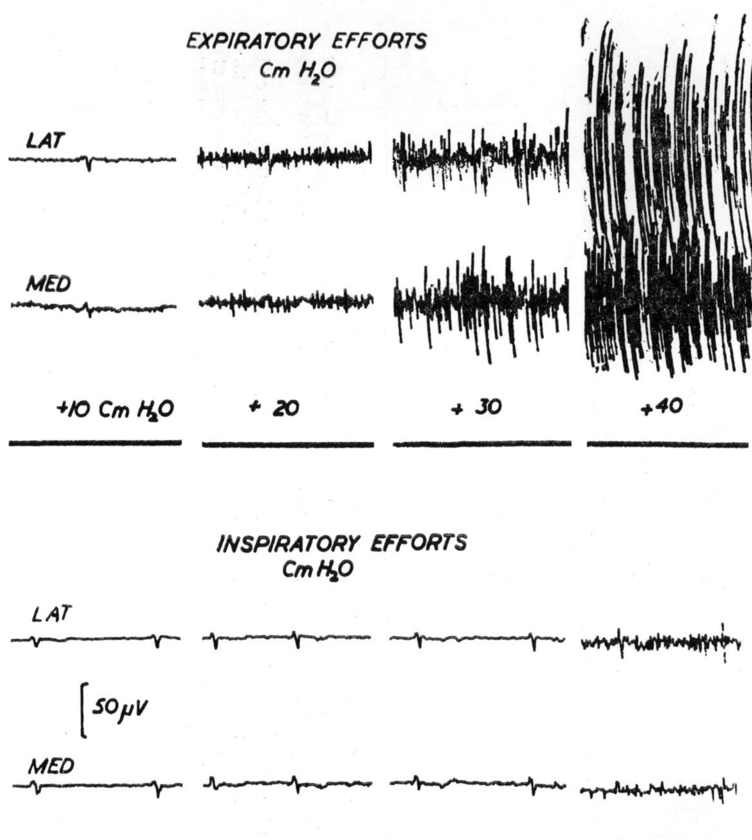


FIG. 70.—THE BACK MUSCLES: BEHAVIOUR DURING GRADED VOLUNTARY EXPIRATORY AND INSPIRATORY EFFORTS

Subject and experimental conditions as in previous figure. Each strip of record was taken during the maintenance of a steady pressure with the chest at the resting respiratory level. The figure corresponding to each strip of record is the alveolar pressure related to atmospheric.

There is much greater activity during expiratory efforts than during inspiratory efforts; also a clear correspondence between the magnitude of the expiratory effort and the intensity of the electrical activity is evident. An inspiratory pressure of —30 cm $H_2O$ can be developed without the use of these muscles. (Campbell, 1954.)

Campbell (1954) examined the sacrospinalis in young male subjects in the course of an electromyographic survey of the respiratory behaviour of other muscles in the lumbar region of the back. He found the same pattern of activity over a wide area from close to the vertebral spine up to 10–12 cm laterally. The same pattern was also found over the erector spinae group in the lower thoracic region. At all these sites other muscles may have contributed to the electromyograms, but most of the activity probably

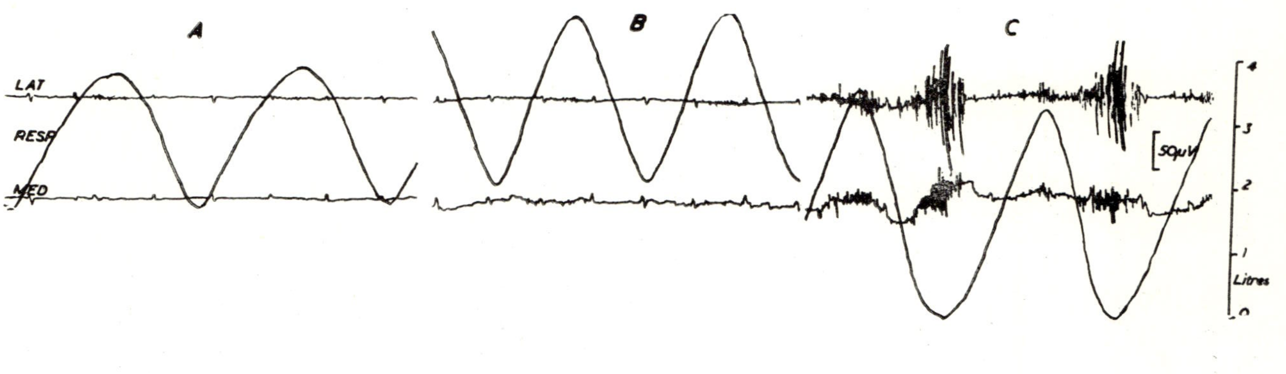


FIG. 71.—THE BACK MUSCLES: BEHAVIOUR DURING INCREASED PULMONARY VENTILATION

Subject and experimental conditions as in previous two figures.

*A* and *B* were recorded while the subject was rebreathing expired air from the spirometer with no $CO_2$ absorber in the circuit. *C* was obtained during voluntarily increased ventilation. During *A* the ventilation rate was 67 l/min.; during *B* 83 l/min; and during *C* 105 l/min. (The level of the spirogram in *C* is not comparable with *A* and *B*.)

There is slight inspiratory activity, but no expiratory activity in *A* and *B*. In *C* there is definite inspiratory activity and marked expiratory activity. Most of the activity recorded at these sites probably arose in the sacrospinalis group. For further interpretation see text. (Campbell, 1954.)

arose in the sacrospinalis. Records were taken both with the subjects supine and erect.

Occasional transient inspiratory activity was found during quiet breathing in one subject, otherwise no respiratory activity was found during quiet breathing.

There was marked activity towards the end of a maximum inspiration and during a maximum expiration (Fig. 69). The expiratory activity was greater than the inspiratory activity.

During graded static voluntary efforts performed with the chest at the resting respiratory level, the activity during expiratory efforts was greater than the activity during inspiratory efforts (Fig. 70). There was a clear gradation of the intensity of activity during progressively increasing expiratory pressures from +10 to +40 cm $H_2O$. During inspiratory efforts there was no activity below pressures of −30 cm $H_2O$.

During increased pulmonary ventilation (Fig. 71) activity first appeared during inspiration at tidal volumes of about 2 litres and ventilation rates of 30–60 l/min. Activity did not usually appear during expiration until the ventilation rate exceeded 75 l/min. At still higher levels, approaching those of maximum voluntary breathing, the expiratory activity became more marked than the inspiratory activity.

The collation of the anatomical and electromyographic data is difficult. Does the similar activity recorded at all areas arise in the same fibres, or are the component muscles so intermingled that electrodes at any site record both vertebro-costal and intervertebral fibres? Anatomically it appears improbable that the more laterally placed electrodes would detect the activity of intervertebral components. Therefore the vertebro-costal components such as the iliocostalis probably contract both during inspiratory and expiratory conditions.

The respiratory function of this group of muscles appears, therefore, to be complex. During expiratory efforts their contraction probably helps to depress the ribs and raise intra-abdominal pressure. The functional significance of their contraction during increased pulmonary ventilation is not so easy to assess. The relatively small participation during voluntary inspiratory efforts when the chest is at the resting respiratory level makes it unlikely that they play any direct part in the development of the force of inspiration when the breathing is increased. It would appear that their contraction during deep breathing is mainly concerned with extension of the spine. This extension may, of course, contribute to the enlargement of the rib cage, and increases the cranio-caudal diameter of the lung (see Chapter II).

## REFERENCES

Beevor, C. E. (1903). The Croonian Lectures on muscular movements and their representation in the central nervous system. London: Adlard (1904). Also Macmillan (1959).

CAMPBELL, E. J. M. (1954). The muscular control of breathing in man. (Ph.D. Thesis, Univ. of London.)

CAMPBELL, E. J. M. (1955). The role of the scalene and sternomastoid muscles in breathing in normal subjects. An electromyographic study. *J. Anat.* (*Lond.*), **89**, 378–386.

CATTON, W. T., and GRAY, J. E. (1951). Electromyographic study of the action of the serratus anterior muscle in respiration. *J. Anat.* (*Lond.*), **85**, 412.

DELHEZ, L. (1963). Méthodes d'examen du comportement électrique des muscles accessibles par électrodes du surface. *Electromyography*, **3**, 165–189.

DELHEZ, L., and PETIT, J. M. (1966). Données actuelles de l'Electromyographie respiratoire chez l'Homme normal. *Electromyography*, **6**, 101–146.

DUCHENNE, G. B. (1867). *Physiology of Motion.* Trans. E. B. KAPLAN, 1949, pp. 443–503. Philadelphia: Lippincott.

FENICHEL, N. M., and EPSTEIN, B. S. (1955). Pulmonary apical herniations. *Arch. intern. Med.*, **96**, 747–751.

FICK, R. (1923). Ueber die Zwischenrippenmuskeln. *S.-B. preuss. Akad. Wiss., Physik-math. Kl.*, 65–72.

GIAUNI, G. (1936). Esplorazione pneumografica della cinematica toracica e della capacita vitale dopo scalenotomia e dopo scalenofrenicoexeresi. *Clin. med. ital.*, **67**, 783–796.

GRONBAEK, P., and SKOUBY, A. P. (1960). The activity pattern of the diaphragm and some muscles of the neck and trunk in chronic asthmatics and normal controls. A comparative electromyographic study. *Acta med. scand.*, **168**, 413–425.

JOLY, H., and VINCENT, PH.-A. (1937). Rôle respiratoire des muscles scalènes et inter-costaux étudié en fonction de la collapsothérapie pulmonaire. *Arch. méd.-chir. Appar. resp.*, **12**, 392–404.

JONES, D. S., BEARGIE, R. J., and PAULY, J. E. (1953). An electromyographic study of some muscles of costal respiration in man. *Anat. Rec.*, **117**, 17–24.

JONES, D. S., and PAULY, J. E. (1957). Further electromyographic studies on muscles of costal respiration in man. *Anat. Rec.*, **128**, 733–746.

LEMON, W. S. (1929). Efficiency of the mechanical factors of respiration. A study of respiratory reserve. *Amer. J. med. Sci.*, **177**, 319–333.

RAPER, A. J., THOMPSON, W. T., JR., SHAPIRO, W., and PATTERSON, J. L., JR. (1966). Scalene and sternomastoid muscle function. *J. appl. Physiol.*, **21**, 497–502.

THOMPSON, W. T., JR., PATTERSON, J. L., JR., and SHAPIRO, W. (1964). Observations on the scalene respiratory muscles. *Arch. intern. Med.*, **113**, 856–865.

TOKIZANE, T., KAWAMATA, K., and TOKIZANE, H. (1952). Electromyographic studies on the human respiratory muscles. *Jap. J. Physiol.*, **2**, 232–247.

VARENE, P., RICHARD, P., and JACQUEMIN, C. (1963). Electromyography of the diaphragm during transverse accelerations in man. *C.R. Acad. Sci.* (*Paris*), **256**, 4975–4976.

# Chapter X

# MUSCLES OF THE LARYNX AND THYROID CARTILAGE

E. J. M. CAMPBELL AND J. NEWSOM DAVIS

## INTRINSIC MUSCLES OF THE LARYNX

THESE muscles will not be discussed individually in detail because their importance in respiration is in their group behaviour. There are two main groups: the abductors (chiefly the posterior crico-arytenoids) and the adductors (chiefly the lateral crico-arytenoids and the thyro-arytenoids). The abductors separate the vocal cords and widen the lumen of the glottis. The adductors approximate the vocal cords and narrow the glottis. The mechanical basis of phonation is summarised on pages 105 to 108. The function of the laryngeal muscles during phonation is considered in detail by several participants in the symposium on "Sound Production in Man" (Bouhuys, 1968).

All the intrinsic laryngeal muscles are innervated by the recurrent laryngeal nerve except the crico-thyroid muscle which is supplied by the external branch of the superior laryngeal nerve. The cells of origin lie in the nucleus ambiguus. The fibre calibre spectrum of the recurrent laryngeal nerve in the cat shows a single peak at 10–12 $\mu$ with no small fibres which might correspond to fusimotor fibres (Murray, 1957). In man, it is also unimodal with a similar peak around 10 $\mu$ (Scheuer, 1964). The innervation ratio of the human laryngeal muscles has been estimated as 100–200 (Faaborg-Anderson, 1957).

In many species (excepting man) these muscles do not contain any receptors morphologically resembling muscle spindles, and neurophysiological attempts to identify such receptors have been unsuccessful (Andrew, 1954; Martensson, 1966, 1968). The function of those receptors which have been identified histologically is uncertain. The proprioceptive control of laryngeal muscles in these species appears to be provided through articular receptors (Martensson, 1964; Kirchner and Wyke, 1965; Martensson, 1968). In man, on the other hand, endings resembling muscle spindles have been shown in the laryngeal muscles (Keene, 1961; Rudolph, 1961).

The contraction times of the laryngeal muscles are in general shorter than those of the limb muscles, with values ranging from 14–35 msec recorded in the dog (Martensson and Skoglund, 1964). Contraction times in the cat and rabbit are similar (Martensson and Skoglund, 1964; Hall-Craggs, 1968). The thyro-arytenoid and lateral crico-arytenoid are "fast" laryngeal muscles (see Chapter I) and the posterior crico-arytenoid and crico-thyroid are "slow". Although the contraction time of the crico-thyroid muscles is fast when compared to limb muscles, its enzyme and microscopic characteristics are those of slow muscle (Hall-Craggs, 1968).

Recording from the recurrent laryngeal nerve in cats shows that most fibres are active only during inspiration, the discharge of these units showing the same general pattern as that of phrenic motoneurones (Green and Neil, 1955; Eyzaguirre and Taylor, 1963; Bianconi and Raschi, 1964; Eyzaguirre *et al.*, 1966). Hyperventilation abolishes all inspiratory activity and causes tonic firing of expiratory units, while hypoxia or breathing 6 per cent $CO_2$ increases inspiratory activity, along with phrenic discharge. Lung inflation may inhibit inspiratory unit discharge and accentuate the discharge of expiratory units. The abductor muscles of the cat are active in inspiration and the adductors in expiration (Green and Neil, 1955).

In resting man, EMG recording reveals tonic activity in the abductors, and in most subjects in the adductors as well. Unexpectedly, in view of the findings in the cat, inspiration is associated with increased activity in the adductors. Thus increased EMG activity occurs first in the thyro-arytenoid followed by discharge half a second later in the crico-thyroid and still later in the vocalis muscle (Faaborg-Anderson, 1957; Buchtal, 1959). Buchtal (1959) has suggested that this activity in the adductors may serve to steady the vocal cords during breathing. The synergistic actions of certain laryngeal muscles have recently been investigated in the dog (Konrad and Rattenborg, 1969), and may in part account for the EMG observations in man. The crico-thyroid muscle adducts the cords when electrically activated alone, and the posterior crico-arytenoid when similarly activated abducts the cords. When the two muscles are stimulated simultaneously, however, there is a greater degree of cord abduction than when the posterior crico-arytenoid is active alone.

During phonation, there is increased activity in all the intrinsic adductor muscles and a decrease in abductor muscles starting about half a second before audible sound. The intensity of the EMG discharge increases with increasing pitch (Faaborg-Anderson, 1957). Activity also increases in the adductors during a cough.

The most important respiratory functions of the adductor muscles are protective. They prevent aspiration of foreign substances into the lungs, and they contract during the compressive phase of a cough, thereby enabling a high intratracheal pressure to be developed. Mills (1950) reported that the glottis is closed at the end of a voluntary maximum inspiration and suggested that this closure is one of the factors limiting the depth of a voluntary maximum inspiration (see p. 70), but Mead and his co-workers (1963) found evidence of this in only one of six normal subjects. Negus (1949) has suggested that the contraction of the adductors during expiration may, by raising the intratracheal, intra-alveolar and intra-thoracic pressures, affect the distribution of air within the lungs and the pulmonary circulation. Rattenborg (1961) reported anatomical, radiographic and spirometric observations which showed a decrease in laryngeal resistance when resistance to expiration was increased artificially or by breathing through the nose. He suggested that this might represent a compensating mechanism which operates to stabilise the time constant (p. 84), the effect of which would be in accord with Negus' suggestion.

The tonic contraction of the abductor muscles is important in maintaining the patency of the glottis. Paralysis of these muscles causes inspiratory stridor.

## MUSCLES ACTING ON THE THYROID CARTILAGE

Few of this large group of muscles are of direct importance in pulmonary ventilation. Their respiratory function is chiefly concerned with the fixation of the larynx.

The sterno-hyoid and the sterno-thyroid may have the mechanical action of raising the sternum, and could therefore be muscles of inspiration. Their small size and the potential mobility of the larynx make them unlikely to be of great importance. There is no evidence of respiratory activity in these muscles in man, but Andrew (1955) gives good evidence from studies in the rat that the sterno-thyroid contracts during inspiration and that this contraction is directly regulated from the respiratory centre.

Mitchinson and Yoffey (1947) pointed out that during a deep inspiration the central tendon of the diaphragm descends by 3–5 cm and that, through the pericardium, this downward traction is applied to the heart, and yet the arch of the aorta does not descend appreciably. The probable explanation is that the aortic arch and the pulmonary arteries are hooked round the main bronchi and that the bronchi are fixed during inspiration. The fixation of the bronchi and the lung root cannot be entirely accounted for by ligamentous attachments. The lung root does descend slightly during a deep inspiration, but Mitchinson and Yoffey showed that in most normal subjects the larynx does not, and they suggested that the larynx is stabilised by the contraction of muscles which elevate it. Mitchinson and Yoffey suggested that the most likely muscles to have this function were the suprahyoid group (mylo-hyoid, stylo-hyoid and posterior belly of digastric). Andrew (1955), as a result of electromyographic studies of these muscles in the rat, agreed with Mitchinson and Yoffey in principle, but suggested that the stylo-glossus is the important elevator.

Whichever are the particular muscles concerned, it seems probable that the elevators of the larynx have the important respiratory function of stabilising the larynx and upper trachea and hence indirectly the lung root. The "tracheal tug" which occurs during inspiration in dyspnoeic patients and in deep anaesthesia may be due to failure of this stabilising mechanism to stand up to the forceful traction of the diaphragm (p. 309).

Andrew also recorded inspiratory activity in the crico-thyroid muscle and suggested that this contraction opposes rotation of the thyroid cartilage at the crico-thyroid joint with consequent narrowing of the glottic aperture during inspiration.

Armstrong and Smith (1955) have reported studies in the dog in which species the behaviour of the sterno-thyroid and crico-thyroid muscles apparently resembles that found in the rat.

## REFERENCES

ANDREW, B. L. (1954). Proprioception at the joint of the epiglottis of the rat. *J. Physiol.* (*Lond.*), **126**, 507–523.

ANDREW, B. L. (1955). The respiratory displacement of the larynx: a study of the innervation of accessory respiratory muscles. *J. Physiol.* (*Lond.*), **130**, 474–487.

ARMSTRONG, B. W., and SMITH, D. J. (1955). Function of certain neck muscles during the respiratory cycle. *Amer. J. Physiol.*, **182**, 599–600.

BIANCONI, R., and RASCHI, F. (1964). Respiratory control of motoneurones of the recurrent laryngeal nerve and hypocapnic apnoea. *Arch. ital. Biol.*, **102**, 56–73.

BOUHUYS, A. Ed. (1968). Symposium on "Sound Production in Man". *Ann. N.Y. Acad. Sci.*, **155**, 1–381.

BUCHTAL, F. (1959). Electromyography of intrinsic laryngeal muscles. *Quart. J. exp. Physiol.*, **44**, 137–148.

EYZAGUIRRE, C., SAMPSON, S., and TAYLOR, J. R. (1966). The motor control of intrinsic laryngeal muscles in the cat. Nobel Symposium I. *Muscular Afferents and Motor Control*, pp. 209–225. Stockholm: Almqvist and Wiksell; New York: John Wiley and Sons.

EYZAGUIRRE, C., and TAYLOR, J. R. (1963). Respiratory discharge of some vagal motoneurones. *J. Neurophysiol.*, **26**, 61–78.

FAABORG-ANDERSON, K. (1957). Electromyographic investigation of intrinsic laryngeal muscles in humans. *Acta physiol. scand.*, **41**, Suppl. 140.

GREEN, J. H., and NEIL, E. (1955). The respiratory function of the laryngeal muscles. *J. Physiol.* (*Lond.*), **129**, 134–141.

HALL-CRAGGS, E. C. B. (1968). The contraction times and enzyme activity of two rabbit laryngeal muscles. *J. Anat.* (*Lond.*), **102**, 241–255.

KEENE, M. F. L. (1961). Muscle spindles in human laryngeal muscles. *J. Anat.* (*Lond.*), **95**, 25–29.

KIRCHNER, J. A., and WYKE, B. (1965). Afferent discharges from laryngeal articular mechanoreceptors. *Nature* (*Lond.*), **205**, 86–87.

KONRAD, H. R., and RATTENBORG, C. C. (1969). Combined action of laryngeal muscles. *Acta oto-laryng.* (*Stockh.*), **67**, 646–649.

MARTENSSON, A. (1964). Proprioceptive impulse patterns during contraction of intrinsic laryngeal muscles. *Acta physiol. scand.*, **62**, 176–194.

MARTENSSON, A. (1966). Testing for doubly innervated fibres in the intrinsic laryngeal muscles of the dog. *Acta physiol. scand.*, **67**, 152–164.

MARTENSSON, A. (1968). The functional organisation of the intrinsic laryngeal muscles. *Ann. N.Y. Acad. Sci.*, **155**, 91–97.

MARTENSSON, A., and SKOGLUND, C. R. (1964). Contraction properties of intrinsic laryngeal muscles. *Acta physiol. scand.*, **60**, 318–336.

MEAD, J., MILIC-EMILI, J., and TURNER, J. M. (1963). Factors limiting depth of a maximal inspiration in human subjects. *J. appl. Physiol.*, **18**, 295–296.

MILLS, J. N. (1950). The nature of the limitation of maximal inspiratory and expiratory efforts, *J. Physiol.* (*Lond*)., **111**, 376–381.

MITCHINSON, A. G., and YOFFEY, J. M. (1947). Respiratory displacement of larynx, hyoid bone and tongue. *J. Anat.* (*Lond.*), **81**, 118–120.

MURRAY, J. G. (1957). Innervation of the intrinsic muscles of the cat's larynx by the recurrent laryngeal nerve: a unimodal nerve. *J. Physiol.* (*Lond.*), **135**, 206–212.

NEGUS, V. E. (1949). The second stage of swallowing. *Acta oto-laryng.* (*Stockh.*), Suppl. **78**, 78–82.

RATTENBORG, C. (1961). Laryngeal regulation of respiration. *Acta anaesth. scand.*, **5**, 129–140.

RUDOLPH, G. (1961). Spiral nerve-endings (proprioceptors) in the human vocal muscle. *Nature* (*Lond.*), **190**, 726–727.

SCHEUER, J. L. (1964). Fibre size frequency distribution in normal human laryngeal nerves. *J. Anat.* (*Lond.*), **98**, 99–104.

# SUMMARY OF RESPIRATORY MUSCLE ACTIVITY

The following account is simplified and represents only the common or basic pattern of behaviour of the more important muscles. For a full account reference must be made to the chapters dealing with the appropriate muscles.

### Quiet Breathing

During *inspiration* the following muscles are in action:

The diaphragm: in all subjects.

The intercostals: the "parasternal" intercostals of the first five spaces, and the external intercostals posteriorly: in all subjects.

The scaleni: in some subjects.

If postural activity is present in the abdominal muscles, it tends to decrease.

Some small muscles in the neck contract to stabilise the thyroid cartilage; and the abductors of the larynx contract to maintain or increase the patency of the glottis.

During *expiration* there is some persistence of the contraction of the inspiratory muscles which decreases progressively in the first part of expiration and disappears at about half expiration. In the last part of expiration the interosseous internal intercostals of the 7th–10th spaces on the lateral aspect of the chest wall are regularly active, the abdominal muscles may exhibit some increase in postural activity, and the adductors of the larynx contract.

### Moderately Increased Pulmonary Ventilation

The pattern of behaviour up to ventilation rates of 50 l/min is fundamentally the same as that found during quiet breathing; the distribution of intercostal activity spreads from above downwards during inspiration and conversely during expiration, the superficial layer in any region being inspiratory and the deeper layer being expiratory.

Between 50 and 100 l/min the sternomastoids and the extensors of the vertebral column come into action towards the end of inspiration. The antero-lateral abdominal and intercostal muscles begin to contract with increasing vigour towards the end of expiration.

### Greatly Increased Pulmonary Ventilation

Above 100 l/min all the accessory muscles of inspiration come into action throughout inspiration and all the muscles of the abdominal wall come into action throughout expiration. The fundamental pattern becomes obscured by the appearance of the generalised activity found under conditions of great physical effort.

### Maximum Inspiration

All the ordinary muscles of inspiration are probably active. Most of the accessory muscles of inspiration are also usually active.

The antero-lateral abdominal muscles contract in most subjects at the end of maximum inspiration. In a few subjects the adductors of the larynx contract at the end of maximum inspiration.

### Maximum Expiration

The muscles of the abdominal wall are the most important muscles of forced expiration. The internal interosseous intercostals and other accessory muscles of expiration are active. The diaphragm and scaleni also contract.

# APPENDIX B

## Electromyography†

*Electrodes.*—The EMG can be recorded either with surface electrodes applied to the skin over the muscle, or with needle or wire electrodes inserted into the muscle itself. With surface electrodes, individual motor units cannot readily be identified, nor can the precise site of the activity be established with certainty in muscles such as the intercostals which are thin and arranged in layers. For these purposes needle or wire electrodes are required.

A concentric needle electrode consists of a central electrode insulated from the shaft of the needle which acts as the other electrode. In order to reduce further the range of the electrode (i.e. the volume of muscle from which spike potentials are recorded) bipolar needles can be employed, which contain a pair of central electrodes. Such electrodes may have a range of less than 0·5 mm and are thus suitable for sampling from intercostal muscles (Taylor, 1960). In some circumstances, for instance when recording for long periods, wire electrodes may be used. A pair of fine insulated wires, bared only at their cut ends, are led down the shaft of a needle and hooked over its bevel.* The needle is then inserted into the muscle and the shaft withdrawn, leaving the wires *in situ.* A recording period of many hours can be achieved in this way. It should be emphasised, however, that long periods of recording are likely to be associated with changes in the impedance of the electrode surface and with local tissue damage, whatever type of electrode is used.

*Quantitative methods.*—In order to be able to relate the EMG to the force of contraction, it is necessary to quantify the electrical activity. Several methods are available (see Rosenfalck, 1960).

1. The electrical activity can be integrated either graphically or electronically. It is now easily achieved using an operational amplifier with capacitive feedback. It is often convenient to have facilities for resetting the integrator at regular predetermined intervals (see Newsom Davis and Sears, 1970). When recording with an oesophageal electrode, special techniques are available for minimising the ECG signal (Lourenço and Mueller, 1967).

2. The potentials can be summed using an RC filter with a time constant greater than the spike duration. With this method, the output voltage deviates from the true mean voltage, the deviation varying at different potential durations and frequencies. If inductance is added to the RC

† A more detailed review of earlier work and, in particular, of the use and limitations of surface electrodes is contained in the Appendix to the first edition of this monograph.

* When recording from intercostal muscles with this technique, it is important that the needle bevel should be short to reduce the risk of pneumothorax.

filter, a close agreement is achieved between the measured output voltage and the true mean voltage (Rosenfalck, 1960).

3. The number of individual motor unit potentials can be counted (Bergström, 1959). This has the disadvantage that it is unsuitable for strong contractions of the muscle because of the difficulty in accurate counting. The impulse frequency has a linear relationship with the integrated EMG (Bergström, 1959).

4. The rate and mean amplitude of potential changes can be counted electronically (Rose and Willison, 1967; Dowling *et al.*, 1968). The potential changes counted with this technique do not correspond directly to individual motor units because of the effects of summation and cancellation.

## REFERENCES

Bergström, R. M. (1959). The relation between the number of impulses and the integrated electric activity in electromyogram. *Acta physiol. scand.*, **45**, 97–101.

Dowling, M. H., Fitch, P., and Willison, R. G. (1968). A special purpose digital computer (Biomac 500) used in the analysis of the human electromyogram. *Electroenceph. clin. Neurophysiol.*, **25**, 570–573.

Lourenço, R. V., and Mueller, E. P. (1967). Quantification of electrical activity in the human diaphragm. *J. appl. Physiol.*, **22**, 598–600.

Newsom Davis, J., and Sears, T. A. (1970). The proprioceptive reflex control of the intercostal muscles during their voluntary activation. *J. Physiol.* (*Lond.*). (In press).

Rose, A. L., and Willison, R. G. (1967). Quantitative electromyography using automatic analysis: studies in healthy subjects and patients with primary muscle disease. *J. Neurol. Neurosurg. Psychiat.*, **30**, 403–410.

Rosenfalck, A. (1960). *Proceedings of the 2nd International Congress on medical electronics, Paris, June 1959*, pp. 9–12. London: Iliffe.

Taylor, A. (1960). The contribution of the intercostal muscles to the effort of respiration in man. *J. Physiol.* (*Lond.*), **151**, 390–402.

# CONTROL AND ORGANISATION

# Chapter XI

# SPINAL CONTROL

J. Newsom Davis

One of the most interesting recent developments in the study of the control of breathing has been the increasing emphasis placed on segmental proprioceptive mechanisms. It has been implicit in most earlier accounts, as for instance in Pitts' (1946) review, that the output from the "respiratory centre" in the brain stem was transmitted unchanged to the respiratory muscles. It has now become clear that the spinal mechanism performs an important function in integrating the segmental input from respiratory muscle proprioceptors with the descending drives concerned with respiration, posture and the other acts in which the respiratory muscles participate.

The general features of proprioceptive innervation of striated muscle have been outlined in Chapter I, and illustrated in Fig. 7. In animals, information about the proprioceptive control of the respiratory muscles has been derived from the effects of dorsal root section, from the discharge characteristics of their receptors and the efferent control of their muscle spindles, and from intracellular recording of their alpha motoneurones. In man, evidence for a segmental proprioceptive control is based upon the effects of dorsal root section and of sudden load changes upon respiratory muscle electrical activity.

## DORSAL ROOT SECTION

The earliest evidence for a segmental influence on the respiratory muscles came from a study of the effects of dorsal root section which interrupts the afferent arc of the gamma loop as well as all other afferent fibres. Coombs (1918) reported that thoracic dorsal root section in the cat caused a diminution of costal respiration although abdominal respiration remained unaltered and the rate very little changed. In further experiments (Coombs and Pike, 1930) it was found that this procedure almost abolished costal respiration in kittens during the first ten days of life, and that cervical dorsal root section much reduced diaphragmatic activity. Although Stella (1938) stated that respiration in the cat was unaltered after cervical and thoracic dorsal root section, in those experiments the animal was only showing diaphragmatic breathing. When the phrenic nerves were sectioned in addition, the costal respiration which resulted was somewhat irregular in rhythm and depth. The unaltered diaphragmatic activity following dorsal root section of the phrenic nerve was confirmed in the cat by Sant'Ambrogio *et al.* (1962) and in the rabbit by Sant'Ambrogio and Widdicombe (1965).

The abdominal muscles are inactive during quiet breathing, but during positive-pressure breathing regular respiratory activity occurs (Bishop, 1963, 1964). This activity is abolished by dorsal root section from T8–L3, demonstrating its dependence upon segmental afferent inflow (Bishop, 1964), but it is not a purely segmental reflex because it is also abolished by acute cord transection and by cervical vagotomy.

Nathan and Sears (1960) examined in man the effects upon certain respiratory muscles of dorsal root section, carried out for the relief of intractable pain. In the two patients who underwent dorsal root section of the phrenic nerve, the diaphragm was shown by fluoroscopy to be temporarily paralysed, some movement returning in one case by the eleventh post-operative day but impaired movement persisting up to three months. In the patient who had dorsal root section of the 4th, 5th and 6th thoracic roots, EMG activity in the affected intercostal muscles was found to be reduced two weeks after the operation when compared to the unoperated side. In their discussion of these findings, the authors suggested that the paralysis caused by dorsal root section followed sudden withdrawal of the habitual afferent inflow to the alpha motoneurone of muscle spindle afferents.

The gain of the dorsal root afferents *in toto* has been compared to that of the vagal afferents by von Euler and Fritts (1963). These authors measured the change in intratracheal pressure developed by the spontaneously breathing anaesthetised cat (output) per unit change in thoracic volume (input) after tracheal closure at different lung volumes before and after vagal section or dorsal root section from C3 to T12. The vagus exerted an inhibitory influence on the force of respiratory muscle contraction, which was almost linearly related to lung volume. The net effect of the dorsal roots, on the other hand, was facilitatory, and decreased with increasing lung volume. The gain of the two systems was similar. This facilitation is presumably attributable to the effects of deafferentation on intercostal activity alone since dorsal root section of the phrenic nerve does not alter the inspiratory pressures developed against a closed airway (Sant'Ambrogio, personal communication; see p. 216).

Thus dorsal root section of the intercostal nerves both in animals and man leads to a temporary impairment of muscle activity. Dorsal root section of the phrenic nerve is without effect on diaphragm activity in some animals (e.g. cat and rabbit) but leads to temporary impairment of activity in man, presumably accounted for by a species difference. These observations establish the importance of a segmental input to the maintenance of excitability in intercostal, and possibly phrenic, alpha motoneurones, but do not specify its nature.

## THORACIC REFLEXES

An intercostal nerve has reflex connections with the intercostal nerves of other segments, the phrenic nerve and the splanchnic nerves. This has been demonstrated by recording the efferent discharge in one nerve evoked by electrical stimulation of another.

### Intercostal-intercostal and Splanchnic-intercostal

Electrical stimulation of an intercostal or a splanchnic nerve in anaesthetised or decerebrate cats evokes polysynaptic reflex motor responses in other intercostal nerves (Downman, 1955; Alderson and Downman, 1966). These reflexes are of smaller amplitude in the decerebrate animal, but become more prominent and of shorter latency after spinal cord transection, indicating release from a tonic inhibition operating in the decerebrate state. The pathway mediating the tonic inhibitory effect lies in the ipsilateral dorsolateral region of the cord (Downman and Hussain, 1958). The inhibition originates from the ventro-medial reticular formation of the medulla and lower pons (Downman and Hussain, 1958), and may be elicited in the anaesthetised cat by electrical stimulation in this region (Alderson and Downman, 1966). It acts upon interneuronal pathways serving these reflexes. The tonic inhibition is unmodified by the rhythm of breathing but may sometimes be abolished by vagotomy, and then enhanced by stimulation of the cut central end of the nerve (Alderson and Downman, 1966). Splanchnic-intercostal reflexes can be modified by electrical stimulation of the orbital cortex (Korn, 1967).

### Intercostal-phrenic and Splanchnic-phrenic

Polysynaptic reflex responses can be evoked in the phrenic nerve of the cat after latency of 10–16 msec by stimulation of the nerves to the intercostal muscles of the lower rib cage, where the costal part of the diaphragm is attached (Downman, 1955; Decima *et al.*, 1967; von Euler, 1968). In a recent detailed study of these responses in spinal animals, von Euler and his colleagues (Decima *et al.*, 1969) have shown that intercostal afferents exert a reflex control over the diaphragm, the afferent fibres concerned arising from spindle secondary endings and from tendon organs. The ascending pathways from the lower thoracic segments to the phrenic motor nucleus which serve these reflexes lie in the ventro-lateral funiculus. A further interesting observation was made that a twitch of the diaphragm produced by stimulation of the peripheral stump of the cut phrenic nerve could evoke single or multiple volleys in the phrenic motoneurones after a latency of about 25 msec (Decima and von Euler, 1969*a*). Thus the phrenic motoneurone can influence its own excitability through a reflex arc which includes a mechanical link between the diaphragm and thoraco-abdominal wall. These authors suggest that this reflex mechanism might "compensate" for the absence of autogenetic facilitation in the diaphragm (see page 214).

The intercostal-phrenic reflex in decerebrate as opposed to spinal animals is generally similar although smaller, and the response waxes and wanes with the animal's own respiratory cycle, being maximal in inspiration and sometimes unobtainable at end-expiration (Decima and von Euler, 1969*b*). This inhibition at end-expiration could be abolished by cerebellar stimulation. Significant differences were noted between the reflex responses recorded from branches of the phrenic nerve to different parts of the diaphragm in the decerebrate animal. The fact that these differences disappeared after spinalisation implies that a supraspinal mechanism determines these selective effects.

The order of recruitment of phrenic motoneurones in these reflexes was always the same as that during spontaneous respiratory activity, small spikes being recruited earlier than large spikes (Decima and von Euler, 1969*a*).

Stimulation of the nerves to the lower thoracic spinal muscles (Decima *et al.*, 1967) and of the splanchnic nerves (Downman, 1955) also evokes a reflex response in the phrenic nerve. In contrast to these effects, stimulation of somatic afferents from the limbs in spinal animals only occasionally causes reflex discharge in the phrenic nerve (Calma, 1952). In the intact preparation, long latency responses can be elicited which are influenced by the respiratory phase; the reflex arcs probably involve brain stem structures.

## INTERCOSTAL MUSCLE PROPRIOCEPTORS

The intercostal muscles contain muscle spindles, tendon organs and Pacinian corpuscles (see Chapter VII). The afferent discharge from these receptors can be recorded either in single fibres of the dorsal roots or from the fine nerve filaments which supply the muscle, and the receptors can be identified physiologically by their response to certain stimuli such as stretch or ventral root stimulation.

### Muscle Spindles

**Spontaneous breathing.**—As indicated in Chapter I, muscle spindles discharge when the main muscle is stretched and, conversely, cease discharging when the main muscle shortens, unless this is offset by fusimotor activity. The discharge pattern of intercostal muscle spindles implies that there is fusimotor driving of the spindle during the phase of respiration in which the main muscle is shortening (Critchlow and von Euler, 1962, 1963). The pattern of activity of a muscle spindle in an inspiratory intercostal is shown in Fig. 72. In trace C, the fusimotor (gamma) fibres have been blocked with lidocaine so that the spindle is responding as a passive stretch receptor, and its discharge is seen to be maximum during expiration when it is being stretched. Its activity is different under "physiological" conditions, however, when its fusimotor fibres are intact (Fig. 72*a*) for although the spindle still fires during expiration it reaches its maximum discharge frequency during inspiration when the muscle in which it lies is shortening.

Thus there must be a rhythmic fusimotor drive to the inspiratory intercostal spindles during inspiration which is sufficient not only to offset the unloading of the spindle caused by main muscle shortening, but actually to increase the spindle discharge during this period. A similar discharge pattern occurs in an expiratory muscle spindle, with maximum discharge frequency in expiration. Some spindles are entirely dependent on fusimotor innervation for their ability to respond to changes in muscle length during spontaneous breathing because their discharge is abolished when their fusimotor fibres are blocked with lidocaine.

Direct evidence for a rhythmic fusimotor drive to the intercostal

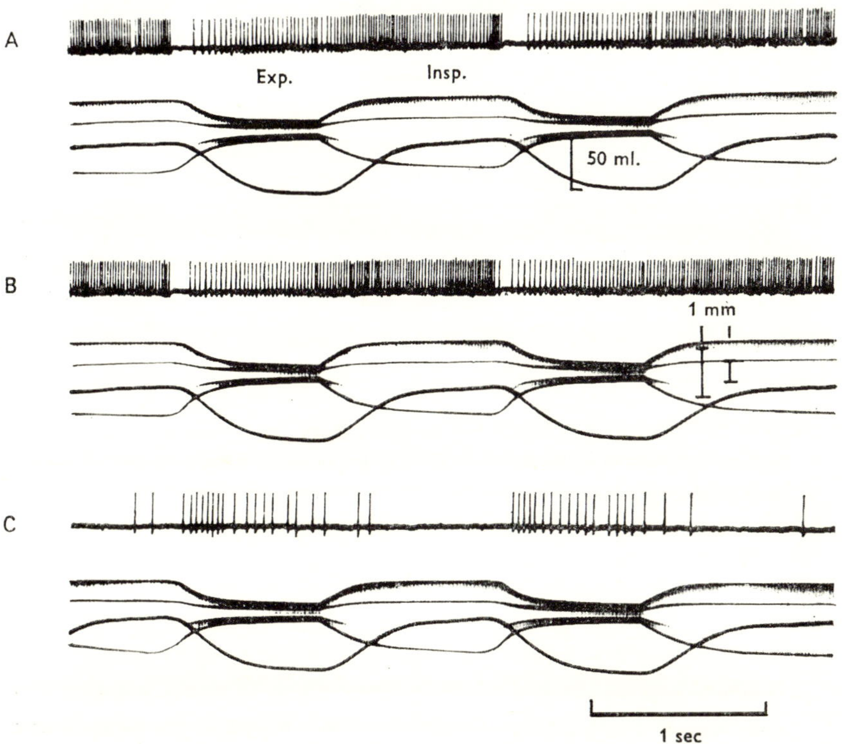


FIG. 72.—AFFERENT DISCHARGE FROM A MUSCLE SPINDLE OF AN INSPIRATORY INTERCOSTAL MUSCLE

A, control. B, 2·5 minutes after application of lidocaine solution 0·25% to the intercostal nerve. C, 3·5 minutes after the lidocaine application. Alpha motoneurone conduction appeared unimpaired at 4·2 minutes after cocainisation as judged from the "myographic" record of the response to ventral root stimulation. Note the "reversal" of the discharge pattern. Intercostal width (middle traces) and tidal volume (lowest trace) recorded together with the neurogram. (From Critchlow and von Euler, 1963.)

muscle spindles was obtained by recording the efferent activity in fine nerve filaments supplying the muscle (Sears, 1962, 1963, 1964*a*; Eklund *et al.*, 1963, 1964). The pattern of discharge recorded from a filament to an expiratory intercostal muscle is illustrated in Fig. 73*a-c*, and from an inspiratory filament in Fig. 73*d*. The spikes of large amplitude arise from alpha motoneurones (which innervate the main muscle) while those of small amplitude arise from fusimotor fibres innervating the muscle spindles. The respiratory phase is indicated by the diaphragm EMG in the lower trace of each pair. In *a-c*, phasic alpha motoneurone activity can be

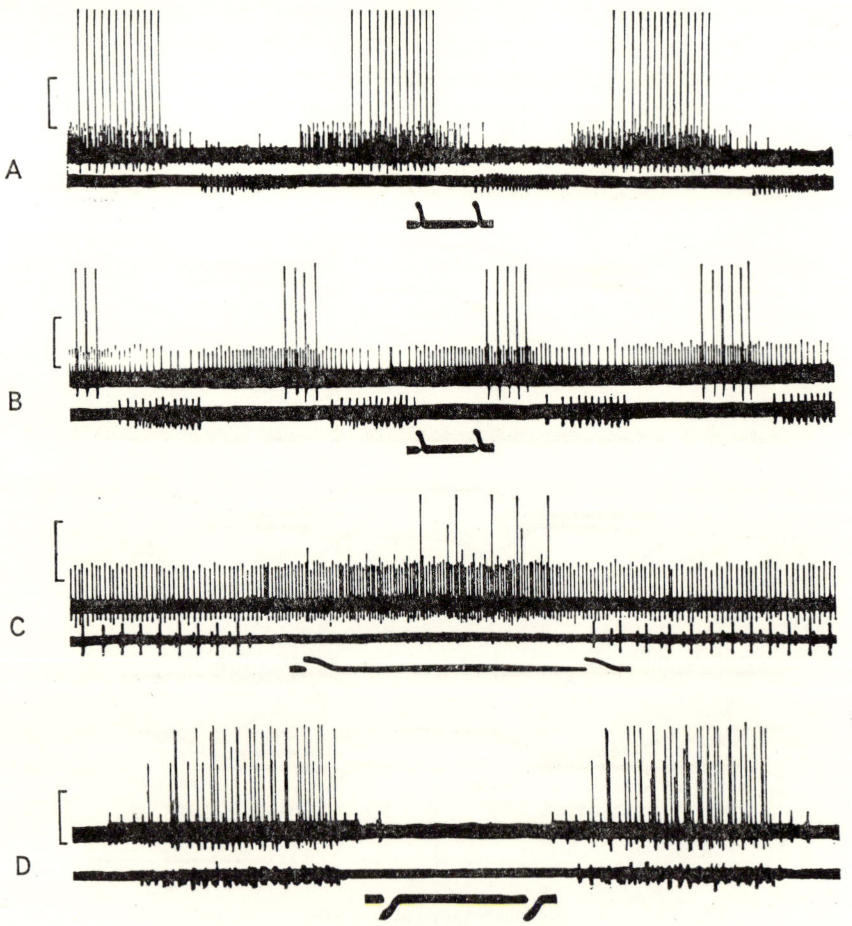


FIG. 73.—ALPHA AND FUSIMOTOR DISCHARGES RECORDED FROM INTERCOSTAL NERVE FILAMENTS

Upper traces show monophasic recording of efferent discharges in intercostal nerve filaments during normal respiration (pentobarbital anaesthesia): "expiratory nerve filament" (A, B, and C); "inspiratory nerve filament" (D). Lower trace EMG diaphragm. Time scale 1 sec. Vertical calibration 100 $\mu$V (alpha spikes retouched). (From Sears, 1963.)

seen during expiration. In *a*, fusimotor discharge is also phasic during expiration, but it precedes the onset of alpha motoneurone activity and persists after it. In *b* and *c*, fusimotor discharge is present throughout the respiratory cycle, but accelerates at the onset of expiration before alpha motoneurone firing. In *d*, rhythmic alpha and fusimotor discharge is present in an inspiratory nerve filament.

These records show examples of "rhythmic" fusimotor activity. Some fusimotor neurones, in contrast to this, are "tonic", showing little or no variation in their discharge frequency (Eklund *et al.*, 1964; Corda *et al.*, 1966). These authors have argued that the two types of fusimotor activity are functionally distinct. Rhythmic fusimotor activity appeared to be closely linked with respiratory alpha discharge, although the relationship could be changed by cerebellar stimulation. Tonic fusimotor activity, on the other hand, was more readily influenced by cerebellar stimulation and was not abolished by cervical cord transection in contrast to rhythmic fusimotor activity (Corda *et al.*, 1966). In particular, tonic fusimotor neurones seemed more responsive to postural and other reflex effects (see below). On the other hand, Sears (1964*a*) considers that it would be unwise to suggest any fundamental difference between rhythmic and tonic fusimotor activity.

Static and dynamic fusimotor fibres (see Chapter I) are represented in both the tonic and rhythmic fusimotor fibres (Corda *et al.*, 1966; von Euler and Peretti, 1966). Only static fibres are present in the rhythmic fusimotor fibres to secondary endings. It is of interest that von Euler and Peretti (1966) noted in most of the cats studied that high frequency vibration caused a significant increase in the rate of inspiration without a compensating decrease in tidal volume.

**Reflex effects.**—An *increase in inhaled* $CO_2$ or a *reduction in* $O_2$ causes an increased depth and rate of breathing with increased activity both in alpha and fusimotor neurones to inspiratory intercostals. Inhibition in the antagonist expiratory intercostals during inspiration is indicated by a marked reduction in its fusimotor neurone activity (Sears, 1964*a*). Artificial *hyperventilation* on the other hand produces apnoea during which all rhythmic alpha and fusimotor activity may be abolished in both inspiratory and expiratory nerve filaments, the rhythmic changes in fusimotor activity being the last to disappear (Sears, 1964*a*). The afferent discharge from some spindles decreases or ceases altogether, whereas in others it may be little affected (Critchlow and von Euler, 1963).

*Vagotomy*, apart from slowing respiratory frequency, does not affect alpha and fusimotor activity, as shown both by direct recording and the spindle afferent discharge (Critchlow and von Euler, 1963; Eklund *et al.*, 1964). Electrical stimulation of the vagus at an adequate strength to inhibit inspiration and initiate expiration does not change the phase of spindle afferent discharge (Critchlow and von Euler, 1963). Alpha motoneurone

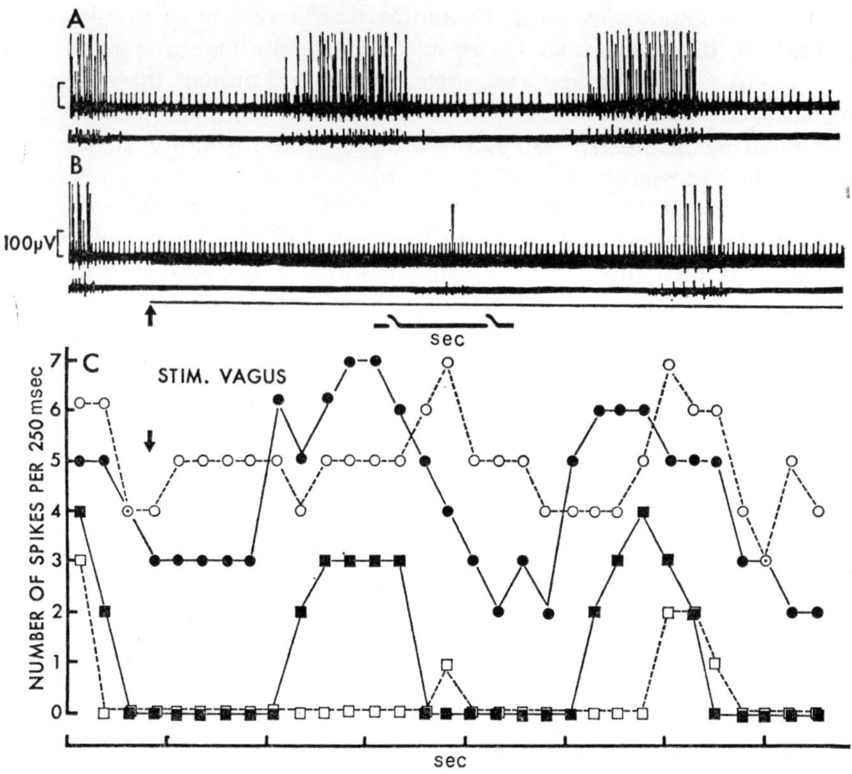


FIG. 74.—EFFECTS OF ELECTRICAL STIMULATION OF THE VAGUS NERVE ON THE ACTIVITY OF INSPIRATORY ALPHA AND FUSIMOTOR NEURONES

Upper traces of A and B, recording from inspiratory nerve filament (T8); lower traces, EMG of diaphragm. A, control; B, stimulation of left vagus nerve (at arrow) at 250 c/s. Note "vagal escape" of alpha and fusimotor neurones. In C, the number of spikes/250 msec are plotted as ordinates against the time at which the measurements were made as abscissae; the ordinates were plotted in the middle of the corresponding periods. Filled symbols measured from A, open circles from B. The fusimotor spike is shown by circles, and the small alpha spike, which showed vagal escape, by squares. (From Sears, 1964*a*.)

discharge to inspiratory intercostals is abolished and the fusimotor discharge loses its rhythmic characteristic (Sears, 1964*a*). If stimulation is maintained, however, alpha motoneurone "escape" occurs, the alpha spikes being preceded by an acceleration in the fusimotor discharge as in Fig. 74.

*Tracheal obstruction with vagus intact* at the height of inspiration elicits, in animals, the Hering-Breuer inspiratory-inhibiting, expiratory-activating reflex, and is associated with an increase in both the alpha and fusimotor discharge to expiratory intercostals. Obstruction at the onset of inspiration causes an initial increase followed by a marked decrease in inspiratory muscle spindle discharge (Critchlow and von Euler, 1963) and there is an increase in alpha and fusimotor discharge to the inspiratory intercostals

(Sears, 1964*a*). The segmental afferent inflow makes a significant contribution to these responses as only a small increase in alpha discharge occurs with tracheal obstruction at the height of inspiration after dorsal root section. The respiratory pattern produced by alteration in the prevailing chemical drive or by the Hering-Breuer reflex is therefore effected through changes in both intercostal fusimotor neurone and alpha motoneurone excitability.

*Cutaneous stimulation* exerts reflex effects upon intercostal alpha and fusimotor activity (Sumi, 1963; Eklund *et al.*, 1964). Inspiratory fusimotor neurones in particular are excited by skin stimulation.

*Effects of added loads.* A stretch reflex, which depends upon spindle primary afferent excitation of the alpha motoneurone, can be elicited in the

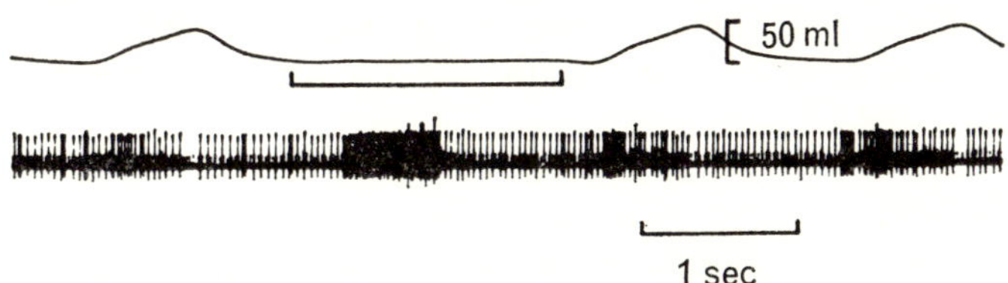


FIG. 75.—AFFERENT DISCHARGE FROM AN INSPIRATORY (EXTERNAL) INTERCOSTAL MUSCLE SPINDLE (THE LARGE SPIKE) TO SHOW THE RESPONSE TO TRACHEAL OCCLUSION

Mark indicates period of tracheal occlusion. Upper tracing represents tidal volume. (From Corda, Eklund and von Euler, 1965.)

intercostal muscles of the cat and rabbit, and is abolished by dorsal root section (Ramos and Mendoza, 1959; Glebovskii, 1965). In Chapter I, it was indicated that the "gamma loop" (fusimotor neurone—spindle primary afferent fibre—alpha motoneurone) might operate during a postural contraction or in the performance of a movement so as to compensate for changes in load. An increase in load, for example, would prevent the "demanded" shortening of the muscle and lead to misalignment between extrafusal and intrafusal length, with consequent reflex excitation round the gamma loop. Experimental evidence from the intercostal muscles of vagotomised animals (to eliminate any vagal responses) provides some support for this. Tracheal obstruction at the onset of inspiration causes an increased discharge from inspiratory intercostal muscle spindles (Fig. 75), and reflex increase in alpha motoneurone discharge to the muscle which is abolished by dorsal root section (Corda, Eklund and von Euler, 1965). Electromyographic recording of single motor units in the parasternal intercostals of the rabbit has confirmed an increased frequency and prolongation of discharge with large resistive loads or tracheal obstruction, which are abolished by appropriate dorsal root section. With smaller resistive loads (e.g. 9·4 cm $H_2O$/l/sec, which is far less than the airways flow resistance of normal rabbits) no augmentation of activity was found (Sant'-Ambrogio and Widdicombe, 1965).

*Postural effects.* Passive movements of the head cause reflex changes in the electrical activity of intercostal muscles in the cat (Massion *et al.*, 1960) and such reflexes have also been demonstrated in human subjects with chronic airways obstruction (Moltke and Skouby, 1963). Direct recording in the cat of intercostal alpha and fusimotor activity and of spindle discharge has confirmed these reflex effects (Corda *et al.*, 1966), the characteristics of the change in spindle discharge suggesting that these effects were achieved mainly by changes in tonic fusimotor activity.

Passive movements of the limbs will evoke fusimotor activity in an animal in which such activity had disappeared (Sears, 1964*a*) and which for the hind limbs appears to be of the tonic type (Corda *et al.*, 1966). Furthermore, von Euler and his co-workers have found evidence for proprioceptive activation of tonic fusimotor neurones by movements of the chest wall (see Corda *et al.*, 1966).

### Tendon Organs

The presence of tendon organs in the intercostal muscles has been identified by the characteristic increase in their discharge frequency during muscle contraction (Critchlow and von Euler, 1963). The discharge pattern of these receptors has not yet been studied in detail.

## DIAPHRAGMATIC PROPRIOCEPTORS

Histological examination of the diaphragm shows that, unlike the intercostal muscles, it contains relatively few proprioceptors (see Chapter VI). The ratio of spindles to tendon organs is low when compared with other skeletal muscles. Other receptors present are free-ending afferents, Pacinian corpuscles and encapsulated peritoneal end organs.

### Muscle Spindles

**Spontaneous breathing.**—Afferent discharges in the phrenic nerve from receptors in the diaphragm sensitive to stretch were first recorded by Cardin (1939, 1944) and later by Cuénod (1961) and Yasargil (1962). The discharge pattern of diaphragmatic proprioceptors has been analysed further by Corda, von Euler and Lennerstrand (1965). The types of muscle spindle discharge which may be recorded in the spontaneously breathing cat are illustrated in Figs. 76 and 77. Most spindles fire maximally during expiration (Fig. 76*a*), behaving in a passive manner in that they discharge only when they are being stretched. The application of lignocaine, which blocks the fusimotor fibres, causes a decrease in spindle discharge frequency but no change in the diaphragmatic EMG or the tidal volume (Fig. 76*b* and *c*). A few spindles on the other hand fire maximally in inspiration (Fig. 77*a*) which implies rhythmic fusimotor driving to the spindle to compensate for the effects of main muscle shortening (Corda, von Euler and Lennerstrand, 1965). Fusimotor block caused these spindles to become "passive" in type (Fig. 77*b* and *c*). A third group are intermediate with respect to the fusimotor drive upon them, their discharge frequency being high in the initial phase of inspiration, with a decrease in firing rate at the time of maximal muscle shortening.

**Reflex effects.**—*Effects of added loads.* Tracheal occlusion in vagotomised animals causes an increase in the discharge frequency of those diaphragmatic spindles which are driven by rhythmic fusimotor activity, being particularly marked in the spindles whose fusimotor driving was insufficient to compensate for the maximum rate of muscle shortening. In the majority of spindles, where no fusimotor driving is apparent, discharge frequency is unaffected by tracheal occlusion. This manoeuvre, however, does not lead to any reflex increase of alpha motoneurone discharge in the cat (Corda, Eklund and von Euler, 1965). This may be accounted for by the weak fusimotor driving to the spindles so that the prevention of muscle shortening causes only a relatively small spindle afferent response, and by the inhibitory effects of tendon organ afferents. Other workers have also failed to show any load-compensating mechanism in the vagotomised cat

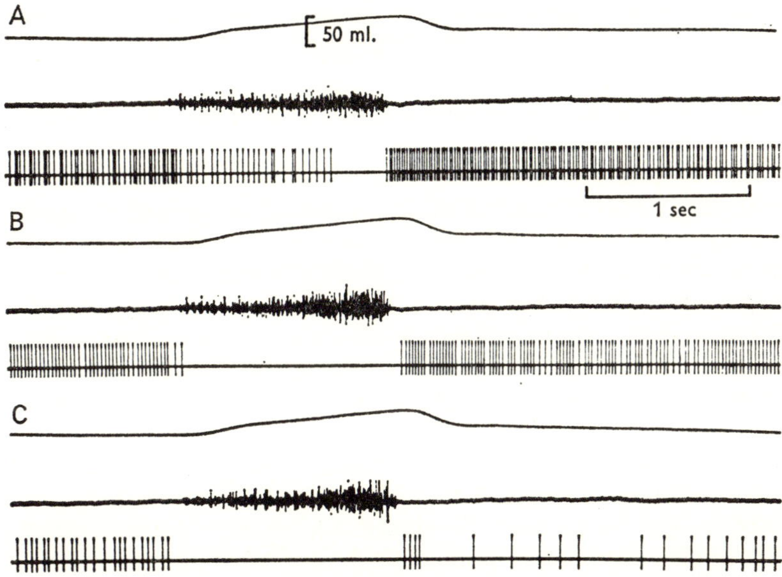


FIG. 76.—DISCHARGE FROM A DIAPHRAGMATIC MUSCLE SPINDLE OF THE "PASSIVE" TYPE

Spindle located in the pars costalis of the diaphragm, and recorded from C5 dorsal root together with tidal volume (top tracing, upward deflection denotes inspiration) and the electromyogram of the diaphragm (middle tracing). Spontaneous breathing of air. A, control; B, 3 minutes; C, 4·5 minutes after application of 0·4% lignocaine solution to the phrenic nerve. Lignocaine blockade of the fusimotor fibres led to a general decrease in firing rate. Note, no effect on EMG or tidal volume. (From Corda, von Euler and Lennerstrand, 1965.)

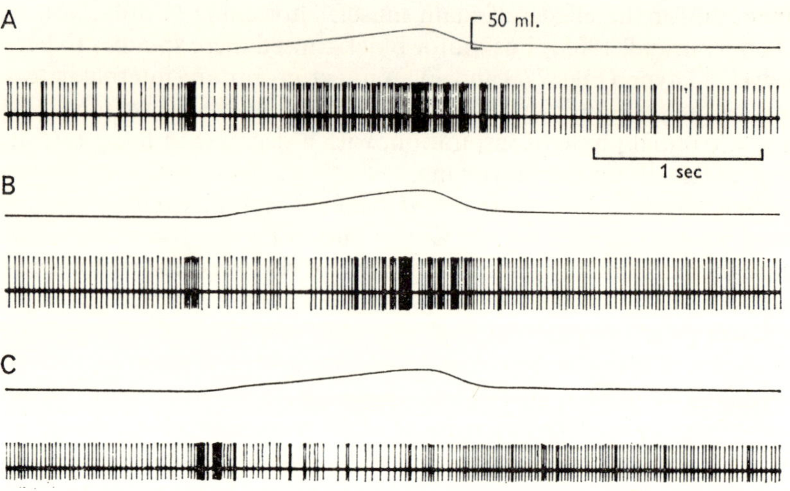


FIG. 77.—DISCHARGE FROM A DIAPHRAGMATIC MUSCLE SPINDLE OF THE "ACTIVE" TYPE

Highest discharge rate in inspiration; recorded from C5 dorsal root together with tidal volume (upper tracing, upward deflection denotes inspiration). This spindle was located in pars cruralis of the diaphragm. Cat spontaneously breathing air. A, control: B, 1 minute; C, 3 minutes 40 seconds after application of 0·4% lignocaine solution to the phrenic nerve. Note the "reversal" of the discharge pattern due to a marked decrease in the discharge rate during the inspiratory phase. Note also the regularisation of the discharge during the expiratory phase. (From Corda, von Euler and Lennerstrand, 1965.)

with electromyographic recording (Fink *et al.*, 1958; Sant'Ambrogio *et al.*, 1962).

In the vagotomised rabbit, the results are conflicting. The integrated diaphragmatic EMG is reduced by lung inflation or manual compression of the abdomen (Ramos, 1959). Small inspiratory loads have been reported to cause an increase of diaphragmatic electrical activity, although less than with vagus intact, but larger loads either produce no change or lead to a decrease in activity (Cuénod, 1961). With an added elastic load to breathing, an increase in tension has been shown in the mechanically isolated diaphragmatic slip (Campbell *et al.*, 1964). However, Sant'Ambrogio and Widdicombe (1965) found no change in the diaphragmatic EMG with an added airflow resistance, while tracheal occlusion in the unanaesthetised animal sometimes inhibited the EMG. In the anaesthetised animal spinalised at T1, the pressure developed during static inspiratory efforts with the trachea clamped is similar before and after phrenic dorsal root section, as is the Hering-Breuer reflex as measured by the ratio of duration of occluded breath to normal breath (Sant'Ambrogio, personal communication).

*Laryngeal stimulation* evokes a considerable increase in diaphragmatic spindle discharge both in the inspiratory and expiratory phase of the cough response it initiates (Corda, von Euler and Lennerstrand, 1965).

### Tendon Organs

Most diaphragmatic tendon organs increase their firing rates during inspiration (Corda, von Euler and Lennerstrand, 1965), with decrease in frequency or silence during the expiratory phase (Fig. 78). Some of them have a rather low mechanical threshold. Tracheal occlusion at the onset of

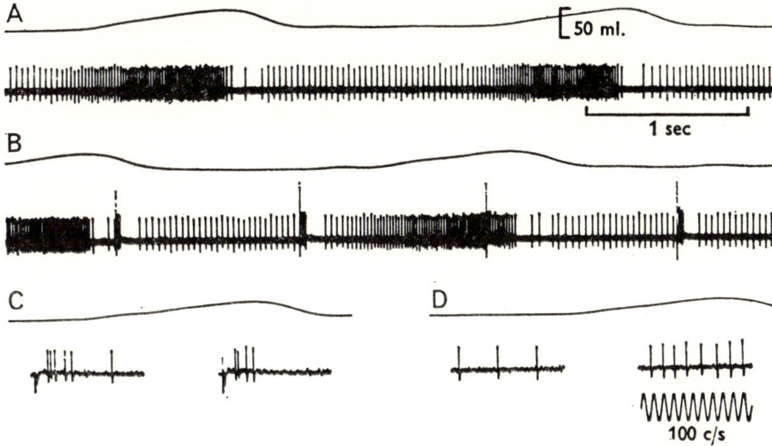


FIG. 78.—DISCHARGE FROM A TENDON ORGAN LOCATED IN THE CRURAL PART OF THE DIAPHRAGM

Recorded from C5 dorsal root, together with tidal volume (upper tracings). A, Spontaneous breathing; B, single shocks to the ventral root of C5 2 db above threshold for diaphragmatic contraction; C, same on sweeps; D, serves as control for C. (From Corda, von Euler and Lennerstrand, 1965.)

inspiration does not result in an increase in firing rate, while occlusion at the end of inspiration, which prevents the stretching of the muscle which would otherwise occur during the subsequent expiration, silences the tendon organ in the expiratory phase.

Tendon organ afferents from the diaphragm are probably responsible for the inhibitory reflex effects described by Dolivo (1953) and Glebovski and Pavlova (1962) and appear to dominate its proprioceptive regulation, at least in the cat (Corda, von Euler and Lennerstrand, 1965).

## ABDOMINAL MUSCLE PROPRIOCEPTORS

Direct recording from abdominal muscle proprioceptors has not been made. Their potential importance in controlling abdominal muscle excitability is indicated by the effects of dorsal root section (see p. 206).

## RESPIRATORY MOTONEURONES

We have seen in the previous sections that there is an important segmental input from respiratory muscle proprioceptors which may be inhibitory or excitatory. This input is integrated with the central drives at the alpha motoneurone, which is the final common pathway to the respiratory muscles. The integrative function of the alpha motoneurone can be examined by studying the changes in its excitability, either directly by intracellular recording, or indirectly by measuring the size of the response of the motoneurone pool to electrical monosynaptic excitation.

In intracellular recording, microelectrodes are inserted into the spinal cord and directed towards the anterior horn until a motoneurone is impaled, as indicated by the occurrence of a negative potential of −40 to −70 mV. When the membrane potential of the motoneurone becomes less negative it is said to be depolarised, and when it becomes more negative it is said to be hyperpolarised. If the depolarisation is sufficient to bring the cell to the critical firing threshold, a further automatic transient change in potential occurs known as the spike potential (see Katz, 1966). In lumbar motoneurones the membrane potential may rise to about +40 mV during the spike potential, but in the thoracic motoneurones the peak value is usually only about −5 mV (Sears, 1964*b*). Repetitive firing will occur for as long as the membrane potential is held above the critical firing threshold (Fig. 79). Changes in membrane potential are brought about by activation of those neurones which make synaptic connections with the cell, the resulting change in potential being called a "postsynaptic potential" (PSP). Some neurones give rise to a PSP which is depolarising, and because this thereby increases the tendency of the cell to discharge, it is termed an "excitatory postsynaptic potential" (EPSP). A muscle spindle primary afferent, for example, gives rise to an EPSP in the alpha moto-

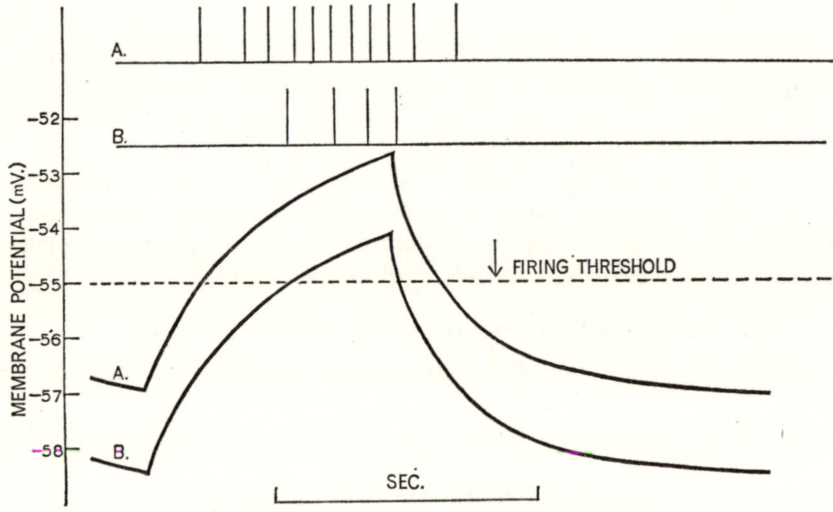


FIG. 79.—DIAGRAMMATIC REPRESENTATION OF THE EFFECTS ON MOTONEURONE DISCHARGE OF CHANGE IN MEMBRANE POTENTIAL

Note repetitive spike potentials for as long as the membrane potential remains above the critical firing threshold.

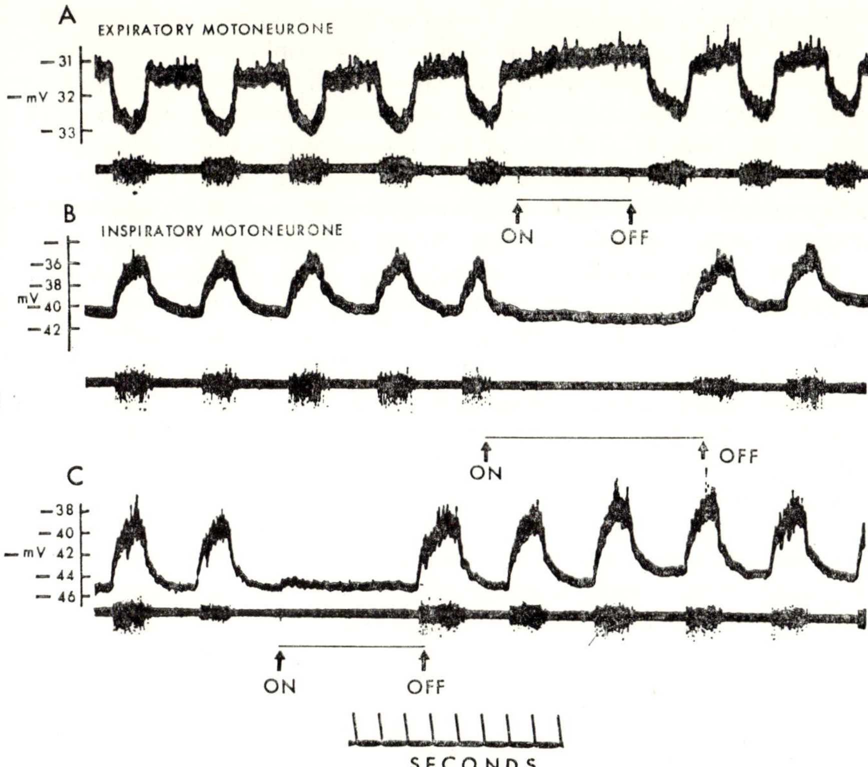


FIG. 80.—CENTRAL RESPIRATORY DRIVE POTENTIALS (CRDPs) OF EXPIRATORY AND INSPIRATORY MOTONEURONES

The effects of stimulating the central end of the left vagus at 300 c/s on the CRDPs of the inspiratory and expiratory motoneurones (T8) are also shown. Upper traces, intracellular d.c. recordings from an expiratory motoneurone in A, and an inspiratory motoneurone in B and C; lower traces, electromyogram of diaphragm. A and B were recorded within 15 minutes of each other and their records have been aligned with respect to the diaphragm electromyogram. C was recorded 10 minutes after B when the membrane potential had increased. (From Sears, 1964*c*.)

neurone. Other neurones act in the opposite manner producing an "inhibitory postsynaptic potential" (IPSP), such as those interneurones activated by group Ib (tendon organ) afferents which give rise to an IPSP in the alpha motoneurone (see Fig. 7).

## Intercostal Motoneurones

Intercostal motoneurones have relatively short after-hyperpolarisations (Sears, 1964*b*), a characteristic of phasic motoneurones innervating limb muscles (see Chapter I). However, intercostal muscle is known to contain both fast and slow motor units, so that the relationship of the after-hyperpolarisation to the motoneurone type in motoneurones supplying limb muscles does not appear to apply to intercostal motoneurones. In addition, no evidence of recurrent (Renshaw) inhibition was found (Sears,

1964*b*). These characteristics are similar to those of phrenic motoneurones.

Intracellular recording from an intercostal alpha motoneurone in a spontaneously breathing animal reveals that the membrane potential is subject to regular fluctuations with a respiratory periodicity (Eccles *et al.*, 1962; Sears, 1964*c*), as shown in Fig. 80 where the respiratory phase is indicated by the diaphragmatic EMG in the lower trace. In inspiratory motoneurones, identified by antidromic stimulation of the appropriate intercostal nerve, the phase of depolarisation is during inspiration while expiratory motoneurones are subject to a phase of depolarisation in expiration. These slow potential changes in the intercostal motoneurones are of central origin, because they are abolished by section of the spinal cord above the recording site. They have been called the central respiratory drive potentials (CRDPs). In addition, superimposed on the CRDPs are slower changes of membrane potential with a periodicity of 2–15 minutes; their origin is unknown.

**Vagal effects.**—Repetitive electrical stimulation of the vagus at low intensity inhibits inspiration and prolongs the expiratory pause, causing associated changes in the CRDPs (Sears, 1964*c*; Fig. 80). The expiratory intercostal motoneurone remains in a progressively depolarised state during the period of stimulation, in contrast to the inspiratory moto-

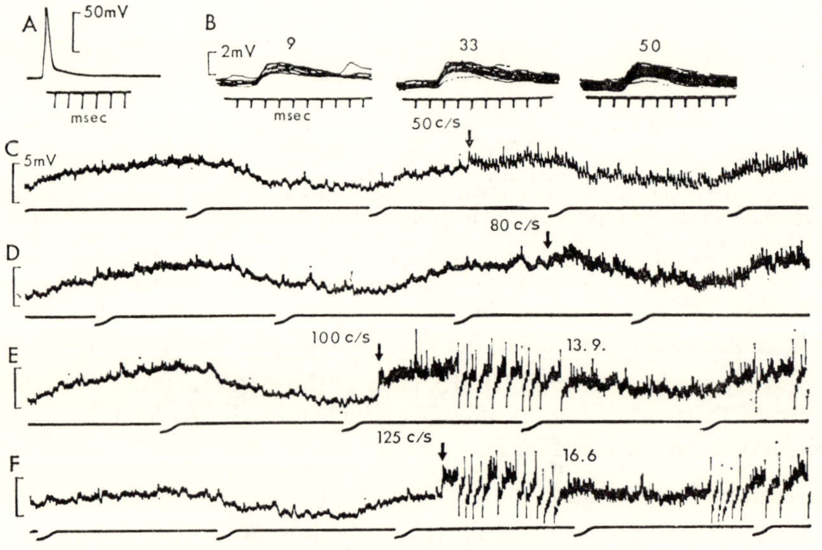


FIG. 81.—INTEGRATION AT THE MOTONEURONE OF THE CENTRAL RESPIRATORY DRIVE WITH A SEGMENTAL INPUT

Summation of repetitively evoked monosynaptic EPSPs with CRDP in an expiratory motoneurone. A, antidromic somadendritic spike; B, superimposed EPSPs evoked at 9, 33 and 50 c/sec to show absence of depression (a.c. recording). C, D, E and F, d.c. intracellular recordings to show at the left side of each trace a CRDP alone. The right side of each trace shows the CRDP in summation with the depolarisation evoked by repetitive stimulation of the monosynaptic pathway at the frequencies indicated in c/sec. Time scale 1 sec. (From Sears, 1964*d*.)

neurone whose depolarising phase was abruptly terminated and transformed into repolarisation. If stimulation is maintained, vagal "escape" of intercostal activity occurs with the reappearance of the CRDPs.

**Segmental effects.**—By low intensity stimulation of intercostal nerves, it is possible to excite selectively group Ia afferent fibres (primary muscle spindle afferents), such stimulation giving rise to an EPSP in the alpha motoneurone. Increasing the stimulus repetition rate causes an increase in the EPSP by temporal summation (Fig. 81*b*) which, when it is of sufficient amplitude, brings the membrane potential to the critical firing threshold.

Monosynaptic excitation of all expiratory intercostal motoneurones and most inspiratory motoneurones can be demonstrated from low threshold (group Ia) stimulation of the nerves supplying the respective muscles. When the stimulus intensity to the intercostal nerves is increased, an inhibitory effect (IPSP) is added to the EPSP, which probably arises from stimulation of tendon organ afferent fibres.

The integration of the central respiratory drive potentials with segmental monosynaptic excitation is clearly shown in Fig. 81*c-f*, recorded from an expiratory motoneurone (Sears, 1964*d*). The CRDPs can be seen as before (note the change in time scale). An EPSP is evoked by group Ia afferent stimulation at the points indicated by an arrow, and a control cycle is shown in the left half of each trace. In *c* and *d*, the EPSPs produced by stimulus repetition rates of 50 and 80 per second were insufficient to cause firing of the motoneurone, although a slight increase in membrane potential occurs. In trace *e*, the increase at the onset of stimulation at 100 c/s is more apparent, but only causes cell firing when it summates with the CRDP towards the peak of its depolarisation. In trace *f*, with a still higher stimulus rate producing a larger EPSP, alpha motoneurone firing almost immediately follows the onset of stimulation.

There is evidence that the CRDPs arise from both excitatory and inhibitory influences of central origin. This has important implications with regard to the organisation of the brain stem respiratory centres for it implies that these centres exert an alternating excitatory and inhibitory influence on spinal respiratory neurones (Sears, 1964*c*).

The influence of the segmental input must therefore be seen against the background of the central respiratory drive potentials. An EPSP occurring at the peak of the depolarising phase of the CRDP is more likely to cause alpha motoneurone firing than an EPSP during the hyperpolarised phase. This is an important means by which the central nervous system exerts control over the segmental proprioceptive mechanism (Sears, 1964*d*).

### Phrenic Motoneurones

Phrenic motoneurones cannot satisfactorily be classified as either tonic or phasic (see Chapter I). The conduction velocity of their axons is com-

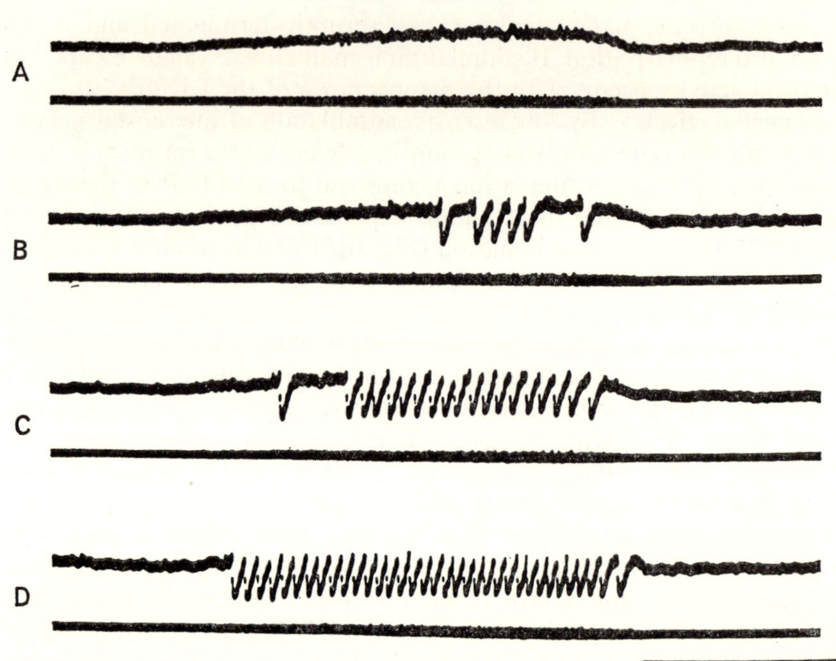


FIG. 82.—INTRACELLULAR RECORDINGS FROM A PHRENIC MOTONEURONE
Intracellular potentials from phrenic motoneurone (upper beams) with discharges from phrenic nerve trunk (lower beams) during the inspiratory phase. Depolarising current through the membrane was increased in intensity from B to D. Voltage calibration, 10 mV for intracellular potential; time 1 second. (From Gill and Kuno, 1963*b*.)

parable to that of tonic motoneurones but the duration of the after-hyperpolarisation is more consistent with phasic motoneurones (Gill and Kuno, 1963*a*). Moreover, phrenic motoneurones are probably not subject to recurrent (Renshaw) inhibition (Gill and Kuno, 1963*a*; but see also Baumgarten *et al.*, 1963), whereas such inhibition is a feature of tonic motoneurones in the limbs. The direct effect of $CO_2$ on phrenic motoneurones is inhibitory (Gill and Kuno, 1963*a*).

Slow changes in membrane potential of central origin occur in phrenic motoneurones of the cat as in intercostal motoneurones, shown in Fig. 82 (Gill and Kuno, 1963*b*). In trace *a*, the motoneurone has not reached firing threshold but in traces *b* to *d*, when a progressive increase in depolarisation was achieved by passing a current through the membrane, an increase in cell firing occurs.

Stimulation of descending tracts in the lateral column of the severed cervical cord produces a depolarising (excitatory) phase followed by a hyperpolarising (inhibitory) phase. Medullary stimulation, however, in an area where inspiratory neurones are commonly found results in a pure EPSP in the phrenic motoneurone, while stimulation of an "expiratory" site produced an IPSP. Gill and Kuno (1963*b*) conclude that phrenic motoneurone excitability is dependent upon both descending excitatory and inhibitory impulses from the medullary centres.

**Segmental effects.**—Stimulation of one root of the phrenic nerve inhibits the motor discharge in another root of the ipsilateral nerve (Rijlant, 1942) and produces hyperpolarisation (inhibition) of contralateral phrenic motoneurones (Gill and Kuno, 1963*b*). These effects may originate from excitation of tendon organ afferents. The failure to demonstrate an EPSP in phrenic motoneurones when stimulating phrenic afferent fibres is not unexpected in view of the characteristics of diaphragmatic proprioceptive innervation. Inhibitory influences appear to be dominant.

### Abdominal Motoneurones

Some of the thoracic expiratory motoneurones sampled by Sears (1964*c*) innervated abdominal muscles; thus abdominal motoneurones are subject to the same rhythmic central respiratory drive potentials as those of expiratory intercostal muscles. Bishop (1968) has studied the excitability of abdominal motoneurones in a different way by measuring the amplitude of the monosynaptic reflex. A fluctuation in excitability with a respiratory rhythm was demonstrated, with peak excitation during early expiration. Thus both direct intracellular recording and monosynaptic testing show that abdominal motoneurones, like intercostal motoneurones, are subject to slow changes in potential with a respiratory periodicity.

During continuous positive pressure breathing, there is an increase in the size of the motor response of the abdominal motoneurone pool to monosynaptic testing, but the overall time course of the change in excitability is unaltered (Bishop, 1968). There is thus an increase in the excitability of the abdominal motoneurone pool throughout the respiratory cycle during positive-pressure breathing, on which the rhythmic changes of the central respiratory drive are superimposed. The increase in excitability with pressure breathing depends both on the vagus and on the segmental afferent inflow. The slow changes in potential in the abdominal motoneurones, as in the intercostal motoneurones, provide a means by which the stretch reflex is inhibited during the "inappropriate" phase of the respiratory cycle, i.e. during inspiration for the abdominal muscles (Bishop, 1968).

## PROPRIOCEPTIVE CONTROL OF RESPIRATORY MUSCLES IN MAN

In man, the evidence for the operation of a proprioceptive control mechanism similar to that demonstrated in animal studies rests mainly upon the reflex effects of alterations in load, although the effects of dorsal root section (see p. 206) indicate that a segmental afferent inflow is normally necessary for the maintenance of alpha motoneurone excitability. The short latency reflex responses to a change in load provide the clearest information about the segmental proprioceptive mechanism, because responses of longer latency are likely to be influenced by the supraspinal effects of the stimulus. The supraspinal control of respiratory muscles, however, operates through the segmental mechanism, so that the short latency responses must not be regarded as the sole contribution of this mechanism to the observed effects of an alteration in load. The reflex responses which depend upon vagal afferents, for example, operate through fusimotor neurones as well as alpha motoneurones (see p. 213).

The effects of alterations in mechanical load to the respiratory muscles by sudden interruptions to airflow were first studied systematically by Fleisch (1928) and his co-workers, who described a series of compensating reflexes which, in animal experiments, were unaffected by vagotomy. However, when the experiments were repeated after dorsal root section from C3–C7 and cord transection at C7 the tracings were shown to be unchanged (Niekerk and ter Braak, 1935) while similar airflow changes occur in dead animals respired by changes in pressure of a plethysmograph (Riedstra and Dirken, 1953). Recently, Delhez and Petit (1966) have failed to demonstrate from airflow records any compensating reflex for the reduction in airflow produced by a sudden increase in airway resistance in healthy man.

The respiratory response over the first few breaths following an added "elastic" or "threshold" load has been examined both in conscious and anaesthetised subjects (Campbell, Dinnick and Howell, 1961; Campbell, Dickinson, Dinnick and Howell, 1961), but the precise contribution of the segmental proprioceptive mechanism to these responses cannot easily be analysed for the reasons outlined above. Their significance in relation to the stabilisation of ventilation is considered in Chapter XIII.

The segmental proprioceptive mechanism controlling the voluntarily activated intercostal muscles has been studied recently by examining the changes in integrated EMG activity in the period immediately following an alteration in load, particular attention being paid to the first 100 msec (Newsom Davis *et al.*, 1966; Sears and Newsom Davis, 1968; Newsom Davis and Sears, 1970). These responses can be regarded as reflex because the minimal voluntary reaction time of the intercostal muscles to an auditory stimulus is 140 msec (Draper *et al.*, 1960), and to a change in load somewhat longer (Newsom Davis, Sears and Taylor, unpublished).

The first important finding was that the muscle spindle afferent discharge appears to be essential for the maintenance of alpha motoneurone excitability. The evidence for this comes from the silent period which develops in the EMG when the load on a contracting muscle is suddenly removed, so that it is allowed to shorten. Shortening of the muscle abolishes spindle afferent discharge and the withdrawal of its excitatory effect

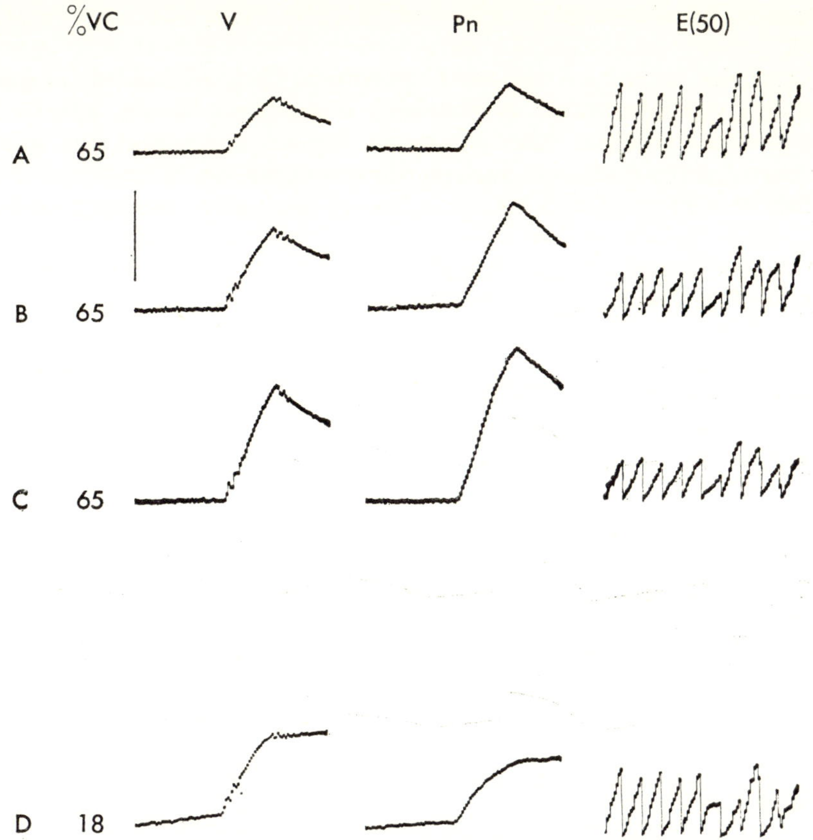


FIG. 83.—IMMEDIATE RESPONSE OF AN INSPIRATORY INTERCOSTAL MUSCLE TO SUDDEN SHORTENING CAUSED BY A DECREASE IN LOAD

Averaged responses (10 runs) from an external intercostal muscle (6 ICS). A to C, unloading under static conditions at 65% VC; unloading pressure stimulus +20, +40 and +60 cm $H_2O$ respectively; onset of unloading indicated by deflection of volume trace. D, resistive unloading; subject maintaining mouth pressure of −40 cm $H_2O$ against large airway resistance, and with low inspiratory air flow rate. Onset of unloading, by connection to a low resistance pathway for a 100 msec period, is indicated by deflection of volume trace. V, lung volume as %VC; Pn, pneumograph; E(50), integrated EMG reset at 50 msec intervals. Calibration, 5% VC. The duration of each trace is 500 msec. (From Newsom Davis and Sears, 1970.)

upon the alpha motoneurone leads to a cessation of alpha motoneurone firing and the "silent period". This is illustrated for an inspiratory muscle in Fig. 83*d*. The subject is inspiring against a very large resistive load, which is suddenly removed at the point indicated by the sharp rise on the volume and pneumograph trace. The integrated EMG shows a steady level of activity up to this point (as shown by the equal heights of the first five epochs). Following the removal of the load, at a latency of about 25 msec, a silent period can be seen as a plateau in the 6th epoch. The extent to which the EMG is reduced depends upon the degree of unloading, as illustrated in Fig. 83*a* to *c* (Newsom Davis and Sears, 1970).

From the proposed function of the gamma loop, an increase in load might be expected to cause an increase in alpha motoneurone activity, as the extrafusal-intrafusal misalignment in this situation would lead to an increased spindle discharge. Such an increase in activity indeed occurs on loading an intercostal muscle, after a latency of 50–60 msec, and is shown

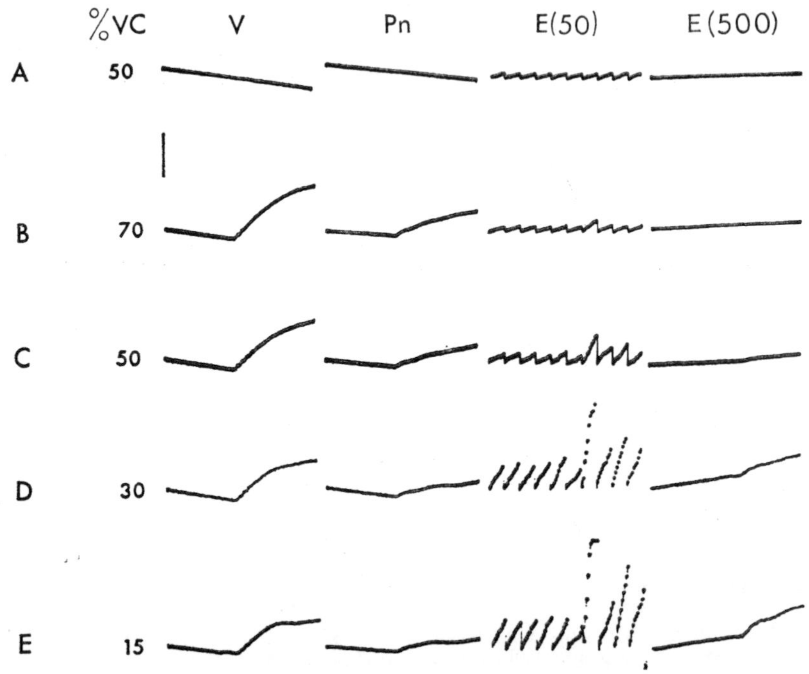


FIG. 84.—IMMEDIATE RESPONSE OF AN EXPIRATORY INTERCOSTAL MUSCLE TO SUDDEN STRETCHING CAUSED BY AN INCREASE IN LOAD DURING VOLUNTARY EXPIRATION

Average responses (10 runs) from internal intercostal muscle (6 ICS), with loading carried out at decreasing lung volume. Expiratory airflow 0·2 l/sec in A to D, and 0·14 l/sec in E. Load +70 cm $H_2O$; onset of load indicated by deflection of volume trace; duration 100 msec. Control experiment in A. Note progressive increase in excitatory response with decreasing lung volume, in contrast to the relatively constant inhibitory response. Symbols as Fig. 83. E(500), integrated EMG reset at 500 msec interval. Calibration 5% VC. The duration of each trace is 500 msec. (From Newsom Davis and Sears, 1970.)

for an expiratory muscle in Fig. 84 (Sears and Newsom Davis, 1968; Newsom Davis and Sears, 1970). This excitatory response (ER) can be seen in the 7th epoch and increases in magnitude as the lung volume at which the loading was carried out is decreased. The ER at 15 per cent VC in this experiment shows a 740 per cent increase in the EMG activity compared to the control (pre-load) level. The size of the ER also tends to increase as the airflow rate at which the load is applied is increased. The ER is particularly prominent during phonation. The dependence of the

ER upon fusimotor innervation was confirmed by the effects of intercostal nerve block with local anaesthetic (Newsom Davis and Sears, 1970).

The experiments also provided evidence for tendon organ ("autogenetic") inhibition of the intercostal muscles following an increase in load. In Fig. 84, an inhibition of EMG activity can be seen in the 6th epoch, its onset occurring about 20–25 msec after the application of the load. This inhibition was sometimes sufficient to silence the EMG (Newsom Davis and Sears, 1970).

## FUNCTIONAL SIGNIFICANCE OF THE PROPRIOCEPTIVE MECHANISM

From animal experiments we know that in the intercostal muscles the characteristics of spindle control are consistent with the operation of the gamma loop as a servo mechanism. Misalignment between extrafusal and intrafusal muscle fibre length leads to appropriate spindle discharge and compensatory changes in alpha motoneurone excitability. The functional effects of tendon organs on the other hand, have not yet been studied in detail. In the diaphragm, in contrast to the intercostal muscles, autogenetic facilitation is weak and autogenetic inhibition is dominant. The reflex effects of intercostal afferents may compensate for this lack of autogenetic facilitation, but the functional significance of these reflexes has not yet been defined.

It is natural to consider the significance of the segmental proprioceptive mechanism in relation to the control of ventilation, and in particular to consider the extent to which the gamma loop might act as a load compensating mechanism, as was discussed in two theoretical papers by Campbell and Howell (1962, 1963). Although the EMG changes characterising the short latency proprioceptive responses of human intercostal muscle are large, the ventilatory effects of the responses are in fact small because of their brief duration. There follow reflex responses of longer latency, occurring later in the first breath or in subsequent breaths (see Chapter XIII), which may be associated with relatively large changes in ventilation. These latter responses are dependent upon the central effects of proprioceptive, vagal and other afferents which lead to an altered demand to the motor mechanism. It must be emphasised, however, that these responses are still effected through the segmental mechanism, so that the properties of the proprioceptive control mechanism remain available throughout the response.

If the proprioceptive control mechanism is not primarily concerned with achieving compensatory changes in ventilation in the face of sudden large changes in load, what is its function? Von Euler (1966) has emphasised the possible importance of the fusimotor—muscle spindle control system in integrating the postural and respiratory movements which are required of the same muscle, particularly in the case of the intercostal muscles. The studies of intercostal activity in man have indicated that alpha motoneurone excitability depends upon spindle afferent inflow, so that during a movement or a held posture the gamma loop would appear to be operating. In the event of an increase in load, a large compensatory change in EMG activity occurs after a latency of 50–60 msec, but the *initial* response at 20–25 msec is one of inhibition, presumed to arise from tendon organs. Thus the operation of the gamma loop seems to be transiently interrupted by autogenetic inhibition when there is an unexpected increase in load. These observations suggest that, in the face of a predictable load, high loop gain is available for the control of the movement or posture, while with unexpected loads the immediate and possibly inappropriate correction by the gamma loop is suppressed (Newsom Davis and Sears, 1970). The gamma loop may therefore be primarily concerned with maintaining the accuracy of a held posture or a movement against predicted loads. It is significant that the proprioceptive responses to changes in load are most prominent during phonation in which an accurate control over airflow is of particular importance.

## Summary

The effects of dorsal root section demonstrate the importance of segmental proprioceptive input to respiratory mononeurone excitability. Dorsal root section of the intercostal muscles both in animals and man leads to an impairment of muscle activity; in the phrenic nerves it is without effect on diaphragmatic activity in spontaneously breathing animals, but in man leads to a temporary impairment of activity. Activity in the abdominal muscles evoked by pressure breathing is abolished by dorsal root section. The gain of the dorsal root system, determined by the inspiratory pressure-volume relations, is similar to that of the vagus, the former being facilitatory and the latter inhibitory. There is evidence for a reflex relationship between the spinal mechanisms controlling intercostal and diaphragmatic muscle activity, which is under supraspinal control.

Intercostal muscle spindles are subject both to rhythmic fusimotor driving in phase with the respiratory cycle and to tonic fusimotor driving. The rhythmic fusimotor activity is closely linked with intercostal alpha motoneurone discharge during spontaneous breathing, but this may be altered by cerebellar stimulation. Changes in the chemical respiratory drive and vagal reflex effects do not alter the alpha-gamma linkage. Tonic activity appears to respond more readily to cerebellar stimulation and to passive changes in posture, and may be activated by movements of the chest wall. Added loads to an intercostal muscle may evoke an increase in its spindle discharge and reflex excitation of the alpha motoneurone, which are not dependent on an intact vagus.

Diaphragmatic muscle spindles are few in number and the majority are not subject to rhythmic fusimotor driving. Some increase their discharge frequency with tracheal occlusion. Tendon organ afferents are more numerous than spindle afferents. The effects of an increase in load indicate the functional absence of autogenetic facilitation in the diaphragm; inhibition appears to be dominant in its proprioceptive control. However, spindle secondary afferents and tendon organs from intercostal muscles of the lower rib cage exert a reflex control over phrenic motoneurones, and phrenic motoneurones may influence their own excitability through a reflex arc involving these afferents which includes a mechanical link between the diaphragm and thoraco-abdominal wall. These reflexes may "compensate" for the lack of autogenetic facilitation.

Intercostal, phrenic and abdominal motoneurones are subject to slow changes in membrane potential of central origin and of respiratory periodicity, the central respiratory drive potentials (CRDPs). The CRDPs in intercostal motoneurones may be modified by vagal stimulation. Both excitatory (from group Ia afferents) and inhibitory (from group Ib afferents) segmental effects on intercostal motoneurone excitability can be demonstrated, while in phrenic motoneurones inhibitory effects alone are apparent. The CRDPs and segmental inflow are integrated at the motoneurone.

In man, muscle spindle afferent discharge is essential for the maintenance of intercostal alpha motoneurone discharge, as shown by the silent period which develops in the EMG when the voluntarily activated muscle is suddenly allowed to shorten. An increase in load typically evokes a short latency (20–25 msec) inhibitory response (IR) attributed to autogenetic inhibition arising from tendon organs, followed by an excitatory response (ER). During the ER there may be a more than sevenfold increase in EMG activity, the size of the ER tending to increase with the airflow rate at which the load is applied and, for an expiratory muscle, as lung volume decreases. The responses are prominent during phonation. The characteristics of the ER are consistent with its being a manifestation of the gamma loop, transiently interrupted by autogenetic inhibition. The function of the proprioceptive mechanism may be primarily to control the accuracy of movement against predicted loads rather than to provide stability of ventilation in the face of unpredicted loads.

## REFERENCES

ALDERSON, A. M., and DOWNMAN, C. B. B. (1966). Supraspinal inhibition of thoracic reflexes of somatic and visceral origin. *Arch. ital. biol.*, **104**, 309–327.

BAUMGARTEN, R. VON, SCHMIEDT, H., and DODICH, N. (1963). Microelectrode studies of phrenic motoneurones. *Ann. N.Y. Acad. Sci.*, **109**, 536–544.

BISHOP, B. (1963). Abdominal muscle and diaphragm activities and cavity pressures in pressure breathing. *J. appl. Physiol.*, **18**, 37–42.

BISHOP, B. (1964). Reflex control of abdominal muscles during positive-pressure breathing. *J. appl. Physiol.*, **19**, 224–232.

BISHOP, B. (1968). Neural regulation of abdominal muscle contractions. *Ann. N.Y. Acad. Sci.*, **155**, 191–200.

CALMA, I. (1952). The reflex activity of the respiratory centre. *J. Physiol.* (*Lond.*), **117**, 9–21.

CAMPBELL, E. J. M., DICKINSON, C. J., DINNICK, O. P., and HOWELL, J. B. L. (1961). The immediate effects of threshold loads on the breathing of men and dogs. *Clin. Sci.*, **21**, 309–320.

CAMPBELL, E. J. M., DICKINSON, C. J., and HOWELL, J. B. L. (1964). The immediate effects of added loads on the inspiratory musculature of the rabbit. *J. Physiol.* (*Lond.*), **172**, 321–331.

CAMPBELL, E. J. M., DINNICK, O. P., and HOWELL, J. B. L. (1961). The immediate effects of elastic loads on the breathing of man. *J. Physiol.* (*Lond.*), **156**, 260–273.

CAMPBELL, E. J. M., and HOWELL, J. B. L. (1963). The sensation of breathlessness. *Brit. med. Bull.*, **19**, 36–40.

CAMPBELL, E. J. M., and HOWELL, J. B. L. (1962). Proprioceptive control of breathing. *Ciba Foundation Symposium: Pulmonary Structure and Function*, pp. 29–45. Eds. A. V. S. de Reuck, and M. O'Connor, London: J. & A. Churchill.

CARDIN, A. (1939). Il muscolo diaframma e le sue componenti cinestesiche. *Arch. Sci. biol.* (*Bologna*), **25**, 51–88.

CARDIN, A. (1944). Recettori di tensione nel diaframma. *Arch. Sci. biol.* (*Bologna*), **30**, 9–22.

COOMBS, H. C. (1918). The relation of the dorsal roots of the spinal nerves and the mesencephalon to the control of the respiratory movements. *Amer. J. Physiol.*, **46**, 549–471.

COOMBS, H. C., and PIKE, F. H. (1930). The nervous control of respiration in kittens. *Amer. J. Physiol.*, **95**, 681–693.

CORDA, M., EKLUND, G., and EULER, C. VON (1965). External intercostal and phrenic α motor responses to changes in respiratory load. *Acta physiol. scand.*, **63**, 391–400.

CORDA, M., EULER, C. VON, and LENNERSTRAND, G. (1965). Proprioceptive innervation of the diaphragm. *J. Physiol.* (*Lond.*), **178**, 161–177.

CORDA, M., EULER, C. VON, and LENNERSTRAND, G. (1966). Reflex and cerebellar influences on α and on "rhythmic" and "tonic" γ activity in the intercostal muscle. *J. Physiol.* (*Lond.*), **184**, 898–923.

CRITCHLOW, V., and EULER, C. VON (1962). Rhythmic control of intercostal muscle spindles. *Experientia* (*Basel*), **18**, 426–427.

CRITCHLOW, V., and EULER, C. VON (1963). Intercostal muscle spindle activity and its γ motor control. *J. Physiol.* (*Lond.*), **168**, 820–847.

CUÉNOD, M. (1961). Réflexes proprioceptifs du diaphragme chez le lapin. *Helv. physiol. pharmacol. Acta.*, **19**, 360–372.

DECIMA, E. E., and EULER, C. VON (1969*a*). Excitability of phrenic motoneurones to afferent input from lower intercostal nerves in the spinal cat. *Acta physiol. scand.*, **75**, 580–591.

DECIMA, E. E., and EULER, C. VON (1969*b*). Intercostal and cerebellar influences on efferent phrenic activity in the decerebrate cat. *Acta physiol. scand.*, **76**, 148–158.

DECIMA, E. E., EULER, C. VON, and THODEN, U. (1967). Spinal intercostal-phrenic reflexes. *Nature* (*Lond.*), **214**, 312–313.

DECIMA, E. E., EULER, C. VON, and THODEN, U. (1969). Intercostal-to-phrenic reflexes in the spinal cat. *Acta physiol. scand.*, **75**, 568–579.

DELHEZ, M., and PETIT, J. M. (1966). Influence de l'application sondaine d'une resistance additionnelle au débit aérien sur le régime ventilatoire de l'homme conscient. *Arch. int. Physiol.*, **74**, 270–275.

DOLIVO, M. (1953). "Crossed phrenic phenomenon" et phénomène phrénique bilatéral. *Helv. physiol. pharmacol. Acta.*, **11**, 251–269.

DOWNMAN, C. B. B. (1955). Skeletal muscle reflexes of splanchnic and intercostal nerve origin in acute spinal and decerebrate cats. *J. Neurophysiol.*, **18**, 217–235.

DOWNMAN, C. B. B., and HUSSAIN, A. (1958). Spinal tracts and supraspinal centres influencing visceromotor and allied reflexes in cats. *J. Physiol.* (*Lond.*), **141**, 489–499.

DRAPER, M. H., LADEFOGED, P., and WHITTERIDGE, D. (1960). Expiratory pressures and airflow during speech. *Brit. med. J.*, **1**, 1837–1843.

ECCLES, R. M., SEARS, T. A., and SHEALY, C. N. (1962). Intracellular recording from respiratory motoneurones of the thoracic spinal cord of the cat. *Nature* (*Lond.*), **193**, 844–846.

EKLUND, G., EULER, C. VON, and RUTKOWSKI, S. (1963). Intercostal $\gamma$-motor activity. *Acta physiol. scand.*, **57**, 481–482.

EKLUND, G., EULER, C. VON, and RUTKOWSKI, S. (1964). Spontaneous and reflex activity of intercostal gamma motoneurones. *J. Physiol.* (*Lond.*), **171**, 139–163.

EULER, C. VON (1966). Proprioceptive control in respiration. Nobel Symposium I: *Muscular Afferents and Motor Control*, pp. 197–207. Stockholm: Almqvist and Wiksell; New York: Wiley.

EULER, C. VON (1968). The proprioceptive control of the diaphragm. *Ann. N.Y. Acad. Sci.*, **155**, 204–205.

EULER, C. VON, and FRITTS, H. W. (1963). Quantitative aspects of respiratory reflexes from the lungs and chest walls of cats. *Acta physiol. scand.*, **57**, 284–300.

EULER, C. VON, and PERETTI, G. (1966). Dynamic and static contributions to the rhythmic $\gamma$ activation of primary and secondary spindle endings in external intercostal muscle. *J. Physiol.* (*Lond.*), **187**, 501–516.

FINK, B. R., NGAI, S. H., and HOLADAY, D. A. (1958). Effect of airflow resistance on ventilation and respiratory muscle activity. *J. Amer. med. Ass.*, **168**, 2245–2249.

FLEISCH, A. (1928). Proprioceptive Atmungsreflexe. *Pflügers Arch. ges. Physiol.*, **219**, 706–725.

GILL, P. K., and KUNO, M. (1963*a*). Properties of phrenic motoneurones. *J. Physiol.* (*Lond.*), **168**, 258–273.

GILL, P. K., and KUNO, M. (1963*b*). Excitatory and inhibitory actions on phrenic motoneurones. *J. Physiol.* (*Lond.*), **168**, 274–289.

GLEBOVSKII, V. D. (1965). Stretch reflexes of intercostal muscles. *Fiziol. Zh.* (*Mosk.*), **51**, 1420. (Transl. in *Fed. Proc.*, 1966, **25**, T937–T942.)

GLEBOVSKII, V. D., and PAVLOVA, N. A. (1962). Diaphragm reflexes in response to adequate stimulation of receptors of lungs and respiratory muscles. *Fiziol. Zh.* (*Mosk.*), **48**, 545. (Transl. in *Fed. Proc.*, 1963, **22**, II T405–T410.)

KATZ, B. (1966). *Nerve, Muscle, and Synapse.* New York: McGraw-Hill Book Co.

KORN, H. (1967). Control of the splanchno-intercostal reflex by orbital cortex in the cat. *J. Physiol.* (*Lond.*), **192**, 33–34P.

MASSION, J., MEULDERS, M., and COLLE, J. (1960). Fonction posturale des muscles respiratoires. *Arch. int. Physiol.*, **68**, 314–326.

MOLTKE, E., and SKOUBY, A. P. (1963). The influence of tonic neck reflexes on the activity of some muscles of the trunk in patients with asthma and emphysema. *Acta med. scand.*, **173**, 299–305.

NATHAN, P. W., and SEARS, T. A. (1960). Effects of posterior root section on the activity of some muscles in man. *J. Neurol., Neurosurg. and Psychiat.*, **23**, 10–22.

NEWSOM DAVIS, J., and SEARS, T. A. (1970). The proprioceptive reflex control of the intercostal muscles during their voluntary activation. *J. Physiol.* (*Lond.*). (In press.)

NEWSOM DAVIS, J., SEARS, T. A., STAGG, D., and TAYLOR, A. (1966). The effects of airway obstruction on the electrical activity of intercostal muscles in conscious man. *J. Physiol.* (*Lond.*), **185**, 19P.

NIEKERK, J. VAN, and BRAAK, J. W. G. ter (1935). Die Anpassung des Atmungsvorganges an Widerstandsänderungen in den Atmungswegen. *Pflügers Arch. ges. Physiol.*, **236**, 44–51.

PITTS, R. F. (1946). Organization of the respiratory centre. *Physiol. Rev.*, **26**, 609–630.

RAMOS, J. G. (1959). On the integration of respiratory movements. *Acta physiol. lat.-amer.*, **9**, 246–256.

RAMOS, J. G., and MENDOZA, E. L. (1959). On the integration of respiratory movements. II. The integration at spinal level. *Acta physiol. lat.-amer.*, **9**, 257–266.

RIEDSTRA, J. W., and DIRKEN, M. N. J. (1953). On proprioceptive respiratory reflexes. *Acta physiol. pharmacol. neerl.*, **3**, 19–26.

RIJLANT, P. (1942). Contribution à l'étude du contrôle réflex de la respiration. *Bull. Acad. roy. Méd. Belg.*, **7**, 58–107.

SANT'AMBROGIO, G., and WIDDICOMBE, J. G. (1965). Respiratory reflexes acting on the diaphragm and inspiratory intercostal muscles of the rabbit. *J. Physiol.* (*Lond.*), **180**, 766–779.

SANT'AMBROGIO, G., WILSON, M. F. and FRAZIER, D. T. (1962). Somatic afferent activity in reflex regulation of diaphragmatic function in the cat. *J. appl. Physiol.*, **17**, 829–832.

SEARS, T. A. (1962). The activity of the small motor fibres system innervating respiratory muscles of the cat. *Aust. J. Sci.*, **25**, 102.

SEARS, T. A. (1963). Activity of fusimotor fibres innervating muscle spindles in the intercostal muscles of the cat. *Nature* (*Lond.*), **197**, 1013–1014.

SEARS, T. A. (1964*a*). Efferent discharges in alpha and fusimotor fibres of intercostal nerves of the cat. *J. Physiol.* (*Lond.*), **174**, 295–315.

SEARS, T. A. (1964*b*). Some properties and reflex connections of respiratory motoneurones of the cat's thoracic spinal cord. *J. Physiol.* (*Lond.*), **175**, 386–403.

SEARS, T. A. (1964*c*). The slow potentials of thoracic respiratory motoneurones and their relation to breathing. *J. Physiol.* (*Lond.*), **175**, 404–424.

SEARS, T. A. (1964*d*). Investigations on respiratory motoneurones of the thoracic spinal cord. *Progr. Brain Res.*, **12**, 259–272.

SEARS, T. A., and NEWSOM DAVIS, J. (1968). The control of respiratory muscles during voluntary breathing. *Ann. N.Y. Acad. Sci.*, **155**, 183–190.

STELLA, G. (1938). On the mechanism of production, and the physiological significance of "apneusis". *J. Physiol.* (*Lond.*), **93**, 10–23.

SUMI, T. (1963). The segmental reflex relations of cutaneous afferent inflow to thoracic respiratory motoneurones. *J. Neurophysiol.*, **26**, 478–493.

YASARGIL, G. M. (1962). Proprioceptive Afferenzen im *N. phrenicus* der Katze. *Helv. physiol. pharmacol. Acta*, **20**, 39–58.

# Chapter XII

# SUPRASPINAL CONTROL

J. NEWSOM DAVIS

EXPERIMENTAL studies of the central mechanism controlling respiration have in the past relied largely upon the experimental techniques of surgical transection or ablation, and electrical stimulation. In recent years, microelectrode recording of neurone activity, both at brain stem and spinal cord level, has allowed a different approach to the study of the respiratory mechanism and has provided information which, in some respects, is not compatible with earlier accounts of the organisation of the central control of breathing. These have usually taken as their starting point a consideration of those "centres" of specified function to which the earlier work, based on transection and stimulation experiments, seemed to point. In the organisation of this chapter the existence of such centres has not been presupposed.

The respiratory mechanism in the medulla and pons will be considered first, the justification for this being that in the experimental animal breathing continues apparently unchanged following transection above the pons. Midbrain and higher structures nevertheless have an important influence on respiratory muscle activity, and the experimental evidence for this will be given later. The final section will deal with the efferent pathways to the respiratory motoneurones in the spinal cord.

## MEDULLA AND PONS

It is proposed to begin with the characteristics of respiratory neurones in the intact medulla and pons because, although the precise function of these neurones has not yet been defined, many workers believe them to be concerned with the genesis of the rhythmic component of breathing. The influence of vagal afferents on breathing will next be described. Then, in outline, the effects on respiratory muscle activity of brain stem transection and electrical stimulation will be given. These longstanding experimental techniques have provided a wealth of experimental data, much of it rather difficult to interpret. It is necessary, however, to be familiar with these basic observations in order to understand how some current views on the control of breathing have arisen. The extent to which these and other experimental observations support the operation of centres in the brain stem serving specific respiratory functions, or whether an alternative interpretation is preferable, will be discussed at the end of the section. It has to be recognised that the detailed function of brain stem structures is far from being fully understood so that it is not altogether surprising that a generally accepted statement cannot yet be given of how the respiratory mechanism in the medulla and pons is organised.

## Respiratory Neurones

Before considering brain stem respiratory neurones in greater detail, several points need to be emphasised. Firstly, a discharge in phase with respiration does not thereby imply that the neurone is concerned directly with the genesis of rhythmic breathing. Secondly, the possibility that these neurones may show periodic changes in their response characteristics has to be borne in mind (*cf.* Scheibel and Scheibel, 1965). Finally, a rigid correlation should not necessarily be expected between the discharge of those neurones concerned with the genesis of the respiratory rhythm and that of spinal respiratory motoneurones, i.e. phrenic, intercostal and abdominal motoneurones. As discussed in the preceding chapter the spinal respiratory motoneurone is the site of integration of segmental inputs and supraspinal drives. The central respiratory drive potentials recorded in these motoneurones only cause cell firing (and thus muscle contraction) if the depolarisation is sufficient to bring the membrane potential to the critical firing threshold. Thus activity in brain stem "expiratory" neurones may give rise to changes in membrane potential of spinal motoneurones to expiratory muscles but not necessarily to expiratory muscle contraction. Excitatory or inhibitory effects of segmental origin or arising from other supraspinal pathways can also profoundly influence the discharge pattern of the motoneurone. Moreover, the central drives on the respiratory motoneurones probably operate through spinal interneurones (Sears, 1964, 1966), which could themselves be under separate central control.

When a micro-electrode is inserted into the intact brain stem, spike potentials can be recorded arising from the cell bodies of individual neurones. Amongst many units which can be recorded in this way, there are some whose discharge has a consistent relationship with the respiratory cycle as monitored, for instance, by changes in pleural pressure, pneumogram or phrenic neurogram (action potentials recorded in the phrenic nerve). Such spike potentials arising from respiratory neurones were first recorded by Gesell *et al.* (1936) in the dog, and later by Woldring and Dirken (1951) in the rabbit. Subsequent studies have shown that respiratory neurones are much more frequently encountered in the medulla than in the pons, and more information is available about them at the former site. The following account, based on observations in the cat, will refer to medullary respiratory neurones unless otherwise stated. A summary of the properties of respiratory neurones is given at the end of this section.

## Site

Units having a respiratory discharge are found in the medulla at the level of the obex (Hukuhara *et al.*, 1954; Haber *et al.*, 1957; Nelson, 1959; Salmoiraghi and Burns, 1960*a*), most of them lying between 2 and 4·5 mm from the midline and between 2 and 4 mm deep to the floor of the fourth ventricle (Fig. 85). In the longitudinal axis, respiratory units can be recorded from the caudal border of the facial nucleus to the level of the pyramidal decussation. Some separation between inspiratory and expiratory units has been demonstrated in this axis, inspiratory units being

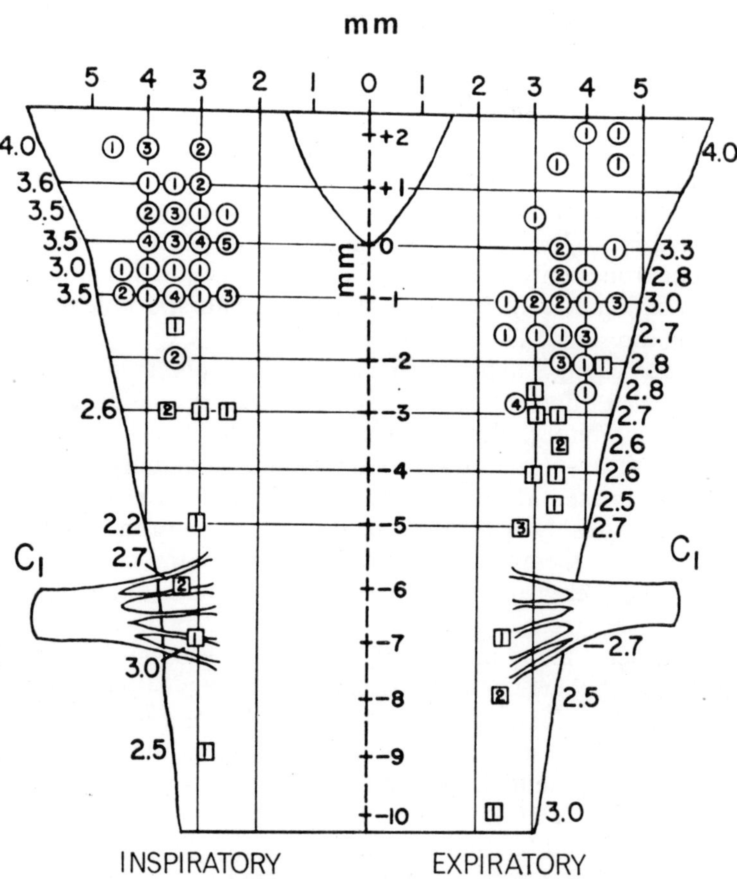


Fig. 85.—Localisation of Respiratory Neurones in the Medulla

Dorsal projection of medulla to represent localisation of unit activity. Stereotaxic coordinates with reference to obex are marked in mm. ○, units with "cell body" wave forms. □, units with "axon" wave forms. Number of units isolated at given point is shown within circle or square. Inspiratory units are grouped on left, expiratory units on right, though both are present bilaterally. Numbers at lateral margins show average depth (mm) of units at that level. (From Nelson, 1959.)

concentrated mainly rostral to the obex, expiratory units caudal to the obex (Haber *et al.*, 1957; Nelson, 1959; Batsel, 1964).

The relationship of these neurones to other brain stem structures was studied by Batsel (1964), who used histological methods to establish that unit discharges recorded with micro-electrodes were not arising from motoneurones innervating muscles which are also known to be active during respiration, such as those of the face, tongue, pharynx

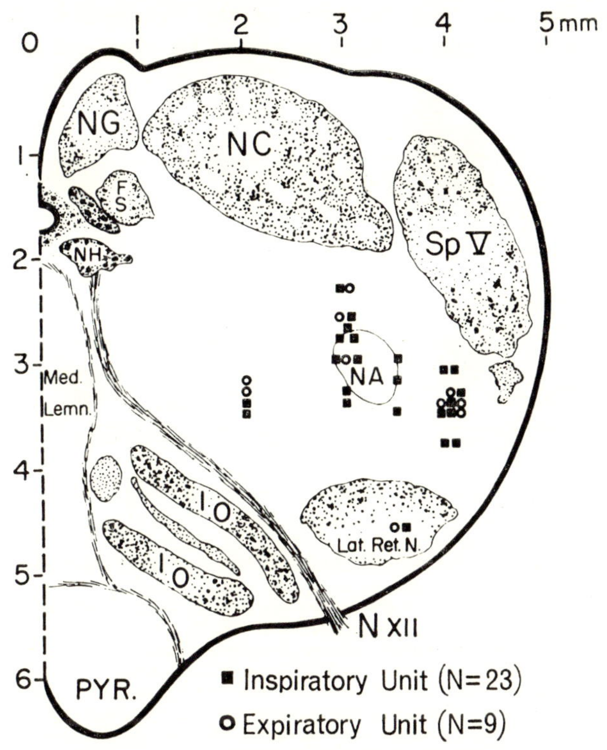


FIG. 86.—LOCALISATION OF RESPIRATORY NEURONES IN THE MEDULLA

Transverse section at the level of the obex. Respiratory units cluster about nucleus ambiguus (NA) at a depth of about 3 mm from the surface of the medulla. Histological confirmation showed that this method, although subject to error, gave a fair representation of the position of respiratory units. NH = nucleus hypoglossus; FS = fasciculus solitarius; NG = nucleus gracilis; NC = nucleus cuneatus; Sp.V. = spinal trigeminal nucleus. (From Batsel, 1964.)

and larynx. Two aggregations of respiratory units were located (Fig. 86). The larger, consisting of inspiratory, expiratory and phase-spanning units (see below) lay in the region of the nucleus ambiguus, dorsal to the lateral reticular nucleus and ventral to the descending trigeminal nucleus. The cross-section of this area considerably exceeded that of the nucleus ambiguus (motor nucleus of the vagus). Batsel (1964) has emphasised the possible teleological significance of this site, which lies along that of the branchial motor nuclei of the glossopharyngeal and the vagus nerves, and has drawn attention to the report by Baker *et al.* (1950) of neuronal damage and focal necrosis in this region in patients who have died of respiratory failure in bulbar poliomyelitis. The second smaller group of respiratory neurones consisted of inspiratory units situated immediately ventro-medial to the fasciculus solitarius, and 1 to 3 mm rostral to the obex (see also von Baumgarten and Kanzow, 1958). No units were recorded in the ventro-medial reticular formation.

It may be recalled that the sites of the "inspiratory centre" and "expiratory centre" in the medulla proposed by Pitts and his colleagues

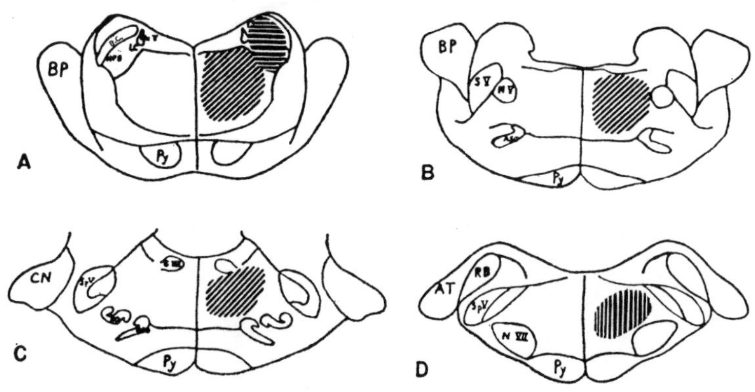


FIG. 87.—LOCALISATION OF RESPIRATORY NEURONES IN THE PONS

Schematic drawings of sections of brain stem at various levels. A: Level of locus coeruleus. B: Level of motor nucleus of trigeminal nerve. C: Level of superior olive. D: Level of acoustic tubercles. Cross hatching in A indicates pneumotaxic region, where predominantly inspiratory discharges are found. Diagonal hatching in A, B, and C indicates regions of pons where predominantly inspiratory-expiratory discharges are found. Vertical hatching in D indicates region where expiratory-inspiratory discharges are found more frequently than in other regions. Abbreviations: ASO: accessory nucleus of superior olive. AT: acoustic tubercle. BC: brachium conjunctivum. BP: brachium pontis. CN: cochlear nuclei. G VII: Genu of seventh nerve. LC: locus coeruleus. M V: motor nucleus of the fifth nerve. Me V: mesencephalic root of fifth nerve. N VII: nucleus of seventh nerve. NPB: nucleus parabrachialis. Py: pyramidal tract. RB: restiform body. S V: sensory nucleus of the fifth nerve. SO: superior olive. Sp V: spinal tract of fifth nerve. (From Cohen and Wang, 1959.)

(1939*a*) do not correspond with those regions outlined above in which respiratory neurones are encountered on microelectrode sounding. Pitts *et al.* placed these "centres" in the medial reticular formation of the medulla on the basis of the changes in respiratory pattern following electrical stimulation there. In retrospect, these changes can probably be attributed to activation of descending reticulo-spinal pathways to the respiratory muscles (see below) and not to direct excitation of rhythmically discharging expiratory or inspiratory neurones. In support of this is the observation of Pitts *et al.* (1939*a*) that the pathway serving the inspiratory response was crossed at medullary level since the response was not affected by an ipsilateral hemisection of the cord at C1; the pathway concerned with rhythmic breathing, on the other hand, appears to be mainly uncrossed since a mid-line sagittal section through the medulla leaves the rhythm of breathing unchanged (Kahn and Wang, 1965; see also p. 246).

In a survey of the pons, using $CO_2$ to stimulate respiration, Cohen and Wang (1959) found that in the lateral part of the upper pons in the region of the locus coeruleus, brachium conjunctivum and nucleus parabrachialis (Fig. 87) most units were inspiratory or inspiratory-expiratory. In the medial part of the upper and middle pons, corresponding approximately to the nucleus pontis oralis and nucleus pontis caudalis, inspiratory-expiratory units predominated. In the lower pons, in the region of nucleus giganto-cellularis, expiratory-inspiratory units were most common.

**Discharge Characteristics**

Different discharge patterns are encountered amongst respiratory neurones with respect to the duration of burst discharge, the timing in relation to the respiratory cycle and the number of spikes per burst, but the characteristics for a particular neurone under steady conditions tend to remain constant over successive bursts (Salmoiraghi and Burns, 1960*a*; Nesland and Plum, 1965). Units may be characterised by the phasing of their discharge as inspiratory or expiratory, and their timing within that phase as early, late or throughout (Fig. 88), (Batsel, 1965; Nesland and Plum, 1965; Cohen, 1968). In addition, phase-spanning units occur (predominantly in the pons rather than the medulla) classed as inspiratory-expiratory or expiratory-inspiratory depending on the phase in which the burst discharge has its onset (Batsel, 1965; Cohen, 1968), or continuous when units with respiratory modulation of frequency show no discharge gap between bursts. The different types of respiratory neurone discharge are shown schematically in Fig. 89.

Although this classification of units in relation to the timing of their discharge is convenient, the units contained in each group are not homogeneous for they do not necessarily respond in a similar way to stimuli such as changes in $P_{CO_2}$ or $P_{O_2}$, and lung inflation.

*Changes in $CO_2$ pressure.*—Increasing inspired $CO_2$ usually produces an increased discharge frequency both of inspiratory and expiratory units (Haber *et al.*, 1957; Batsel, 1965; Nesland *et al.*, 1966). This increase correlates with the increase in tidal volume, but since the burst duration also shortens, the total number of spikes per burst is usually unchanged (Nesland *et al.*, 1966). In the vagotomised animal, discharge frequency of respiratory neurones also increases with $CO_2$, but the burst duration does not shorten and the respiratory rate fails to increase, in contrast to the intact animal, so that the number of spikes per burst increases (Nesland *et al.*, 1966). An exception to this response occurs in some early expiratory units which are silenced by increased $CO_2$ and may be vagal motoneurones (Batsel, 1965).

Hypocapnic apnoea is associated with silence in some units but in others produces a continuous discharge which is usually modulated but sometimes aperiodic (Nesland and Plum, 1965; Batsel, 1967). These effects in both pontine and medullary units have been subject to a further

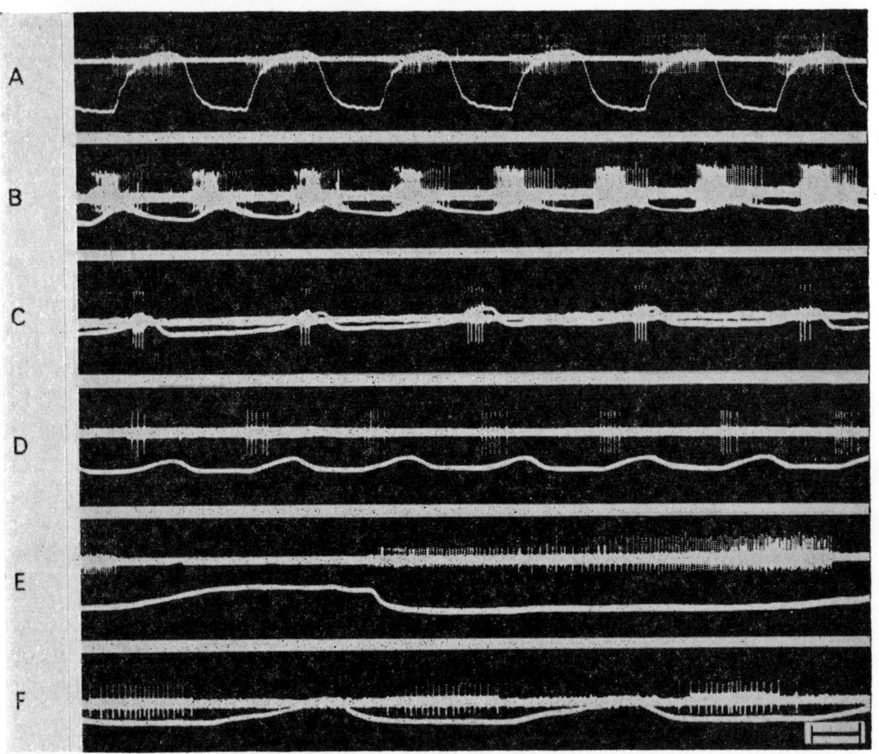


FIG. 88.—DISCHARGE PATTERNS OF RESPIRATORY NEURONES

In this figure, the lower trace is the abdominal pneumograph, an upward deflection being inspiration. The mark equals 1 sec. A and B. I-throughout neurones showing differences in cessation of activity. C. I-late neurone. D. I-early neurone. E. Two expiratory neurones recorded at the same location, demonstrating phase difference between an E-early neurone and an E-late neurone. F. E-late neurone having constant frequency throughout the burst. Records A, B, C and D were from non-vagotomised, E and F from vagotomised cats. (From Nesland and Plum, 1965.)

analysis by Cohen (1968) in vagotomised, paralysed animals maintained on artificial ventilation. In this study, the phrenic neurogram was used to indicate central respiratory activity. When alveolar $Pco_2$ was lowered from relatively high levels (e.g. 50 mm Hg), units maintained their phase relationship with the overall respiratory cycle. Neurone responses were of three types. In type 1, shown mainly by inspiratory, expiratory or inspiratory-expiratory neurones, discharge frequency diminished in all portions of the cycle, reaching zero at sufficiently low levels of alveolar $CO_2$ (e.g.

10 mm Hg). In type 2, occurring predominantly in expiratory or expiratory-inspiratory neurones, the discharge became continuous, and at very low levels of $CO_2$ lost its respiratory modulation. In type 3, occurring only in units characterised by continuous discharge with respiratory modulation, the discharge frequency was reduced in all parts of the cycle and the respiratory modulation was lost, but activity still continued, even at the lowest $CO_2$ levels.

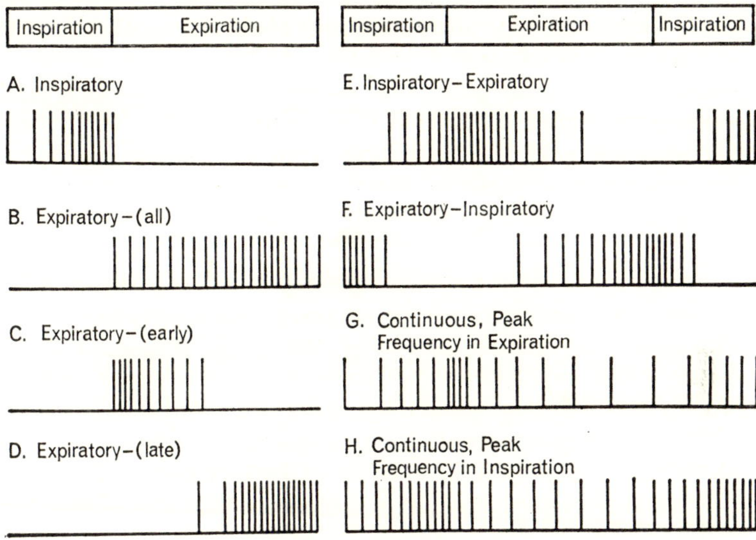


FIG. 89.—SCHEMATIC DIAGRAM OF MAJOR PATTERNS OF RESPIRATORY UNIT DISCHARGE

Based on spike counts for a representative neurone of each type. (From Cohen, 1968.)

*Changes in $O_2$ pressure.*—The changes in discharge pattern with hypoxia are less consistent. Discharge frequency sometimes increases in inspiratory units and decreases in expiratory units (Batsel, 1965; Nesland *et al.*, 1966). If 100 per cent $O_2$ is substituted for room air, little change in respiratory unit discharge is found. When 100 per cent $O_2$ is given to hypoxic animals, on the other hand, inspiratory units and most expiratory units showed a decrease in discharge frequency and spikes per burst, some units ceasing activity altogether (Nesland *et al.*, 1966). A few expiratory units increased their frequency.

*Vagal effects.*—Bilateral vagal section gives rise to a brief irregularity of inspiratory discharges, but the only permanent change during quiet breathing is an increase in burst duration and a delay in reaching the peak frequency (Batsel, 1965). With a rise in inspired $CO_2$, the normal shortening of burst discharge no longer occurs, as noted above. Electrical stimulation of the vagus at a frequency which inhibits inspiration causes inhibition of inspiratory neurones and prolongation of activity in expiratory neurones (Haber *et al.*, 1957). In a study of the effects of lung inflation (vagi intact) in 10 inspiratory units and 24 expiratory units, Nesland and Plum (1965) found that most of the inspiratory units and half of the expiratory units were inhibited, the remaining units being facilitated. Cohen (1969) has also reported that although some inspiratory neurones, as might be expected, are inhibited by lung inflation, other neurones with otherwise similar discharge characteristics are facilitated by the inflation. Similarly, expiratory neurones could show either facilitation or inhibition with lung inflation. Expiratory-inhibitory neurones, which are found predominantly in the pons, were usually depressed by lung inflation, substantiating an earlier suggestion (Cohen and Wang, 1959) that they may be concerned with the transition from inspiration to expiration. Breathing against an expiratory pressure increases expiratory neuronal activity, the response being abolished by vagotomy (Price and Batsel, 1969).

*Somatic afferents.*—It is known that stimulation of somatic afferents, whether from skin, muscle or mixed nerves, produces changes in ventilation. Stimulation of muscle nerves increases the discharge frequency of inspiratory units and causes an earlier onset of activity in the inspiratory phase. In contrast to this, expiratory units are relatively unaffected, a slight inhibition being apparent. In some cases, stimulation of a muscle nerve which gave rise to inhibition of an expiratory unit had no effect on inspiratory units (Koizumi *et al.*, 1961).

**Correlation with Ventilation**

Although some authors have found evidence for recruitment of units with hypercapnia or hypoxia, a specific search for this phenomenon, while measuring pulmonary ventilation, indicates that it plays only a minor part in the graded response of medullary neurones to stimulation (Nesland *et al.*, 1966). The units quiescent in apnoea became active as breathing resumed, few further units being recruited with further increase in ventilation. Thus changes in ventilation are associated with changes in discharge frequency of medullary respiratory neurones rather than with recruitment of units.

**Mechanism of Chemosensitivity**

Chemosensitive structures in the brain stem, which include respiratory neurones responding to changes in blood gases, are distinct from those mechanisms which subserve certain respiratory reflexes (von Euler and Söderberg, 1952*a*). Chloralose depresses these chemosensitive structures in vagotomised animals as shown by the absent response in the phrenic neurogram after $CO_2$ had been added to the inspired air, but leaves unchanged the response to superior laryngeal nerve stimulation and intravenous lobeline. Slow changes in electrical potential can be recorded with a unipolar electrode in the medulla, which seem to relate specifically to alterations of inspired $CO_2$ and which are depressed by chloralose (von Euler and Söderberg, 1952*b*). These "chemopotentials" thus appear to arise from the activity of chemosensitive receptors.

Direct injection of chemical substances into the medulla has been used to investigate its chemosensitivity (Comroe, 1943; Liljestrand, 1952). The latter found that respiratory responses followed injection of a few $\mu$l of sodium bicarbonate into the lateral reticular formation, an area which corresponds more closely with the site of respiratory neurones than with that area in the medial reticular formation believed by Pitts *et al.* (1939*a*) to be the site of inspiratory and expiratory centres. The specificity of bicarbonate as the stimulus has since been questioned. Citrate produces a similar response and it is suggested that these substances exert their effects through the formation of un-ionised calcium complexes which prevent the usual stabilising properties of calcium ions at the axonal membrane (Kim and Carpenter, 1961).

Areas of respiratory chemosensitivity have since been demonstrated in anaesthetised cats on the ventro-lateral surface of the medulla (Mitchell *et al.*, 1963). Appropriate chemical stimuli applied to this region give rise to large changes in ventilation. The authors suggest that the chemoreceptors lie within, or just below, the pia mater at this site, and respond to changes in CSF hydrogen ion concentration. However, rather smaller changes in ventilation are produced by chemical stimulation applied to the medullary surface if the depressant effect of anaesthesia is avoided by employing mid-collicular preparations (Cozine and Ngai, 1967). Furthermore, Pappenheimer and his co-workers (1965) believe the chemoreceptor mechanism to lie within the brain substance, on the basis of their perfusion studies in the conscious goat. The role of the CSF and brain extracellular fluid in the control of pulmonary ventilation has been the subject of a recent review by Cameron (1969) and will not be considered further here.

### Genesis of Rhythmic Discharge

It has been proposed that the rhythmic characteristic of respiration depends upon the activities of two groups of neurones in the brain stem, the neurones of each group being self re-exciting and the groups mutually inhibitory (Burns and Salmoiraghi, 1960; Salmoiraghi and Burns, 1960*b*; Burns, 1963). Although it is clear now that respiratory neurones consist of a relatively large number of sub-groups rather than two functionally homogeneous groups of inspiratory and expiratory units, these studies still indicate the principles on which rhythmic genesis might depend.

The most likely alternative mechanism of respiratory rhythmicity had been the supposed property of certain neurones to discharge spontaneously in a manner comparable with that of the cardiac pacemaker. If such pacemaker neurones exist, progressive neurological isolation of the brain stem would be expected to leave the number of active respiratory units in the medulla relatively unchanged. Experimental results do not confirm this expectation (von Euler and Söderberg, 1952*a*; Salmoiraghi and Burns, 1960*b*). A midline sagittal section of the brain stem in a mid-collicular decerebrate animal abolished spontaneous respiratory movements (*cf.* p. 246), but 60 per cent of micro-electrode penetrations in the medulla revealed units with the discharge characteristics of respiratory neurones (Salmoiraghi and Burns, 1960*b*). Following spinal cord section, this figure fell to 26 per cent while with total isolation of one-half of the brain stem only about 10 per cent of micro-electrode penetrations were successful. Following stimulation of the cut caudal end of the neurologically isolated medulla and pons, some units were caused to discharge for several bursts in a manner typical of respiratory neurones. In addition, a micro-electrode study of the isolated upper pons showed no units firing with a respiratory rhythm (Cohen, 1958). These observations thus make it rather unlikely that the respiratory rhythmicity of certain brain stem neurones depends upon the activity of pacemaker units.

That there were two groups of reciprocally innervated inspiratory and expiratory neurones was indicated by the observation that, when units in both populations were considered, the probability of firing remained a constant (Salmoiraghi and Burns, 1960*a*). Moreover, if such a reciprocal pattern of innervation links these two groups, it might be expected that under certain circumstances rhythmic activity could be halted with continuous activity in one of the groups. This was observed in some inspiratory and expiratory units using an incremental dosage of intravenous barbiturates (Robson *et al.*, 1963). Similar observations following hypocapnia had been reported earlier by Cohen (1959: see also 1968).

The repetitive discharge which the neurones of each group display during the period of their burst activity seems to result from a self-excitatory process within each group (Burns and Salmoiraghi, 1960). Firstly, these neurones share the characteristic of irregularity of discharge frequency with certain cells in isolated cerebral cortex where this is believed to be due to a self re-exciting process. Secondly, self re-exciting circuits would be expected to respond in a predictable manner to an electrical stimulus. Thus a shock sufficient to excite all members of the population should abruptly terminate activity in all the constituent neurones since they would pass simultaneously into an absolute refractory state (see Fig. 90, fifth pair of traces). In contrast to this, a submaximal stimulus would leave some neurones undischarged and still available for re-excitation of those other units which had been caused to discharge and which were thus temporarily rendered refractory. Because there would then be a relatively large number of units undergoing re-excitation, the burst discharge

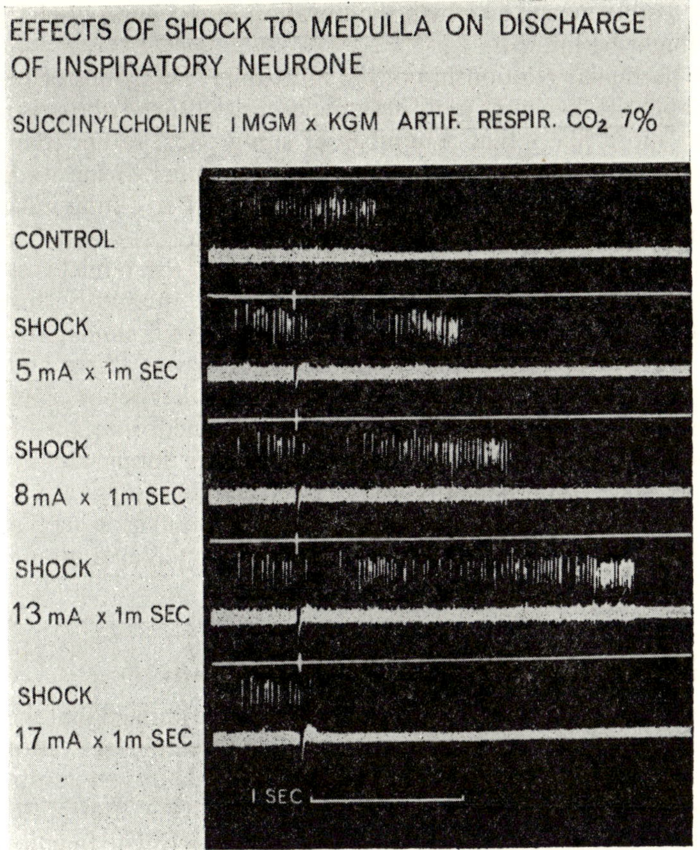


FIG. 90.—EFFECTS OF SHOCKS OF PROGRESSIVELY INCREASING STRENGTH GIVEN TO MEDULLA UPON ACTIVITY OF INSPIRATORY NEURONE

Anaesthetised, decerebellate, mid-collicularly decerebrate, vagotomised cat paralysed with intravenous succinylcholine (1 mg/kg). Artificial ventilation with 7 per cent $CO_2$ in oxygen. In each pair of records: upper trace monitors timing of shock; lower trace: inspiratory neurone. First pair of records from top: control (no shock). Second pair: shock of strength 5 mA; third pair: shock of strength 8 mA; fourth pair: shock of strength 13 mA; fifth pair: shock of strength 17 mA. All shocks of 1 msec duration. In all instances, stimulus triggered by first action potential of inspiratory burst after identical delays. (Burns and Salmoiraghi, 1960.)

would be prolonged. Experimental observations on respiratory neurones were consistent with these predictions, as illustrated in Fig. 90 (Burns and Salmoiraghi, 1960).

Intracellular recording from respiratory neurones, which is technically difficult, reveals that during an inspiratory burst discharge the neurone becomes progressively less excitable because of a shift in a depolarising direction of the membrane potential threshold at which the cell fires (Salmoiraghi and von Baumgarten, 1961). Although such a change in excitability during repetitive firing is not confined to respiratory neurones, it is a possible means by which the duration of the burst discharge of the group might be limited.

The particular relationship of the individual sub-groups of respiratory neurones to the mechanism of rhythmic genesis outlined above is not clear. It seems more likely that a number of sub-groups, rather than a single pair, interact with each other in the proposed manner. From studies of the responses of respiratory neurones to changes in $P_{CO_2}$, lung inflation and brain stem electrical stimulation, Cohen (1970) has proposed a model for the genesis of respiratory rhythm. At the centre of this model is a "master-oscillatory" loop, which consists of mutually inhibitory inspiratory-facilitating and expiratory-facilitating networks (in a similar relationship to that of the inspiratory and expiratory neurones of Burns and Salmoiraghi). The master loop is influenced by three other loops. One of these initiates inspiration, and expiratory-inspiratory neurones are thought to be concerned with this function. The two other loops are inspiratory inhibiting and expiratory-inhibiting respectively. It is suggested that those inspiratory neurones which are excited by lung inflation (see page 242) might form part of the inspiratory-inhibiting loop. Vagal inputs are also incorporated into the model.

## Rhythmic Discharge and Respiratory Muscle Activity

Most accounts of the control of breathing have implied that the respiratory neurones believed to give rise to the rhythmic activation of the respiratory muscles relay directly to the phrenic or intercostal motoneurones in the spinal cord, and that their activity is transmitted thence unchanged to the respiratory muscles. The integrative function of the spinal respiratory motoneurone, however, is incompatible with such a view (see Chapter XI). There is also no anatomical evidence that the axons of rhythmically discharging respiratory neurones synapse directly with respiratory motoneurones, their exact projection being unknown.

The neural pathway subserving rhythmic breathing, through which some of these neurones may be presumed to act, appears to be predominantly uncrossed in the medulla, for a mid-sagittal section extending 3–4 mm caudal and 2–6 mm rostral to the obex does not abolish rhythmic respiratory movements or rhythmic discharge in the phrenic nerve (Kahn and Wang, 1965). Some changes were seen in these experiments in the rate and depth of breathing and in the integrated pattern of the phrenic neurogram which were attributed to disturbance of the neural structures responsible for generating the rhythmic discharge rather than to interruption of the descending pathways. That some workers (e.g. Salmoiraghi and Burns, 1960*b*) have noted cessation of rhythmic respiration on sagittal section of the medulla is perhaps attributable to differences in technique, but variation between individual animals may also play a part. The efferent pathways in the spinal cord to the respiratory motoneurones are discussed at the end of this chapter.

**Summary**

Most neurones discharging with respiratory periodicity are found in the lateral reticular formation of the medulla at the level of the obex. Some units are also found close to tractus solitarius and in the pons. Respiratory neurones differ with respect to the phasing of their discharge (the main groups being inspiratory, expiratory, phase-spanning and "continuous" with respiratory modulation), and in their response to chemical stimuli, and lung inflation or deflation. Changes in ventilation are associated with changes in discharge frequency of respiratory neurones rather than with recruitment of units; neurones serving respiratory reflexes are distinct from those having a chemoreceptor function. The genesis of rhythmic discharge may depend upon the activities of two groups of these neurones, the neurones of each group being self re-exciting and the groups mutually inhibitory. The functional role of the individual groups of neurones so far recognised, however, has not yet been clearly defined. In order to characterise these neurones further, information is needed about their axonal projections, the precise relationship of their discharge to changes in excitability of respiratory motoneurones and their sensory responsiveness to non-respiratory as well as to other respiratory stimuli.

## Vagal Afferents

It has long been recognised that vagal afferents have an important influence on respiratory muscle activity, although they are not essential to the rhythmic component of breathing. Many reflex effects can be elicited through vagal afferents (for review, see Widdicombe, 1964), but here we shall be concerned primarily with the inflation and deflation reflexes, and the role of vagal afferents in the response of the respiratory mechanism to increasing $P_{CO_2}$. The central connections of the vagal pulmonary afferent fibres will also be outlined to provide the anatomical background to the experimental observations which relate to them, described in the next section.

The influence of the vagal afferents *in toto* on the force of contraction of respiratory muscles was investigated quantitatively by von Euler and Fritts (1963) by comparing, before and after vagotomy, the pressure changes developed during the respiratory efforts which follow closure of the trachea at different lung volumes. The vagi exerted an inhibitory effect which was almost linearly related to lung volume. Thus after vagotomy greater inspiratory pressures were developed, this being particularly marked at large volumes of inflation. In some animals, the pressure-volume curves obtained before and after vagotomy crossed at a volume close to the mid-position, suggesting that the vagi inhibited contraction at lung volumes above this volume and facilitated contraction at lung volumes below it. The gain of the inspiratory-inhibiting component of the Hering-Breuer reflex increases with increasing $P_{CO_2}$ (Woldring 1965).

**Reflex Effects**

*The inflation reflex*, first described by Hering and Breuer (1868; and Breuer, 1868), is characterised by an inhibition of inspiratory muscles and prolongation of the expiratory phase elicited either by closure of the trachea at end-inspiration or by lung inflation, the degree of inhibition being related to the lung volume change. There is associated expiratory muscle activation. The reflex is abolished by vagal block. In quietly breathing animals of most species, the inflation reflex probably participates in the termination of each inspiration and is a means by which the frequency of breathing can be controlled. Both cooling of the vagus nerve to block the inflation reflex and total vagal section cause slowing and deepening of respiratory movements (Widdicombe, 1964). There is also some evidence for a tonic action of these afferents upon breathing (see Widdicombe, 1964). However, in healthy man bilateral vagal block causes no change in the pattern of normal breathing either when conscious or anaesthetised (Guz *et al.*, 1964; Guz, Noble, Widdicombe, Trenchard, Mushin and Makey, 1966). An inflation reflex in man can be elicited, however, with large volumes and is abolished by vagal block; inflation volumes in the tidal volume range produce little or no inhibitory response (Widdicombe, 1961; Guz *et al.*, 1964).

The balance of inspiratory to expiratory muscle activity appears to be dependent upon vagal afferents. In normal circumstances, breathing is mainly an inspiratory act, expiration being largely achieved by the net recoil forces of lungs and chest wall. Under conditions of positive pressure breathing, however, in anaesthetised animals the diaphragm is inhibited and breathing becomes an expiratory act with rhythmic activity in the abdominal muscles (Bishop, 1964). This change no longer occurs after cervical vagotomy. The vagal afferents responsible for this effect may not be those serving the Hering-Breuer inflation reflex.

*The deflation reflex* is characterised by increased frequency and force of inspiratory muscle contraction after closure of the trachea at the end of expiration or deflation of the lungs; the reflex is abolished by vagal block. In normal breathing, the reflex is inactive and its physiological role is not clear (Widdicombe, 1964).

*The ventilatory response to $CO_2$* appears to depend in part upon pulmonary afferents in the vagus. In conscious man, breathing high oxygen concentrations to eliminate carotid and aortic chemoreceptor stimulation, combined glossopharyngeal and vagal block markedly diminishes the ventilatory response, the effect being due predominantly to a failure of the respiratory rate to increase (Guz, Noble, Widdicombe, Trenchard and Mushin, 1966). Similar changes occur in the anaesthetised or conscious rabbit following vagal block (Scott, 1908; Richardson and Widdicombe, 1969).

### Central Connections

The primary visceral afferent fibres of the vagus (which include those from pulmonary receptors) pass dorso-medially on entering the medulla. Most of them then turn caudally in the tractus solitarius and synapse in the intermediate or caudal portion of its nucleus (Cottle, 1964). From tractus solitarius, there are collaterals to the commisural nucleus (Foley and Du Bois, 1934), and some vagal fibres reach the contralateral nucleus solitarius (Kerr, 1962).

Further information about the central vagal connections has been obtained by looking for the evoked responses when stimulating the nerve (Harrison and Bruesch, 1945; Anderson and Berry, 1956; Porter, 1963). A negative potential change can be recorded in the region of the nucleus of tractus solitarius, and the nucleus ambiguus. Out of these mass responses, which probably represent the summed potentials from many cells or fibres, individual units can be identified (Porter, 1963). Units were recorded in the region of nucleus ambiguus (motor nucleus of the vagus) with short constant latencies and were consistent with antidromic excitation of the motor fibres in the recurrent laryngeal nerve. In a small region which included the dorsal motor nucleus of the vagus and the nucleus and tractus solitarius, most units had variable latencies and could not follow high rates of stimulation, indicating that they were being excited synaptically by stimulation of afferent fibres. Many of these units discharged physiologically with a respiratory periodicity. Some units in the posterior part of nucleus tractus solitarius could be activated by stimulation of either vagus. No evoked responses were recorded when stimulating the abdominal vagus. Over the orbital surface of the frontal lobes in the cat, evoked potentials with a latency of 15–25 msec can be recorded on vagal stimulation (Bailey and Bremer, 1938; Siegfried, 1962).

## Effects of Transection or Ablation

The lesion produced by these techniques is crude in neuroanatomical terms, and unavoidable tissue damage and haemorrhage may extend beyond the immediate site of the lesion. This leads to difficulties in interpreting the data derived from such experiments. The effects of these procedures on breathing are presented in a simplified form in Fig. 91.

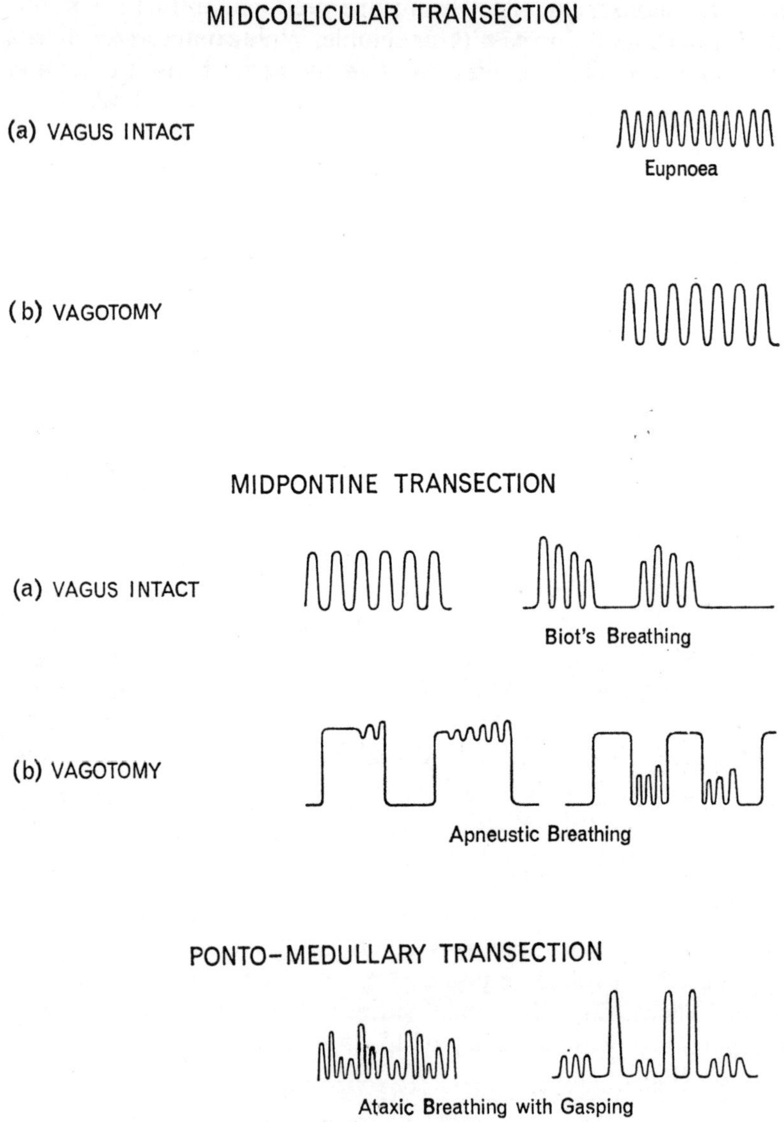

MIDCOLLICULAR TRANSECTION

(a) VAGUS INTACT

Eupnoea

(b) VAGOTOMY

MIDPONTINE TRANSECTION

(a) VAGUS INTACT

Biot's Breathing

(b) VAGOTOMY

Apneustic Breathing

PONTO-MEDULLARY TRANSECTION

Ataxic Breathing with Gasping

FIG. 91.—DIAGRAMMATIC REPRESENTATION OF THE EFFECTS ON BREATHING OF BRAIN STEM TRANSECTION AND VAGOTOMY

Vagotomy does not alter breathing in the ponto-medullary transected animal. Biot's breathing may also be seen in this preparation. Some variation in these responses occurs between species.

1. Transection at the extreme rostral pons or above.
   (*a*) With vagi intact, breathing is eupnoeic.
   (*b*) With vagotomy, breathing is eupnoeic, but deeper and slower than in (*a*).

2. Mid-pontine transection.

(*a*) With vagi intact, breathing is rhythmic but slower than in 1(*a*) (Wang *et al.*, 1957). Important differences become apparent in the phrenic neurogram when lung inflation is artificially controlled (Kahn and Wang, 1966). The smooth and steep increase in phrenic activity of the animal with intact pons, which ends abruptly either spontaneously or with lung inflation, is replaced by a pattern in which activity increases less rapidly, reaching a plateau at a lower intensity and tending to be maintained at this level until terminated by lung inflation. The inflation pressures necessary to achieve this inhibition are greater than in the intact animal (Fadiga *et al.*, 1965). If artificial ventilation is carried out with only small inflation pressures, a continuous phrenic discharge occurs which is modulated by the inflation; if the respirator is stopped in the deflation position, activity continues in the phrenic nerve for two minutes or more, before ending spontaneously (Kahn and Wang, 1966).

(*b*) With vagotomy, apneustic breathing occurs. This phenomenon can be produced in the vagotomised animal by transection of the pons at any level below 2 mm from its upper border (Wang *et al.*, 1957). The critical region in the upper pons lies in the dorso-lateral reticular formation of the isthmus of the pons, bilateral electrolytic lesions here giving rise to apneusis (Ngai and Wang, 1957). Apneusis is abolished when section is made between pons and medulla. The pattern is that of inspiratory spasm or cramp ("apneusis") which if interrupted often enough by expiration is termed apneustic breathing. The state of apneusis commonly lasts one or two minutes and is followed by a single expiration or several phasic breaths before a further apneustic period, which usually becomes progressively shorter (Breckenridge and Hoff, 1950). Small phasic breaths are commonly superimposed on the inspiratory spasm.

3. Transection at the ponto-medullary junction.

Sometimes breathing remains eupnoeic. More commonly, the breathing pattern is ataxic, the tidal volume varying in amplitude and rate, with the frequent occurrence of gasps which occasionally may be the sole type of respiratory movement (Wang *et al.*, 1957). The gasp starts from a resting expiratory position and, after a spasmodic inspiratory movement, terminates abruptly and is followed by an expiratory pause. Biot's breathing may also be seen. Analysis of the phrenic neurogram shows that the normal pattern of the intact animal is replaced by a short, high amplitude discharge (Kahn and Wang, 1967).

Although vagal primary afferent fibres synapse within the medulla (see p. 249), vagotomy is without effect on the pontomedullary transected animal. Moreover, when the vagus nerves are intact, the phrenic nerve discharge is not consistently affected by lung inflation (Kahn and Wang, 1967). These authors conclude that pontine structures play an essential role in the integration of the Hering-Breuer reflex.

## Effects of Electrical Stimulation

Many investigators have electrically stimulated the brain stem and observed the altered pattern of respiratory movements, notably Pitts *et al.* (1939*a*), Comroe (1943), Liljestrand (1952), Baxter and Olszewski (1955), Ngai and Wang (1957). The main limitations of this technique are that excitation of neurones cannot be distinguished from excitation of neural pathways (Wang and Ngai, 1964), alterations in the characteristics of the stimulus may give rise to different responses from the same site (Liljestrand, 1952), and prolonged stimulation may reverse the initial responses (Baxter and Olszewski, 1955).

The more readily classifiable responses evoked by electrical stimulation include inspiratory spasm (apneusis), expiratory spasm (sometimes called expiratory apneusis), apnoea and increased breathing frequency, but many variations of these responses occur. A strong inspiratory response is elicited from stimulation of the ventro-medial reticular formation in the medulla just rostral to the obex and the apneusis can be maintained until the death of the animal; a strong expiratory response follows stimulation in a slightly more rostral region (Pitts *et al.*, 1939*a*). Less marked inspiratory responses can be elicited by stimulation caudal to the obex close to the midline, while strong expiratory effects are produced by stimulation in the midline at the level of the pyramidal decussation (Pitts *et al.*, 1939*a*; see also, Sears, 1966). Inspiratory responses can also be evoked by stimulation in the dorso-lateral region of the middle and upper pontine tegmentum, particularly around the locus coeruleus (Baxter and Olszewski, 1955; Ngai and Wang, 1957), and pronounced postural adjustments may be associated with the more intense responses (Baxter and Olszewski, 1955). Increased frequency of breathing and reduction in tidal volume are also common responses with stimulation in this region of the upper pons (Ngai and Wang, 1957). Apnoea is produced from stimulation at scattered points throughout the pontine tegmentum (Ngai and Wang, 1957). Breathing can become locked to the rhythm of electrical stimulation with bursts of modified square wave pulses in the anterior hypothalamus, dorsal midbrain and lateral medulla (Hyde and Tan, 1966), and the rate of gasping of the medullary preparation can be increased by stimulation of a localised area on the floor of the fourth ventricle (Brodie and Borison, 1957).

It may also be mentioned here that stimulation of the anterior cerebellum depresses normal respiration, having a more marked effect after reflex respiratory excitation via the carotid sinus (Moruzzi, 1940; see also Decima and von Euler, 1969).

It is difficult to draw any firm conclusions about the *overall* organisation of the mechanism controlling breathing from the gross changes in the pattern of breathing caused by electrical stimulation or by brain stem transection, although it is natural that these observations should have formed a basis for theories about the control of breathing at a time when other techniques were not available. But electrical stimulation in particular can provide important information about certain aspects of the control mechanism, especially when the effects are analysed at neuronal level. The concept of centres serving specific respiratory functions, which arose from the earlier experiments of transection and electrical stimulation, will be critically reviewed in the next section and the implications of more recent experimental findings will be considered.

## Functional Organisation

### Evidence for Centres of Specific Function

From the breathing patterns which develop following transection, particular areas of the brain stem were thought by Lumsden (1923) to contain "centres" subserving specific functions. Thus the lower pons was believed to contain an "apneustic centre" which "when uncontrolled sends out impulses causing the prolonged inspirations or apneuses", the medulla to contain a "gasping centre", and the upper pons a "pneumotaxic centre" which "inhibits the activity of the apneustic centre and produces normal breathing". Stella (1938) studied the surgical conditions under which apneusis occurred and concluded that there was a mechanism in the upper pons (Lumsden's "pneumotaxic centre") which, in addition to vagal afferents, rhythmically inhibited inspiration.

In a series of papers describing experiments in which electrical stimulation was used in addition to the established surgical techniques of ablation and transection, Pitts and his co-workers (Pitts, Magoun and Ranson, 1939*a*, *b* and *c*; see also Pitts, 1946) formulated the origin of respiratory rhythm in the following manner. The results of the stimulation experiments substantiated the earlier view of Marckwald (1888) that there were two groups of neurones in the reticular formation of the medulla, the inspiratory and expiratory centres, and indicated that there were inhibitory connections between them. The neurones of the inspiratory centre were thought to be tonically active and to have a lower threshold for activity than those of the expiratory centre. This tonic activity could be interrupted periodically by two inhibitory mechanisms, the vagal afferents relating to lung inflation, and the pneumotaxic centre in the upper pons which was activated by the inspiratory centre itself. These mechanisms exerted their inhibitory effects on inspiration by excitation of the expiratory centre and thereby caused reciprocal inhibition of the inspiratory centre.

Subsequently, while still accepting that it was periodic inhibition of apneusis by the pneumotaxic centre and pulmonary stretch afferents which gave rise to normal rhythmicity, Wang *et al.* (1957) related the tonic inspiratory discharge to the activities of an apneustic centre in the middle and lower pons because apneusis is abolished by precise transection between the pons and medulla. The observation that apneusis could be evoked by electrical stimulation in the mid-pontine reticular formation was also regarded as evidence of a pontine apneustic centre (Ngai and Wang, 1957). The pontine centres were thought to dominate the rhythmic mechanism in the medulla (Wang *et al.*, 1957; Kahn and Wang, 1967).

Kahn and Wang (1967) looked further for evidence of an apneustic centre by recording respiratory neurone activity in the lower pons in mid-pontine transected animals (transection 3 mm from the rostral border of the pons). They were able to record from several inspiratory pontine units discharging in phase with the phrenic nerve which, in contrast to medullary neurones, were progressively inhibited during repeated weak activation of the Hering-Breuer inflation reflex. After vagotomy (which evoked apneusis), two of the four pontine neurones studied showed a pattern of continuous discharge in phase with phrenic activity similar to that of medullary neurones. The authors concluded that these neurones satisfied the criteria of site and activity for pontine units whose integrity would be required for producing and maintaining apneusis.

A basic difficulty with the hypothesis outlined above relates to the concept that there are centres in the pons and medulla serving specific respiratory functions, a view which stems from the work of Lumsden (1923) on the effects of brain stem transection. The word "centre" used in this way assigns function to a particular region in the intact animal specifically in terms of the disturbance produced by the destruction or partial isolation of that region, and implies the existence there of a spatially discrete mechanism to subserve this function. While other experimental methods might show such centres to exist, the transection experiments of themselves do not. A careful definition of "centre" overcomes some of the limitations of the term (see Wang and Ngai, 1964), but the concept of centres serving specific functions nevertheless takes its origin in this context from phenomena which are primarily experimental artefacts and which may relate in rather a complex manner to the operation of the control mechanism in the intact animal.

The other experimental evidence presented in favour of brain stem centres of specific function has not yet established with certainty that such centres exist. The difficulty in reconciling the inspiratory and expiratory centres as proposed by Pitts *et al.* (1939*a*) with the distribution and complex characteristics of medullary respiratory neurones has already been emphasised. With regard to pontine centres, the observation that electrical stimulation of the dorso-lateral reticular formation of the mid-pons gives rise to a sustained inspiratory activation of the respiratory muscles may indicate no more than that there is a powerful projection from this site to the respiratory motoneurones in the spinal cord. Stimulation in the region of the putative pneumotaxic centre may evoke not only an increase in frequency and decreased amplitude of respiratory movements (Ngai and Wang, 1957), but also an inspiratory response (Baxter and Olszewski, 1955), so that it is not easy to draw conclusions from this about the function of this region. Microelectrode recording in the pons reveals units which are markedly affected by weak activation of the Hering-Breuer reflex as mentioned above (Kahn and Wang, 1967), but, as the authors point out, these units do not by themselves provide evidence for an apneustic centre; the presupposition of a specific facilitatory drive to the respiratory muscles operating in the intact animal has to be made first. The characteristics of respiratory neurones in the upper pons, where phase-spanning neurones are relatively common, are consistent with an involvement in the control of rhythmic respiration, but do not yet provide definite evidence for the specific function assigned to the pneumotaxic centre.

Thus it may be questioned whether the "centre" terminology is a useful way at the present time in which to consider the organisation of the brain stem structures controlling respiratory muscle activity. The term, used as it has been in the context of the control of breathing, is perhaps only properly acceptable when the specific function assigned to a centre is based on a more precise knowledge of its input-output characteristics than is at present available.

**Alternative Interpretation**

Recent work makes it possible to reinterpret some of the observations made in the earlier experiments of brain stem transection and electrical stimulation. Let us first, however, summarise the evidence that *medullary structures* are primarily responsible for the rhythmic component of breathing. Rhythmic breathing can continue in the ponto-medullary transected animal although the smooth pattern of eupnoea is usually lost. Most of the neurones in the brain stem which discharge with respiratory phasing lie in the lateral reticular formation of the medulla around nucleus ambiguus or near tractus solitarius, and there is evidence that a property of at least some of these neurones is that of rhythmic discharge when subjected to an excitatory stimulus. It may also be significant that, in the autopsy studies by Baker *et al.* (1950) of patients with acute poliomyelitis who died from central respiratory failure, focal areas of necrosis were seen around nucleus ambiguus.

It may be asked what the significance is of the powerful and sustained respiratory muscle activation evoked by electrical stimulation of certain regions of the medulla, which led Pitts *et al.* (1939*a*) to believe that these were the sites of "inspiratory" and "expiratory" centres. It now appears that the reciprocal actions upon inspiratory and expiratory motoneurones which characterise these responses (i.e. inspiratory muscle activation is associated with expiratory muscle inhibition, and *vice versa*) may be effected at spinal cord level rather than through inspiratory and expiratory neurones in the medulla, and that the pathway may not be the same as that upon which rhythmic breathing depends. The evidence for this comes from an analysis of the intercostal alpha motoneurone and fusimotor neurone

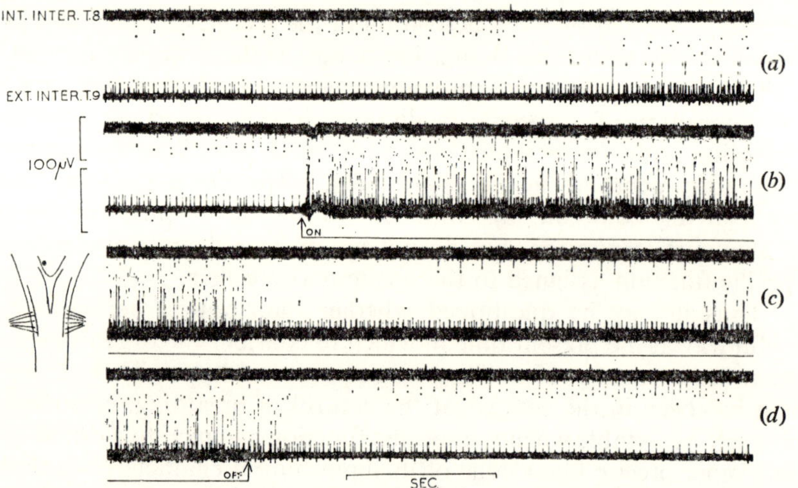


FIG. 92.—INSPIRATORY APNEUSTIC RESPONSE EVOKED BY REPETITIVE STIMULATION IN THE MEDIAL RETICULAR FORMATION OF THE MEDULLA

Site of stimulation: 2 mm rostral and 1·5 mm lateral to the obex. Upper traces, expiratory nerve filament; only fusimotor spikes present. Lower traces, inspiratory nerve filament; alpha spikes (large) and fusimotor spike (small) both present. (*a*), control. (*b*), (*c*) and (*d*), continuous recording during apneustic responses. (From Sears, 1966.)

discharge during the electrically induced apneusis (Andersen and Sears, 1965; Sears, 1966). Stimulation in the medial reticular formation just rostral to the obex causes after a short latency (3–6 msec) an abrupt increase in both fusimotor neurone and alpha motoneurone discharges in the inspiratory intercostal nerves, the response of the fusimotor neurones usually having a shorter latency and lower threshold; simultaneous recording from an expiratory intercostal nerve shows an equally abrupt inhibition of alpha motoneurone and fusimotor activity (Fig. 92). When electrical stimulation is maintained, an expiratory breakthrough may occur with the re-appearance of activity in the expiratory intercostal nerve, while the higher level of fusimotor discharge continues in the inspiratory intercostal nerve. The latency and intensity of these responses to a brief period of electrical stimulation are strongly influenced by the respiratory rhythm, the latency diminishing and the magnitude increasing in the inspiratory intercostal nerve during inspiration and in the expiratory intercostal nerve during expiration. At the site of stimulation, there are nuclei whose axons, as the reticulo-spinal tracts, may descend through the whole length of the spinal cord (Torvik and Brodal, 1957; Nyberg-Hansen, 1965). Thus it was concluded that, in this electrically induced respiratory muscle activation, impulses descend fast conducting reticulo-spinal tracts originating from medullary neurones in the medial reticular formation, to activate interneurones in the spinal cord which, in the case of the pathway serving the inspiratory response, distribute excitation to inspiratory alpha motoneurones and fusimotor neurones and inhibition to expiratory motoneurones. Von Euler (1966) has also shown that alpha-fusimotor linkage is probably effected at spinal level in the case of the rhythmic fusimotor fibres, whose discharge is closely linked with that of alpha motoneurones. Stimulation in the medial reticular formation and in the descending respiratory pathways in the cord did not give rise to a "pause" in the discharge of an inspiratory intercostal muscle spindle which might have occurred if the pathway to alpha motoneurones had been selectively excited.

It was also found that after ablation of one side of the medulla, from a level just below the obex down to $C_2$, sustained inspiratory activity could still be evoked bilaterally with stimulation just rostral to the obex on the same side as the ablation (Sears, 1966). The ablation had abolished all spontaneous rhythmic activity but the sustained inspiratory response was unchanged, suggesting that the integrity of the structures upon which rhythmic breathing depends is not essential to the response.

Andersen and Sears (1970) believe that these pathways activated by electrical stimulation in the medial reticular formation of the medulla are not those subserving the rhythmic component of breathing, and that activity in these two separate pathways may summate at spinal cord level. They suggest that, in addition to a possible postural function, the reticulo-spinal tracts may be specifically involved in the control of breathing by exerting a facilitatory or inhibitory bias upon the spinal respiratory motoneurone, thus determining the operating point about which the central respiratory drive potentials oscillate, through a control of the mean membrane potential (see Chapter XI).

The pathways may also be concerned with mediating the influence of the cerebral cortex upon the respiratory muscles, for stimulation of the orbital gyrus of the frontal lobe evokes short latency electrical potentials in the medial reticular formation of the medulla (Newman and Wolstencroft, 1959) in those areas which, when electrically stimulated, give rise to apneusis. There is also neuro-anatomical evidence that some corticoreticular fibres may originate from the orbital cortex (Rossi and Brodal, 1956). Finally, it is interesting to note the similarity of the electrically maintained inspiratory spasm and subsequent breakthrough of rhythmic breathing in the experimental animal to the pattern of respiratory muscle activity during breath holding in man. In a breath hold, after an initial period of silence rhythmic activity develops in the diaphragm, increasing in amplitude and frequency (Agostoni, 1963). Rhythmic activity also develops in the inspiratory intercostals and, in some subjects, reciprocally in the expiratory intercostal and abdominal muscles (Newsom Davis and Sears, to be published.) If a maximal inspiratory effort is then maintained against a closed airway, the continuous "willed" activity of the inspiratory intercostal muscles is subject to rhythmic modulation, and reciprocating bursts of rhythmic activity continue in expiratory muscles.

The importance of *pontine structures* in respiratory muscle control is not in doubt, but their organisation and their influence on the medullary mechanism is far from clear. The pattern of breathing in the medullary preparation is very different from that of the animal with intact pons, although the rhythmic characteristic remains. The studies of Kahn and Wang (1966, 1967) have shown that the integrity of the pons is necessary for the normal pattern of phrenic nerve discharge. Even in mid-pontine animals with vagi intact, in which apparently eupnoeic breathing continues, the phrenic neurogram indicates that there is a significant disturbance of the respiratory control mechanism. Phase-spanning respiratory neurones are relatively more common in the pons than in the medulla, and may modify the basic rhythmicity of the medulla, perhaps being concerned with the transition between the two phases of breathing (see page 242).

The lower pons is also concerned with the Hering-Breuer inflation reflex, for the response is absent in ponto-medullary transected animals (Kahn and Wang, 1967). It will also be recalled that these authors were able to record from pontine neurones which, in contrast to medullary neurones, responded to small degrees of lung inflation with a marked inhibition and a dissociation from phrenic nerve activity, indicating the effectiveness of vagal afferent inhibition at pontine level.

Two groups of observations indicate the powerful projections which exist from the pons to the respiratory muscles: firstly, the apneusis which develops in mid-pontine transected animals after vagotomy, which is abolished by ponto-medullary section; and secondly, the pronounced inspiratory muscle activation which follows electrical stimulation of certain areas of the mid-pons. The intensity of these responses in these two experimental situations is an indication of the importance of the pathways.

There is still some uncertainty, however, about whether a specific facilitatory drive originating in the pons is a feature of the intact animal, as it is of the mid-pontine preparation. Hoff and Breckenridge (1949; and Breckenridge and Hoff, 1950) noted that apneusis in the experimental animal was neither permanent nor complete and that rhythmic breathing could continue with ponto-medullary transection. In their view, apneusis represented a type of decerebrate rigidity of the inspiratory muscles, as had been proposed earlier by Henderson and Sweet (1930). The evidence against this view was that apneusis was not dependent on proprioceptive reflexes from the respiratory muscles and from the labyrinthine nerves since de-afferentation did not abolish the phenomenon (Stella, 1938), and that apneusis and decerebrate rigidity were not necessarily coexistent (Wang and Ngai, 1964). Apneusis in the transected animal is probably best regarded at the present time as a release phenomenon, following damage to structures in the upper pons, in a motor mechanism concerned both with postural and respiratory functions.

The projection to the spinal cord from the dorso-lateral region of the mid-pons, where electrical stimulation evokes inspiratory spasm, is presumably indirect since pontine reticulo-spinal fibres do not take their origin from this region of the lateral pons (Torvik and Brodal, 1957). Kahn and Wang (1967) believe that apneusis is effected through appropriate driving of medullary respiratory neurones by the pontine mechanism, the characteristics of some pontine units suggesting that they may act as "starter" cells for medullary neurones.

## SUPRAPONTINE STRUCTURES

### Mesencephalon and Diencephalon

Electrical stimulation of the central grey matter and the lateral tegmentum of the mesencephalon elicits an increased amplitude and rate of breathing. At other points in the tegmentum, respiratory movements may be inhibited (Kabat, 1936) or expiratory movements evoked (Baxter and Olszewski, 1955). From observations in transected animals it has been suggested that the posterior diencephalon and upper mid-brain contain structures, possibly the reticular activating system, whose tonic activity tends to increase the respiratory frequency (Fink *et al.*, 1962; Cohen, 1964).

The role of the *reticular activating system* has been further studied in the cat by electrical stimulation of the mesencephalic and diencephalic reticular formation (Hugelin and Cohen, 1963; Cohen and Hugelin, 1965). Analysis of phrenic nerve discharge showed the response to be a complex reaction rather than alteration of a single characteristic of breathing. Thus there was a decrease in the duration of the expiratory pause, an increase in the slope of the integrated phrenic neurogram and an increase in the end-inspiratory level of phrenic discharge. In each individual animal, these responses were related in a relatively stereotyped pattern, although differing between animals, and this stereotype could be produced from extensive stimulation in these regions of the reticular formation (Cohen and Hugelin, 1965). The respiratory responses are interpreted as part of a general "arousal" or "activation" reaction, serving to increase ventilation in preparation for action by the organism (Cohen and Hugelin, 1965).

The *hypothalamus* also has an influence on breathing. Bilateral micro-injections of thiopental into the hypothalamus of cats wakening from barbiturate anaesthesia intensifies the Hering-Breuer inflation reflex, and diminishes the inspiratory response to electrical stimulation of the medullary medial reticular formation, the inspiratory apneusis being interrupted by expirations. The inspiratory activity which follows tracheal closure at end-expiration was reduced (Redgate, 1963). With bilateral destructive lesions in the hypothalamus, the inspiratory response to medullary reticular stimulation is almost abolished. These effects indicate that the hypothalamus normally has a facilitatory influence upon inspiratory neurones (Redgate, 1963).

## Cerebral Cortex

Stimulation of certain regions of the cerebral cortex has long been known to give rise to changes in respiratory movements. In man, as in the cat, respiratory movements are inhibited by stimulation of the anterior cingulate region, the ventro-medial aspect of the temporal lobe, the posterior part of the orbital cortex and the anterior insula (Chapman *et al.*, 1949; Kaada and Jasper, 1952; Liberson *et al.*, 1951; Whitty, 1955; Kaada, 1960). In the monkey, stimulation of the amygdala has been shown to inhibit breathing (Reis and McHugh, 1968). Acceleration of respiration has been demonstrated in man with simultaneous bilateral cingulate stimulation (Pool and Ransohoff, 1949). In the cat, stimulation of the motor cortex, the anterior ectosylvian and sylvian gyri, and the rostral part of the pyriform cortex as well as the middle and anterior part of the cingulate gyrus gives rise to increased frequency of breathing (see Kaada, 1960). The response of respiratory neurones to cortical stimulation has not yet been studied.

A limited amount of information is available about the neural pathways involved in these effects. Using Marchi methods in the monkey after bilateral cingulate gyrus ablations, Ward (1948) found degenerative changes extending through the internal capsule and the ventral and medial portion of the cerebral peduncles, terminating in the medial reticular formation of the brain stem from just below the superior olive to just above the pyramidal decussation. Nyberg-Hansen (1969) has recently demonstrated corticospinal fibres arising from the anterior cingulate region which descend in the lateral funiculi of the cord to sacral levels. In the cat, short latency (about 1 msec) potentials can be recorded ipsilaterally from the medial reticular formation in the caudal pons and medulla following electrical stimulation of the orbital cortex (Newman and Wolstencroft, 1959).

## EFFERENT PATHWAYS IN SPINAL CORD

Cord transection experiments in animals have demonstrated that the pathways subserving rhythmic breathing lie in the anterior part of the lateral column and in the anterior column of white matter (Porter, 1895; Allen, 1927; Tosatti, 1938; Pitts, 1940). Evidence from post-mortem studies of patients who have undergone high cervical cordotomies also shows that, at the level of C1–C3, most of the fibres concerned with rhythmic respiration lie in a relatively discrete area in the most anterior part of the lateral column (Fig. 93; Nathan, 1963).

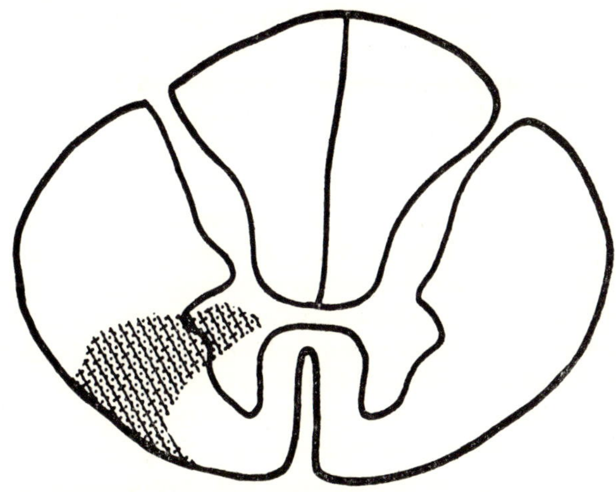

FIG. 93.—DESCENDING RESPIRATORY PATHWAYS

Area in common in two cases with ipsilateral respiratory muscle paralysis following high cervical cordotomy. This region was implicated in the cordotomy lesion in all cases developing ipsilateral respiratory muscle paralysis. (From Nathan, 1963.)

Electrical potentials in phase with respiration have been recorded from the upper cervical cord in the cat in the reticular formation just lateral to the lateral horn (Hukuhara *et al.*, 1954) and, in the dog, anterior and antero-lateral to the anterior horn (Belmusto *et al.*, 1965). Detailed histology of the recording sites is not available, however, and the possibility that some of the unit activity was afferent has not been completely excluded.

The reticulo-spinal pathways mediating the sustained respiratory muscle responses to electrical stimulation in the medial reticular formation of the medulla (see p. 255) probably lie in a similar region of the cord, that is in the anterior portion of the lateral column and the lateral portion of the ventral column (Torvik and Brodal, 1957; Sears, 1966).

Neural pathways subserving respiratory movements may not be confined only to the anterior quadrants of the cervical cord. There is some evidence in the cat for a cortico-spinal pathway which can influence respiratory muscle activity because stimulation in the pyramidal decussation may induce a powerful and sustained expiratory response (Sears, 1966). In man, it has been observed that, following a bilateral high antero-lateral cervical cordotomy, ventilation may be impaired, particularly during sleep, although vital capacity and maximum breathing capacity may be relatively unaffected (Belmusto *et al.*, 1963; see also Plum, 1966). This suggests that the pathways subserving the rhythmic drive to the respiratory muscles may descend the cord separately from those mediating the "willed" movements of these muscles. Furthermore, high cervical cordotomy may almost completely abolish both rhythmic and voluntary activation of the respiratory muscles while the cough reflex remains relatively intact (Newsom Davis and Plum, unpublished observations).

The pathway for abdominal visceral reflexes also seems to be distinct from that serving rhythmic breathing. The powerful coincident activation of diaphragm and abdominal muscles evoked by pelvic nerve stimulation (a reflex response which appears to be closely related to that of urination) is mediated by a pathway lying superficially in the anterior part of the lateral funiculus, which can be totally interrupted without disturbing spontaneous breathing (Yamamoto *et al.*, 1961). The properties of hiccup also indicate that its descending pathway is likely to be independent of that serving rhythmic breathing (Newsom Davis, 1970).

The nature of the *crossed phrenic phenomenon* first became clear from the experimental observations of Porter (1895). Hemisection of the upper cervical cord produces apparent paralysis of the diaphragm on that side, but when the contralateral phrenic nerve is cut, activity returns to the diaphragm on the side of the hemisection. From experimental studies, Porter put forward the hypothesis that some of the longer dendrites of the phrenic motoneurones crossed the midline to synapse with the terminal arborisations of the descending respiratory fibres of the opposite side. This explanation was subsequently doubted but, as Grundfest (1963) has pointed out, the essential correctness of the earlier belief that there were crossed connections at the level of the phrenic nuclei was established by Rosenbaum and Renshaw (1949), who used electromyographic methods to record diaphragmatic activity. These authors showed that, following a hemisection at C2, some activity could still be recorded in the diaphragm of that side, and was augmented by stimuli which increased respiratory drive (e.g. rebreathing or negative pressure breathing) and abolished by hyperventilation. This indicates that there are crossed connections to the phrenic nerve which are quantitatively less powerful than the uncrossed connections (Rosenbaum and Renshaw, 1949). From studies in which recording was made of spike potentials in isolated phrenic nerve filaments, Dolivo (1953) showed that the augmentation of spike activity occurred in the inspiration immediately following electrotonic block of the contra-lateral nerve before any change in the chemical drive could have become effective. From further experiments (Dolivo *et al.*, 1955) it was concluded that the phenomenon depends upon interruption of inhibitory afferents predominantly from the diaphragm and vagus, and to a lesser extent from the chest wall.

## SUMMARY

The rhythmic component of breathing appears to be generated within the medulla, for breathing continues in the animal transected between medulla and pons but not after transection below the medulla. The breathing pattern in the medullary animal, however, usually lacks the smooth and regular characteristics seen when the pons is intact. Most neurones in the brain stem firing with respiratory periodicity are found in the lateral reticular formation of the medulla, although units are also found around tractus solitarius and in the pons. Respiratory neurones differ in the phasing of their discharge and in their responses to chemical stimuli and to lung inflation or deflation, so that a number of sub-groups can be defined. The functional role of these different groups is not yet clear. Two groups of mutually-inhibitory inspiratory and expiratory neurones are probably fundamental to the genesis of rhythmic discharge, but other neuronal groups are almost certainly involved, perhaps as outlined on page 246.

The vagus nerves are not essential for rhythmic breathing either in animals or man, but afferents serving the Hering-Breuer inflation reflex (inspiratory-inhibiting, expiratory-activating) are a means by which the frequency of breathing can be controlled. In many animals, but not in healthy man, vagal section causes slower and deeper breathing, and both in animals and man it markedly diminishes the increase in respiratory frequency on breathing $CO_2$. The central pathway for the inflation reflex appears to involve pontine structures. Vagal afferents are also concerned with determining the balance between inspiratory and expiratory activity, as for example during positive-pressure breathing in animals.

Powerful and sustained inspiratory or expiratory muscle activation can be evoked by electrical stimulation in the medial reticular formation of the medulla (i.e. not in the region where respiratory neurones are encountered). The relationship of the reticulo-spinal pathways which serve this response with the mechanism concerned with the rhythmic component of breathing has not yet been clearly established, but the two systems seem to have functionally separate pathways with interaction probably occurring at segmental level. The former pathway may serve a postural function or exert a facilitatory or inhibitory bias upon spinal respiratory motoneurones. Both pathways lie in the anterior and antero-lateral funiculus. In man, the pathway serving rhythmic breathing lies in a relatively discrete area in the most anterior part of the lateral column. The coincident activation of rhythmic fusimotor neurones and alpha motoneurones through these pathways (see Chapter XI) is probably achieved at spinal level, either directly or through a system of interneurones, rather than through two different descending systems.

The function of pontine structures remains uncertain despite considerable data derived from the techniques of transection or ablation, electrical stimulation and microelectrode recording. Destruction of localised areas in the upper pons alters the pattern of phrenic discharge and, if the vagi are sectioned, leads to sustained inspiratory spasm (apneusis). The role of the mechanism which is being interfered with in this experimental situation is uncertain; some have argued from this and other evidence that it is concerned with rhythmic inhibition of a tonic facilitatory drive believed to originate in the lower pons, and the existence of pontine centres responsible for these specific functions has been proposed. This interpretation is questionable, however, and the observed effects are probably best regarded as a release phenomenon in a motor mechanism serving both postural and respiratory functions. The predominance of "phase-spanning" respiratory neurones in the pons suggests that one of the functions of this region may be to smooth the phase transition of breathing.

The reticular activating system and hypothalamus have a facilitatory influence on respiration. Electrical stimulation of certain areas of cerebral cortex causes inhibition or acceleration of respiratory movements. The pathways concerned with voluntary activation of the respiratory muscles, and those mediating abdominal visceral reflexes such as urination, appear to project to respiratory motoneurones through pathways independent of those serving rhythmic breathing.

*In conclusion*, the increasing evidence of a significant role for spinal mechanisms in the control of respiratory muscle activity (see Chapter XI) justifies a change from the usual emphasis on the integrative functions of specific centres in the brain stem. Motoneurones to respiratory muscles are subject to drives serving respiratory, postural, voluntary, visceral reflex and other functions, and they receive in addition a segmental input which is in part supraspinally determined through the fusimotor system. Transmission through the spinal interneuronal mechanism mediating the segmental proprioceptive input is subject to supraspinal control, and the same may also be true for the central drives. There is evidence now suggesting that the neural mechanisms concerned with some of these drives project to respiratory motoneurones through separate descending pathways. This implies that a principal site of their interaction is at spinal level, and is a further indication of the importance of the spinal mechanism in the control of respiratory muscle activity.

## REFERENCES

AGOSTONI, E. (1963). Diaphragm activity during breath holding: factors related to its onset. *J. appl. Physiol.*, **18**, 30–36.

ALLEN, W. F. (1927). Experimental anatomical studies on the visceral bulbospinal pathway in the cat and guinea-pig. *J. comp. Neurol.*, **42**, 393–456.

ANDERSEN, P., and SEARS, T. A. (1965). Efferent discharges evoked in intercostal nerves by electrical stimulation in the medulla of the anaesthetised cat. *J. Physiol.* (*Lond.*), **178**, 57P–58P.

ANDERSEN, P., and SEARS, T. A. (1970). Medullary activation of intercostal fusimotor and alpha motoneurones. *J. Physiol.* (*Lond.*) (In press).

ANDERSON, F. D., and BERRY, C. M. (1956). An oscillographic study of the central pathways of the vagus nerve in the cat. *J. comp. Neurol.*, **106**, 163–181.

BAILEY, P., and BREMER, F. (1938). A sensory cortical representation of the vagus nerve. *J. Neurophysiol.*, **1**, 405–412.

BAKER, A. B., MATZKE, H. A., and BROWN, J. R. (1950). Poliomyelitis. III. Bulbar poliomyelitis; a study of medullary function. *Arch. Neurol. Psychiat.* (*Chic.*), **63**, 257–281.

BATSEL, H. L. (1964). Localisation of bulbar respiratory centre by microelectrode sounding. *Exp. Neurol.*, **9**, 410–426.

BATSEL, H. L. (1965). Some functional properties of bulbar respiratory units. *Exp. Neurol.*, **11**, 341–366.

BATSEL, H. L. (1967). Activity of bulbar respiratory neurones during passive hyperventilation. *Exp. Neurol.*, **19**, 357–374.

BAUMGARTEN, R. VON, and KANZOW, E. (1958). The interaction of two types of inspiratory neurones in the region of the tractus solitarius of the cat. *Arch. ital. Biol.*, **96**, 361–373.

BAXTER, D. W., and OLSZEWSKI, J. (1955). Respiratory responses evoked by electrical stimulation of pons and mesencephalon. *J. Neurophysiol.*, **18**, 276–287.

BELMUSTO, L., BROWN, E., and OWENS, G. (1963). Clinical observations on respiratory and vasomotor disturbance as related to cervical cordotomies. *J. Neurosurg.*, **20**, 225–232.

BELMUSTO, L., WOLDRING, S., and OWENS, G. (1965). Localisation and patterns of potentials of the respiratory pathway in the cervical spinal cord in the dog. *J. Neurosurg.*, **22**, 277–283.

BISHOP, B. (1964). Reflex control of abdominal muscles during postive-pressure breathing. *J. appl. Physiol.*, **19**, 224–232.

BRECKENRIDGE, C. G., and HOFF, H. E. (1950). Pontine and medullary regulation of respiration in the cat. *Amer. J. Physiol.*, **160**, 385–394.

BREUER, J. (1868). Die Selbststeuerung der Atmung durch den Nervus vagus. *S.-B. Akad. Wiss. Wien, math.-nat. Kl.*, **58**, 909–937.

BRODIE, D. A., and BORISON, H. L. (1957). Evidence for a medullary inspiratory pacemaker. *Amer. J. Physiol.*, **188**, 347–354.

BURNS, B. D. (1963). The central control of respiratory movements. *Brit. med. Bull.*, **19**, 7–9.

BURNS, B. D., and SALMOIRAGHI, G. C. (1960). Repetitive firing of respiratory neurones during their burst activity. *J. Neurophysiol.*, **23**, 27–46.

CAMERON, I. R. (1969). Acid-base changes in cerebrospinal fluid. *Brit. J. Anaesth.*, **41**, 213–221.

CHAPMAN, W. P., LIVINGSTON, R. B., and LIVINGSTON, K. E. (1949). Frontal lobotomy and electrical stimulation of orbital surface of frontal lobes. *Arch. Neurol. Psychiat.* (*Chic.*), **62**, 701–716.

COHEN, M. I. (1958). Intrinsic periodicity of the pontile pneumotaxic mechanism. *Amer. J. Physiol.*, **195**, 23–27.

COHEN, M. I. (1959). Effects of carbon dioxide on discharge patterns of respiratory neurones. *Fed. Proc.*, **18**, 28.

COHEN, M. I. (1964). Respiratory periodicity in the paralysed, vagotomised cat: hypocapnic polypnoea. *Amer. J. Physiol.*, **206**, 845–854.

COHEN, M. I. (1968). Discharge patterns of brain stem respiratory neurones in relation to carbon dioxide tension. *J. Neurophysiol.*, **31**, 142–165.

COHEN, M. I. (1969). Discharge patterns of brain stem respiratory neurones during Hering-Breuer reflex evoked by lung inflation. *J. Neurophysiol.*, **32**, 356–374.

COHEN, M. I. (1970). In: *Ciba Foundation Symposium on Breathing: Hering-Breuer Centenary Symposium*, pp. 125–150. London: Churchill.

COHEN, M. I., and HUGELIN, A. (1965). Suprapontine reticular control of intrinsic respiratory mechanisms. *Arch. ital. Biol.*, **103**, 317–334.

COHEN, M. I., and WANG, S. C. (1959). Respiratory neuronal activity in pons of cat. *J. Neurophysiol.*, **22**, 33–50.

COMROE, J. H. (1943). The effects of direct chemical and electrical stimulation of the respiratory centre in the cat. *Amer. J. Physiol.*, **139**, 490–498.

COTTLE, M. K. (1964). Degeneration studies of primary afferents of IXth and Xth cranial nerves in the cat. *J. comp. Neurol.*, **122**, 329–345.

COZINE, R. A., and NGAI, S. H. (1967). Medullary surface chemoreceptors and regulation of respiration in the cat. *J. appl. Physiol.*, **22**, 117–121.

DECIMA, E. E., and EULER, C. VON (1969). Intercostal and cerebellar influences on efferent phrenic actiivty in the decerebrate cat. *Acta physiol. scand.*, **76**, 148–158.

DOLIVO, M. (1953). "Crossed phrenic phenomenon" et phénomène phrénique bilateral. *Helv. physiol. pharmacol. Acta*, **11**, 251–269.

DOLIVO, M., MEGIRIAN, D., and FLEISCH, A. (1955). L'importance relative des voies nerveuses afférentes dans le "crossed phrenic phenomenon". *Helv. physiol. pharmacol. Acta*, **13**, 300–305.

EULER, C. VON (1966). Proprioceptive control in respiration. Nobel Symposium I: *Muscular Afferents and Motor Control*, pp. 197–207. Stockholm: Almqvist and Wiksell; New York: Wiley.

EULER, C. VON, and FRITTS, H. W. (1963). Quantitative aspects of respiratory reflexes from the lungs and chest walls of cats. *Acta physiol. scand.*, **57**, 284–300.

EULER, C. VON, and SÖDERBERG, U. (1952*a*). Medullary chemosensitive receptors. *J. Physiol.* (*Lond.*), **118**, 545–554.

EULER, C. VON, and SÖDERBERG, U. (1952*b*). Slow potentials in the respiratory centres. *J. Physiol.* (*Lond.*), **118**, 555–564.

FADIGA, E., GESSI, T., and MANZONI, T. (1965). Electrophysiological investigations on the central relays for the vagal reflex inhibiting inspiration. *Arch. Sci. biol.* (*Bologna*), **49**, 291–308.

FINK, B. R., KATZ, R., REINHOLD, H., and SCHOOLMAN, A. (1962). Suprapontine mechanisms in regulation of respiration. *Amer. J. Physiol.*, **202**, 217–220.

FOLEY, J. O., and DUBOIS, F. S. (1934). An experimental study of the rootlets of the vagus nerve in the cat. *J. comp. Neurol.*, **60**, 137–159.

GESELL, R., BRICKER, J., and MAGEE, C. (1936). Structural and functional organisation of the central mechanism controlling breathing. *Amer. J. Physiol.*, **117**, 423–452.

GRUNDFEST, H. (1963). Elementary properties of neurones. *Ann. N.Y. Acad. Sci.*, **109**, 418–435.

GUZ, A. NOBLE, M. I. M., TRENCHARD, D., COCHRANE, H. L., and MAKEY, A. R. (1964). Studies on the vagus nerves in man: their role in respiratory and circulatory control. *Clin. Sci.*, **27**, 293–304.

GUZ, A., NOBLE, M. I. M., WIDDICOMBE, J. G., TRENCHARD, D., and MUSHIN, W. W. (1966). The effect of bilateral block of vagus and glossopharyngeal nerves on the ventilatory response to $CO_2$ of conscious man. *Resp. Physiol.*, **1**, 206–210.

GUZ, A., NOBLE, M. I. M., WIDDICOMBE, J. G., TRENCHARD, D., MUSHIN, W. W., and MAKEY, A. R. (1966). The role of vagal and glossopharyngeal afferent nerves in respiratory sensation, control of breathing and arterial pressure regulation in conscious man. *Clin. Sci.*, **30**, 161–170.

HABER, E., KOHN, K. W., NGAI, S. H., HOLADAY, D. A., and WANG, S. C. (1957). Localisation of spontaneous respiratory neuronal activities in the medulla oblongata of the cat: a new location of the expiratory centre. *Amer. J. Physiol.*, **190**, 350–355.

HARRISON, F., and BRUESCH, S. R. (1945). Intramedullary potentials following stimulation of the cervical vagus. *Anat. Rec.*, **91**, 280.

HENDERSON, V. E., and SWEET, T. A. (1929). On the respiratory centre. *Amer. J. Physiol.*, **91**, 94–102.

HERING, E., and BREUER, J. (1868). Die Selbststeuerung der Atmung durch den Nervus vagus. *S.-B. Akad. Wiss. Wien, math.-nat. Kl.*, **57**, (ii), 672–677.

HOFF, H. E., and BRECKENRIDGE, C. G. (1949). The medullary origin of respiratory periodicity in the dog. *Amer. J. Physiol.*, **158**, 157–172.

HUGELIN, A., and COHEN, M. I. (1963). The reticular activating system and respiratory regulation in the cat. *Ann. N.Y. Acad. Sci.*, **109**, 586–603.

HUKUHARA, T., NAKAYAMA, S., and OKADA, H. (1954). Action potentials in the normal respiratory centres and its centrifugal pathways in the medulla oblongata and spinal cord. *Jap. J. Physiol.*, **4**, 145–153.

HYDE, J. E., and TAN, E. S. (1966). Characteristics of brainstem-evoked stimulus-bound respiration rate in anaesthetised cats. *Exp. Neurol.*, **14**, 396–407.

KAADA, B. R. (1960). Cingulate, posterior orbital, anterior insular and temporal pole cortex. *Handbook of Physiology*, Sect. 1. Neurophysiology, Vol. 2, pp. 1345–1372. Ed. H. W. Magoun. Washington, D.C.: American Physiological Soc.

KAADA, B. R., and JASPER, H. (1952). Respiratory responses to stimulation of temporal pole, insula and hippocampal and limbic gyri in man. *Arch. Neurol. Psychiat.* (*Chic.*), **68**, 609–619.

KABAT, H. (1936). Electrical stimulation of points in the forebrain and midbrain: the resultant alterations in respiration. *J. comp. Neurol.*, **64**, 187–208.

KAHN, N., and WANG, S. C. (1965). Descending respiratory pathways in the medulla oblongata of the cat. *Amer. J. Physiol.*, **209**, 599–603.

KAHN, N. and WANG, S. C. (1966). Pontine pneumotaxic centre and central respiratory rhythm. *Amer. J. Physiol.*, **211**, 520–524.

KAHN, N., and WANG, S. C. (1967). Electrophysiologic basis for pontine apneustic centre and its role in integration of the Hering-Breuer reflex. *J. Neurophysiol.*, **30**, 301–318.

KERR, F. W. L. (1962). Facial, vagal and glossopharyngeal nerves in the cat. *Arch. Neurol.* (*Chic.*), **6**, 264–281.

KIM, J. K., and CARPENTER, F. G. (1961). Excitation of medullary neurones by chemical agents. *Amer. J. Physiol.*, **201**, 1187–1191.

KOIZUMI, K., USHIYAMA, J., and BROOKS, C. MC. C. (1961). Muscle afferents and activity of respiratory neurones. *Amer. J. Physiol.*, **200**, 679–684.

LIBERSON, W. T., SCOVILLE, W. B., and DUNSMORE, R. H. (1951). Stimulation studies of the prefrontal lobe and uncus in man. *Electroenceph. clin. Neurophysiol.*, **3**, 1–8.

LILJESTRAND, A. (1952). Respiratory reactions elicited from medulla oblongata of the cat. *Acta physiol. scand. Suppl.*, **106**, 321–393.

LUMSDEN, T. (1923). Observations on the respiratory centres in the cat. *J. Physiol.* (*Lond.*), **57**, 153–160.

MARCKWALD, M. (1888). *The Movements of Respiration and their Innervation in the Rabbit.* Translated by Haig, T. A. London: Blackie.

MITCHELL, R. A., LOESCHCKE, H. H., MASSION, W. H., and SEVERINGHAUS, J. W. (1963). Respiratory responses mediated through superficial chemo-sensitive areas on the medulla. *J. appl. Physiol.*, **18**, 523–533.

MORUZZI, G. (1940). Paleocerebellar inhibition of vasomotor and respiratory carotid sinus reflexes. *J. Neurophysiol.*, **3**, 20–32.

NATHAN, P. W. (1963). The descending respiratory pathway in man. *J. Neurol. Neurosurg. Psychiat.*, **26**, 487–499.

NELSON, J. R. (1959). Single unit activity in medullary respiratory centres of cat. *J. Neurophysiol.*, **22**, 590–598.

NESLAND, R., and PLUM, F. (1965). Subtypes of medullary respiratory neurones. *Exp. Neurol.*, **12**, 337–348.

NESLAND, R. S., PLUM, F., NELSON, J. R., and SIEDLER, H. D. (1966). The graded response to stimulation of medullary respiratory neurones. *Exp. Neurol.*, **14**, 57–76.

NEWMAN, P. P., and WOLSTENCROFT, J. H. (1959). Medullary responses to stimulation of orbital cortex. *J. Neurophysiol.*, **22**, 516–523.

NEWSOM DAVIS, J. (1970). In: *Ciba Foundation Symposium on Breathing: Hering-Breuer Centenary Symposium*, pp. 201–203. London: Churchill.

NGAI, S. H., and WANG, S. C. (1957). Organisation of central respiratory mechanisms in the brain stem of the cat: localisation by stimulation and destruction. *Amer. J. Physiol.*, **190**, 343–349.

NYBERG-HANSEN, R. (1965). Sites and mode of termination of reticulo-spinal fibres in the cat. An experimental study with silver impregnation methods. *J. comp. Neurol.*, **124**, 71–93.

NYBERG-HANSEN, R. (1969). Corticospinal fibres from the medial aspect of the cerebral hemisphere in the cat. An experimental study with the Nauta method. *Exper. Brain Res.*, **7**, 120–132.

PAPPENHEIMER, J. R., FENCL, V., HEISEY, S. R., and HELD, D. (1965). Role of cerebellar fluids in control of respiration as studied in unanaesthetised goats. *Amer. J. Physiol.*, **208**, 436–450.

PITTS, R. F. (1940). The respiratory centre and its descending pathways. *J. comp. Neurol.*, **72**, 605–625.

PITTS, R. F. (1946). Organisation of the respiratory centre. *Physiol. Rev.*, **26**, 609–630.

PITTS, R. F., MAGOUN, H. W., and RANSON, S. W. (1939*a*). Localisation of the medullary respiratory centres in the cat. *Amer. J. Physiol.*, **126**, 673–688.

PITTS, R. F., MAGOUN, H. W., and RANSON, S. W. (1939*b*). Interrelations of the respiratory centres in the cat. *Amer. J. Physiol.*, **126**, 689–707.

PITTS, R. F., MAGOUN, H. W., and RANSON, S. W. (1939*c*). The origin of respiratory rhythmicity. *Amer. J. Physiol.*, **127**, 654–670.

PLUM, F. (1966). Breathlessness in neurological disease: the effects of neurological disease on the act of breathing. *Boeringer-Ingelheim Symposium on Breathlessness*, pp. 203–222. Eds. E. J. M. Campbell and J. B. L. Howell. Oxford: Blackwell.

POOL, J. L., and RANSOHOFF, J. (1949). Autonomic effects on stimulating rostral portion of cingulate gyri in man. *J. Neurophysiol.*, **12**, 385–392.

PORTER, R. (1963). Unit responses evoked in the medulla oblongata by vagus nerve stimulation. *J. Physiol.* (*Lond.*), **168**, 717–735.

PORTER, W. T. (1895). The path of the respiratory impulse from the bulb to the phrenic nuclei. *J. Physiol.* (*Lond.*), **17**, 455–485.

PRICE, W. M., and BATSEL, H. L. (1969). Behaviour of bulbar expiratory neurones during production of active respiration. *Fed. Proc.*, **28**, 337.

REDGATE, E. S. (1963). Hypothalamic influence on respiration. *Ann. N.Y. Acad. Sci.*, **109**, 606–618.

REIS, D. J., and MCHUGH, P. R. (1968). Hypoxia as a cause of bradycardia during amygdala stimulation in monkey. *Amer. J. Physiol.*, **214**, 601–610.

RICHARDSON, P. S., and WIDDICOMBE, J. G. (1969). The role of the vagus nerves in the ventilatory responses to hypercapnia and hypoxia in anaesthetised and unanaesthetised rabbits. *Resp. Physiol.*, **7**, 122–135.

ROBSON, J. G., HOUSELEY, M. A., and SOLIS-QUIROGA, O. H. (1963). The mechanism of respiratory arrest with sodium pentobarbital and sodium thiopental. *Ann. N.Y. Acad. Sci.*, **109**, 494–502.

ROSENBAUM, H., and RENSHAW, B. (1949). Descending respiratory pathways in the cervical spinal cord. *Amer. J. Physiol.*, **157**, 468–476.

ROSSI, G. F., and BRODAL, A. (1956). Corticofugal fibres to the brain-stem reticular formation. An experimental study in the cat. *J. Anat.* (*Lond.*), **90**, 42–62.

SALMOIRAGHI, G. C., and BAUMGARTEN, R. VON (1961). Intracellular potentials from respiratory neurones in brain stem of cat and mechanism of rhythmic respiration. *J. Neurophysiol.*, **24**, 203–218.

SALMOIRAGHI, G. C., and BURNS, B. D. (1960*a*). Localisation and patterns of discharge of respiratory neurones in brain stem of cat. *J. Neurophysiol.*, **23**, 2–13.

SALMOIRAGHI, G. C., and BURNS, B. D. (1960*b*). Notes on mechanism of rhythmic respiration. *J. Neurophysiol.*, **23**, 14–26.

SCHEIBEL, M. E., and SCHEIBEL, A. B. (1965). Periodic sensory nonresponsiveness in reticular neurones. *Arch. ital. Biol.*, **103**, 300–316.

SCOTT, F. H. (1908). On the relative parts played by nervous and chemical factors in the regulation of respiration. *J. Physiol.* (*Lond.*), **37**, 301–326.

SEARS, T. A. (1964). The slow potentials of thoracic respiratory motoneurones and their relation to breathing. *J. Physiol.* (*Lond.*), **175**, 404–424.

SEARS, T. A. (1966). Pathways of supraspinal origin regulating the activity of respiratory motoneurones. Nobel Symposium I: *Muscular Afferents and Motor Control*, pp. 187–196. Stockholm: Almqvist and Wiksell; New York: Wiley.

SIEGFRIED, J. (1962). Les projections corticales du nerf vague chez le bouc. *Electroenceph. clin. Neurophysiol.*, **14**, 535–539.

STELLA, G. (1938). On the mechanism of production, and the physiological significance of "apneusis". *J. Physiol.* (*Lond.*), **93**, 10–23.

TORVIK, A., and BRODAL, A. (1957). The origin of reticulospinal fibres in the cat. *Anat. Rec.*, **128**, 113–137.

TOSATTI, E. (1939). L'incrociamento del respiro diaframmatico dopo frenicotomia ed emisezione del midollo cervicale. *Arch. Fisiol.*, **38**, 533–564.

WANG, S. C., and NGAI, S. H. (1964). General organisation of central respiratory mechanisms. *Handbook of Physiology*, Sect. 3, Respiration, Vol. 1, pp. 487–505. Eds. W. O. Fenn and H. Rahn. Washington, D.C.: American Physiological Soc.

WANG, S. C., NGAI, S. H., and FRUMIN, M. J. (1957). Organization of central respiratory mechanisms in the brain stem of the cat: genesis of normal respiratory rhythmicity. *Amer. J. Physiol.*, **190**, 333–342.

WARD, A. A. (1948). The cingular gyrus: area 24. *J. Neurophysiol.*, **11**, 13–23.

WHITTY, C. W. M. (1955). Effects of anterior cingulectomy in man. *Proc. roy. Soc. Med.*, **48**, 463–469.

WIDDICOMBE, J. G. (1961). Respiratory reflexes in man and other mammalian species. *Clin. Sci.*, **21**, 163–170.

WIDDICOMBE, J. G. (1964). Respiratory reflexes. *Handbook of Physiology*, Sect. 3, Respiration. Vol. 1, pp. 585–630. Eds. W. O. Fenn and H. Rahn. Washington, D.C.: American Physiological Soc.

WOLDRING, S. (1965). Interrelation between lung volume, arterial $CO_2$ tension, and respiratory activity. *J. appl. Physiol.*, **20**, 647–652.

WOLDRING, S., and DIRKEN, M. N. J. (1951). Site and extension of bulbar respiratory centre. *J. Neurophysiol.*, **14**, 227–241.

YAMAMOTO, S., ARAKI, K., and KIKUCHI, K. M. (1961). Abdominal muscle reflexes of pelvic nerve origin in cats. *Exp. Neurol.*, **4**, 345–357.

# Chapter XIII

# VENTILATORY EFFECTS OF MECHANICAL LOADING

J. MILIC-EMILI* AND L. D. PENGELLY*

THE respiratory muscles are subjected to a wide range of natural loading. For example, switching from mouth to nose breathing approximately doubles the resistance to breathing. Similarly, postural changes in the state of contraction of the thoracic and abdominal muscles, and external constraints from clothing and furniture, may also result in a changed load to breathing. Yet, most studies on regulation of ventilation have centred on chemical aspects, and relatively little attention has been paid to mechanical loading. During the past ten years, however, there have been an increasing number of investigations into the effects of mechanical loading of respiration, and the factors which provide ventilatory stability in the face of changing loads are now beginning to be understood.

In this chapter we describe first the effects of external mechanical loads on respiration and the related load-compensatory mechanisms.† Next, the effects of changes in internal respiratory load are compared briefly with those obtained during external mechanical loading. Finally, the effects of positive and negative pressure breathing are described. Only general features will be discussed, with emphasis on human behaviour. References at the end of this chapter include reviews (Campbell and Howell, 1962 and 1963; Campbell, 1964; Cherniack, 1965), and monographs (Monzein, 1962; Howell and Campbell, 1966; Freedman, 1969), which provide more detailed information.

## EFFECTS OF EXTERNAL MECHANICAL LOADING ON RESPIRATION

When a mechanical load is added to respiration, there is usually an immediate reduction in ventilation. If the load is maintained, ventilation tends to return gradually towards the control values. In this connection, it should be noted that the respiratory system is remarkable in its capacity to maintain adequate ventilation in the face of changed mechanical loads. For example, under steady state conditions, mechanical loading equivalent to about twice the intrinsic load of the respiratory system results in a

* Department of Physiology, McGill University, Montreal.

† Mechanical loading of the respiratory muscles can be produced either by devices external to the body or by increased mechanical impedance within the respiratory system itself. For convenience, these two types of changes will be referred to, respectively, as external and internal loading of breathing.

reduction of ventilation of less than 10 per cent (Zechman *et al.*, 1957; Milic-Emili and Tyler, 1963). Similarly, Kellogg and his colleagues (1970) have shown that when breathing is mechanically assisted by a respirator which takes over the work of ventilating the lungs, steady state ventilation increases by less than 10 per cent. Although this stability of ventilation depends partly on changes in chemoreceptor drive, there is good reason to believe that important compensatory adjustments take place immediately following loading or unloading (i.e., before there are any appreciable changes in blood gas composition).

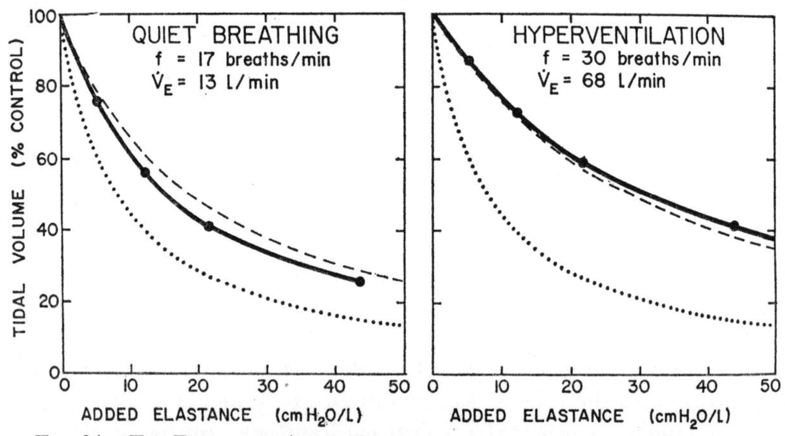


FIG. 94.—THE EFFECTS OF ADDED INSPIRATORY ELASTIC LOADS ON THE TIDAL VOLUME OF THE FIRST LOADED BREATH DURING QUIET BREATHING AND INCREASED VENTILATION

*Filled circles* show the average results on four seated subjects. *Dotted curves* show the changes that would be expected were there no increased end-inspiratory pressure following loading. These curves were predicted according to equation *1*, using the average value of total respiratory elastance of the four subjects, namely 8 cm $H_2O$/1. *Broken curves* show the relationships predicted according to equation *2*, using the values of "effective" respiratory resistance and elastance previously reported (Pengelly *et al.*, 1970), together with the respiratory frequencies indicated in each panel. Note that respiratory frequency did not change during the first loaded breath, i.e. remained the same as under control conditions. (After Pengelly, Bowmer, Greener, Luterman and Milic-Emili (unpublished observations).)

**Immediate response to mechanical loading.**—The immediate ventilatory responses to external mechanical loading have been studied in both man and a variety of animals. The loads used include elastic (Campbell, Dinnick and Howell, 1961; Freedman, 1969; Pope *et al.*, 1968); flow resistive (Davies *et al.*, 1919; Killick, 1935; Hall and Zechman, 1957; Wiley and Zechman, 1965 and 1968; Bennett *et al.*, 1962; Freedman and Weinstein, 1965; Freedman, 1969) and "threshold" loads (Campbell, Dickinson, Dinnick and Howell 1961; Freedman, 1969).* When any of these loads is

* Added elastic loads are usually provided by rigid containers (Campbell, Dinnick and Howell, 1961) and flow resistive loads by narrow tubes. "Threshold" loads are provided by a rigid tube dipping a pre-determined distance beneath the water level in a closed jar (Campbell, Dickinson, Dinnick and Howell, 1961). A "threshold" load requires that a given pressure be developed before air will flow.

added to inspiration, there is almost invariably an immediate reduction in ventilation. In anaesthetised animals, such as cats and dogs, this reduction is due both to decreased tidal volume and respiratory frequency. In man, both conscious and anaesthetised, there are also marked changes in tidal volume but the changes in respiratory frequency are usually small. In both man and animals there is generally little or no change in the end-expiratory lung volume following inspiratory loading.

The immediate reduction in tidal volume following inspiratory loading is invariably less than that expected where there is no increase in end-inspiratory pressure, indicating the existence of mechanisms which stabilise tidal volume in the face of changing loads. This load compensation is present from the first loaded breath, as shown in Fig. 94 which illustrates the effects of added elastic loads on the tidal volume during the first loaded breath of four normal seated subjects, both during quiet breathing and increased ventilation. The dotted curves show the changes that would be expected were there no increase in end-inspiratory pressure following the addition of the elastic loads (added elastances). These curves were computed according to Campbell, Dinnick and Howell (1961), namely:

$$V_T \text{ (per cent control)} = \frac{E}{E + \Delta E} \times 100 \qquad (1)$$

where $V_T$ is the tidal volume expressed as a percentage of its control (unloaded) value, $\Delta E$ is the added elastance, and $E$ is the elastance of the total respiratory system. In the four subjects whose data are shown in Fig. 94, $E$ averaged 8 cm $H_2O$/l. Compensation is relatively slight when the load is added during quiet breathing, but it is more pronounced during increased ventilation. Indeed, during increased ventilation the tidal volumes are roughly twice the values predicted assuming unchanged end-inspiratory pressure.

Various factors have been proposed to explain the tidal volume compensation to added mechanical loads which occurs within the first loaded breath. Broadly speaking, these factors may be divided into two main groups: (*a*) extrinsic mechanisms involving nervous mediation, and (*b*) mechanisms intrinsic to the respiratory pump (including the respiratory muscles). Historically, the former mechanisms were the first to be considered. Indeed, more than a century ago Hering and Breuer (1868) described vagally mediated reflexes originating from stretch receptors in the lungs. These reflexes play an important role in providing immediate load compensation in most animals. In man, however, they are probably of little or no consequence (Guz *et al.*, 1964; Guz, Noble, Widdicombe, Trenchard, Mushin and Makey, 1966). Load compensating reflexes originating from the chest wall proprioceptors have also been described. The nature and pathways of these reflexes, which are mediated by intercostal muscle spindles and by tendon organs, are described in detail in Chapter XI. Briefly, there is evidence that, in man, intercostal muscle proprioceptors elicit within the first loaded breath segmental reflex responses in the intercostal muscles (Sears and Newsom Davis, 1968). These reflex responses to a sudden change in load are of such brief duration that their immediate ventilatory effects are usually negligible. In anaesthetised animals, the functional importance of this rapid reflex is probably somewhat greater. However, according to Lynne-Davies *et al.* (1970), in anaesthetised cats, with added inspiratory elastic loads, compensation within the first loaded breath is provided chiefly by the intrinsic properties of the respiratory pump (including the respiratory muscles) and by the Hering-Breuer reflexes mediated by the vagi. In man, the relative importance of extrinsic (reflex) mechanisms in the adjustments to mechanical loading (first loaded breath) appears to be rather small (Delhez and Petit, 1966), but this is not the case for mechanisms intrinsic to the ventilatory pump. Since the latter mechanisms provide considerably more mechanical stability to ventilation than hitherto supposed, we will describe their mode of operation in some detail.

In a completely relaxed subject, artificially ventilated by a respirator generating a sinusoidal pressure, the effects of added inspiratory elastances ($\triangle E$) and resistances ($\triangle R$) on tidal volume can be predicted from the following expression:*

$$V_T \text{ (per cent control)} = \frac{\sqrt{R^2\, 4\pi^2 f^2 + E^2}}{\sqrt{(R + \triangle R)^2\, 4\pi^2 f^2 + (E + \triangle E)^2}} \times 100 \quad (2)$$

where $V_T$ is the tidal volume (including changes in thoracic volume due to gas compression and expansion) expressed as a percentage of its control (unloaded) value; $f$ is the respiratory frequency; and $E$ and $R$ are the elastance and flow-resistance of the total respiratory system, respectively. Such predictions are illustrated by the solid curves in Fig. 95, which were computed using values of total respiratory elastance and resistance in the normal range for seated men, namely 8 cm $H_2O$/l and 2 cm $H_2O$/l/sec, respectively. It can be seen that, with increasing respiratory frequency, the

* If a sinusoidal pressure is applied to the respiratory system, tidal volume can be approximated by the following expression (ref. Otis *et al.*, 1956, equation *7*):

$$V_T \simeq \frac{2\triangle P}{\sqrt{R^2 + \left(\frac{1}{2\pi f C}\right)^2}} \quad (3)$$

where $\triangle P$ is one-half the amplitude of the pressure, and C = 1/E. This can be rearranged to (ref. Mead, 1960, equation *5*):

$$V_T = \frac{2\triangle P}{\sqrt{R^2 4\pi^2 f^2 + E^2}} \quad (4)$$

If $\triangle E$ and/or $\triangle R$ are added, it follows:

$$V_T = \frac{2\triangle P}{\sqrt{(R + \triangle R)^2 4\pi^2 f^2 + (E + \triangle E)^2}} \quad (5)$$

If $\triangle P$ is constant, by dividing *5* by *4*, and multiplying the resulting equation by 100, equation *2* is obtained. (Note that at $f = 0$, equation *2* becomes equal to equation *1*.)

effect of $\triangle E$ on tidal volume decreases progressively, while the opposite is true for $\triangle R$. Furthermore, as shown by the shape of the iso-frequency curves, the reduction in tidal volume produced by a given added elastance or resistance decreases progressively as their initial value increases, i.e. the decrease in tidal volume is greater if a given elastance or resistance is added to the unloaded rather than to the pre-loaded respiratory system. The price of a high initial elastance and/or resistance is greater work of breathing, but from the standpoint of stability of tidal volume, a high impedance of the respiratory system is clearly advantageous (Mead, 1966). In this connection it should be noted that, during active breathing, the respiratory system behaves *as if* its impedance were about double that commonly thought. Indeed, using electrophrenic respiration, Pengelly and his co-workers (1970) have recently shown that during contraction of the diaphragm the "effective" elastance of the respiratory system in man amounts to about 13 cm $H_2O$/l and the "effective" resistance to about 6 cm $H_2O$/l/sec. (In the same subjects, the total elastance and flow-resistance of the passive respiratory system amounted to about 8 cm $H_2O$/l and 2 cm $H_2O$/l/sec respectively.) The difference between these values and those obtaining when the chest wall is relaxed is due to the intrinsic properties of the diaphragm, i.e. its mechanical arrangement and its force-length and force-velocity relationships (see Chapter VI), and to distortion of the respiratory system from its passive configuration (Mead, 1966; Konno and Mead, 1968; see also Chapter II). Contraction of the intercostal muscles probably has a similar effect.

This extra "impedance" during active breathing provides considerably increased stability of tidal volume in the face of changing mechanical loads. This is shown by the broken curves in Fig. 95, which illustrate the predicted relationship between tidal volume and added mechanical loads (elastic and flow-resistive) during active breathing. The predicted curves were calculated according to equation (2), using the values of "effective" elastance and resistance reported by Pengelly *et al.* (1970). Implicit in these calculations are the assumptions (*a*) that the potential pressure which can be applied by the respiratory muscles during the breathing cycle approximates a sinusoidal pattern and (*b*) that the number of contracting motor units and their frequency of discharge does not change when the load is added (i.e., that there is no reflex recruitment and/or summation of contractions). The good agreement between predicted and experimental data supports the validity of these assumptions. Indeed, as shown in Fig. 94, the broken curves (which were predicted according to equation (2) using the values of "effective" elastance and resistance) are in good agreement with the experimental data (solid curves) during both quiet breathing and increased ventilation. In these predictions, the fact was taken into account that control respiratory frequency increased from an average value of 17 breaths per minute during quiet breathing to 30 breaths per minute during

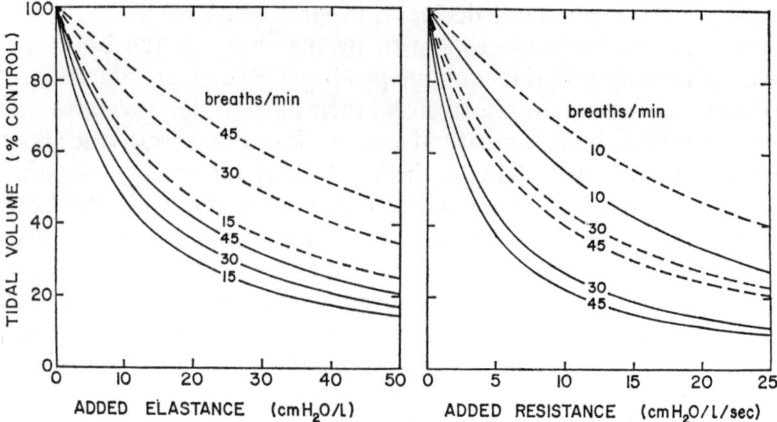


FIG. 95.— RELATIONSHIP BETWEEN TIDAL VOLUME, EXPRESSED AS PER CENT OF CONTROL, AND ADDED INSPIRATORY ELASTANCE AND RESISTANCE PREDICTED ACCORDING TO EQUATION *2* DURING ARTIFICIAL VENTILATION (SOLID CURVES) AND ACTIVE BREATHING (BROKEN CURVES)

Respiratory frequency is indicated on each curve. Note that the iso-frequency isopleths are independent of the magnitude of the tidal volume (and hence also of ventilation). For explanation see text.

increased ventilation. Thus, the increased load compensation during increased ventilation (Fig. 94) appears to be caused chiefly by the increased respiratory frequency. Evidence supporting this conclusion was obtained from a recent study in our laboratory. This showed that in subjects in whom control ventilation increased at the expense of the tidal volume alone, the mechanical stability to added inspiratory elastic loads increased relatively little with increasing ventilation.

It seems logical to conclude that in man the adjustments which take place within the first loaded breath depend largely on the intrinsic mechan-

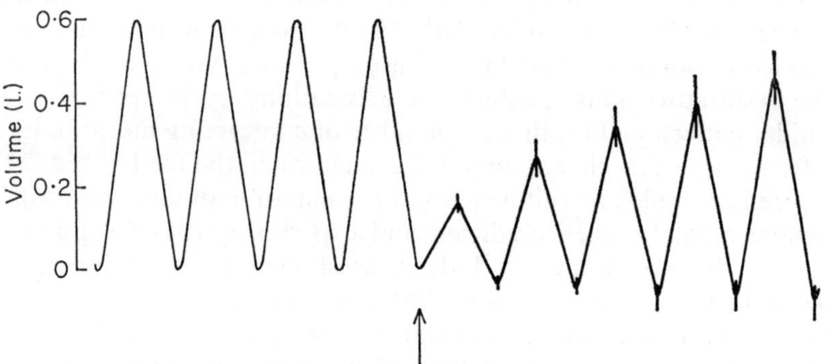


FIG. 96.—THE EFFECTS OF AN ADDED INSPIRATORY ELASTIC LOAD ON THE TIDAL VOLUME OF THE FIRST FIVE LOADED BREATHS OF A CONSCIOUS MAN

Note the progressive restoration of tidal volume from the first to the fifth loaded breath. (After Campbell, Dinnick and Howell, 1961.)

ical properties of the ventilatory pump and on the breathing pattern. As Fenn (1963) stated of an analogous situation: "this at least is the simplest and most obvious interpretation of the results and the burden of proof of the contrary is on those who would explain the results in neurological terms".

The probability of more complex load-compensating mechanisms is suggested by the behaviour of successive breaths after the addition of a

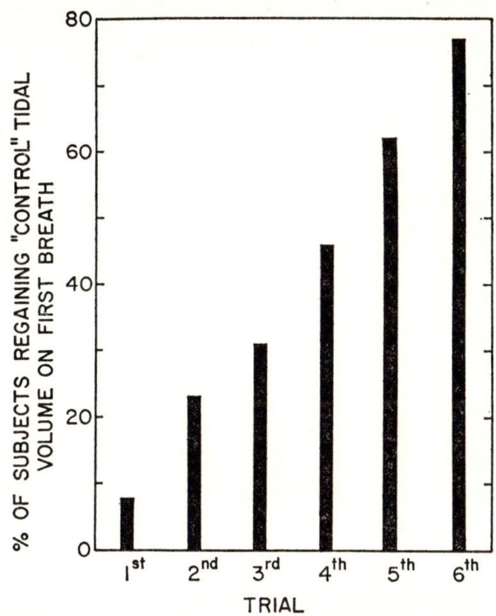


FIG. 97.—RESULTS FROM 13 SEATED SUBJECTS WITH INSPIRATORY "THRESHOLD" LOADING (LOAD OF −6 CM $H_2O$)

Subjects who exceeded control tidal volume on the first breath are included in the designation regaining "control" tidal volume. (From Freedman and Weinstein, 1965.)

load. Figure 96 indicates that after the addition of an elastic load in conscious man there is often a progressive increase in tidal volume for several breaths. This response has also been observed in anaesthetised animals (Campbell, Dickinson, Dinnick and Howell, 1961), and, although considerably reduced, in anaesthetised human subjects (Campbell, Dickinson, Dinnick and Howell, 1961; Campbell, Dinnick and Howell, 1961). The nature of this response is not clearly understood. Its time course is clearly too long to be attributable to segmental spinal reflexes. On the other hand, the response is too prompt to be caused by changes in blood gas composition, and hence in chemical drive. That the reflex in animals involves vagal reflex pathways is shown by its reduction following bilateral vagotomy. The central effects of respiratory muscle proprioreceptors may also be important in determining, at least in part, the additional load compensation in breaths subsequent to the first loaded breath (Campbell *et al.*, 1964; Bishop, 1967). This complex response is greatly accelerated in conscious subjects by repeated loading. Indeed, according to Freedman and Weinstein (1965), repeated presentation of loads may lead to a change in response in a manner "analogous to the learning of voluntary motor acts". They also reported that, under some conditions, these learned responses may provide a complete load-compensation within the first loaded breath. Fig. 97 summarises this point, indicating that at the sixth presentation (trial) of a "threshold" load, 10 out of a group of 13 subjects regained control tidal volume or greater on the first loaded breath, as compared to 1 in the first trial. On the other hand, according to Freedman and Weinstein (1965) the ability to regain control tidal volume on the first loaded breath was virtually nil with elastic loading; at the sixth trial none of the nine subjects studied had regained control tidal volume on the first loaded breath. However, with repeated presentations of an elastic load, most subjects increased respiratory frequency. These findings emphasise that the response to respiratory loading may be greatly modified by a learned process, which in turn can vary greatly with the physical characteristics of the load. Such learned responses probably depend upon consciousness, but, provided they are shown to be reproducible, they must be accepted as representing the behaviour of the organism and not dismissed as the voluntary idiosyncrasies of the experimental subjects.

Expiratory loading results in somewhat different responses. Small resistive loads which would produce a reduction in ventilation if added to inspiration may have no effect if added to expiration. Larger expiratory loads cause an immediate reduction in ventilation and an increase in end-expiratory level and, in some conscious subjects, may at times evoke active expiration. In general, however, the expiratory muscles do not participate in the human response to moderate expiratory loading. Mechanical loading of both phases of respiration results in immediate responses which are a combination of those obtained by loading inspiration and expiration separately (Campbell, Dickinson, Dinnick and Howell, 1961).

**Steady state response to mechanical loading.**—The steady state response to mechanical loading involves changes in chemoreceptor drive. Indeed, following the application of an external mechanical load there is usually an increase in arterial $CO_2$ tension and a reduction in arterial $O_2$ tension (Zechman *et al.*, 1957). Similarly, following unloading of respiration there is an increase in ventilation and a decrease in arterial $CO_2$ tension (Kellogg *et al.*, 1970). This indicates that ventilatory responses to chemical stimuli are affected by mechanical loading (or unloading). It is in fact well established that as a result of loading, the steady-state ventilatory response to exercise (Silverman *et al.*, 1943, 1945*a*, 1945*b* and 1951) and $CO_2$ inhalation (Cherniack and Snidal 1956; Eldridge and Davis, 1959) decreases. In a further study (Milic-Emili and Tyler, 1963), by combining added resistive loads with $CO_2$ inhalation, it has been found that the mechanical power developed by the inspiratory muscles is related linearly to changes in end-tidal $P_{CO_2}$, a 1 mm Hg rise in $P_{CO_2}$ causing an increase in inspiratory muscle work rate (power) averaging about 0·6 kpm/min in normal subjects.

That there is a linear relationship between end-tidal $P_{CO_2}$ and inspiratory power for added elastic loads has also been established (Howell, 1966). Kellogg and his co-workers (1970) have recently studied the effects of decreasing the mechanical load on the respiratory muscles by assisting respiration with a servomechanism which takes over the work of ventilating the lungs. Their results indicate that the generalisation that the inspiratory muscle power is a linear function of alveolar $CO_2$ tension can also be extended to a situation in which both elastic and resistive loads are decreased.

The observation that at any given value of alveolar $CO_2$ tension inspiratory work rate is substantially independent of added (or subtracted) loads has been attributed to the force-velocity relationship of the respiratory muscles (Milic-Emili and Tyler, 1963). Their force-length relationship and mechanical arrangement (Marshall, 1962; Pengelly *et al.*, 1970) might have a similar effect. In addition, nervous reflex mechanisms may also contribute to keeping the inspiratory muscle power a constant function of alveolar $P_{CO_2}$; however, direct evidence of this is lacking. It should be noted that there is no consistent relationship between power developed by the expiratory muscles and alveolar $P_{CO_2}$ (Milic-Emili and Tyler, 1963). As Campbell (1958) suggested in the first edition of this book, the activity of the expiratory muscles may not be directly regulated by the chemoreceptor-respiratory centre control system, and the latter may control only the activity of the inspiratory muscles (but see page 223). The mechanism which regulates the activity of expiratory muscles in man is not clear at present, although there is evidence that in cats a reflex mediated by vagal afferents from the lung is involved (Bishop, 1967). These afferents, however, appear to be distinct from those mediating the diaphragmatic inhibition of the Hering-Breuer inflation reflex (Bishop, 1967 and 1968)

**Tolerance of mechanical loads.**—There are few studies on this topic. Freedman (1969) investigated the ability of normal subjects to tolerate added inspiratory loads for two minute periods. He found that they could tolerate added elastic loads of 60–200 cm $H_2O$/l (mean 150 cm $H_2O$/l), representing about 6–20 times the normal thoracic elastance; flow-resistive loads of 140–380 cm $H_2O$/l/sec (mean 270 cm $H_2O$/l/sec), representing about 50–150 times the normal thoracic resistance; and "threshold" loads of 30–130 cm $H_2O$ (mean 80 cm $H_2O$). The maximum tolerable "threshold" load was closely correlated with maximum static inspiratory pressure, but the ability to tolerate added elastic and flow-resistive loads could not be explained in terms of limitation of respiratory muscle force or power, nor by changes in total ventilation or pattern of breathing; subjective factors appeared to be the main determinant. Similar results were obtained by Silverman *et al.* (1943) who found that inspiratory flow-resistive loads of about 6 cm $H_2O$/l/sec made exercise (at 830 kpm/min) intolerable, with very little objective change in measured respiratory variables. Progressive addition of flow-resistive loads in Freedman's experiments caused a significant fall in ventilation with loads in excess of 70 cm $H_2O$/l/sec, but changes in ventilation with elastic and "threshold" loads were very variable. More striking and reproducible were changes in pattern of breathing. With elastic loads, frequency increased and tidal volume fell, whereas with flow-resistive loads frequency fell and the duration of inspiration increased. These changes are mechanically appropriate, enabling ventilation to be maintained with a smaller inspiratory pressure than if frequency were unaltered (see Chapter V). Similar findings were reported by McIlroy *et al.* (1956) and Bland *et al.* (1967), but are not found in anaesthetised subjects (Freedman, 1969; Nunn and Ezi-Ashi, 1961). This suggests that changes in pattern of breathing with added loads are, at least in part, dependent upon consciousness.

As stated above, during external loading alveolar $CO_2$ tension usually increases, sometimes by as much as 10 mm Hg. This emphasises the fact that in many of the situations which occur in life, and in which there is a change in mechanical load, maintenance of $P_{CO_2}$ may be relegated to secondary importance.

## Effects of Internal Mechanical Loading on Respiration

It has long been recognised that in daily life the respiratory muscles are subjected to a wide range of internal loading. For example, during speech and singing there is a marked increase in expiratory resistance. Changes in position and posture, as well as muscular efforts such as lifting, walking, and straining all presumably vary the mechanical load on respiration. Although measurements in these various conditions are few and furthermore have been made with apparatus which necessarily distorts the natural pattern of breathing, it seems safe to say that ventilation is maintained within relatively narrow limits compared with the changes in mechanical loading (Mead, 1966). Presumably, the response to internal mechanical loading involves the same load compensatory mechanisms which operate during external mechanical loading of respiration. There is, however, some evidence to suggest that, in animals at least, the ventilatory response to internal loading (altered pulmonary compliance and resistance) may not be the same as during external loading (Mead, 1960). These observations are consistent with the lung being an important site of the receptors responsible for the load-compensatory reflexes occurring in experimental animals. External loads can influence intrapulmonary stresses (and hence the lung receptors) only to the extent that the pattern of volume change is altered. Mechanical changes in the lung itself, however, would have direct effects on intrapulmonary stresses (Mead, 1960).

Accordingly, the response to internal and external mechanical loading may well differ.

Finally, it should be emphasised once more that so far all measurements of ventilation have been made with external mechanical devices, which add a mechanical load to respiration and hence necessarily distort the pattern of breathing. Furthermore, these devices may also provide a psychological load. It can, therefore, be said (Mead, 1966) that the patterns of breathing during normal daily activities have still to be determined. Recently, however, Mead *et al.* (1967) have developed a device with which ventilation can be measured without recourse to a mouthpiece or face mask, other than for calibration, and with minimal encumbrance to the subject. Although this device has not been tested in different postures or conditions, continuous measurement of the natural pattern of breathing now appears feasible.

## EFFECT OF POSITIVE AND NEGATIVE PRESSURE BREATHING ON RESPIRATION

The effects of pressure breathing have been studied extensively both in man and animals. The results of these investigations indicate that the response to pressure breathing varies considerably from one animal species to another and, within a given species, from one individual to another. Furthermore, the response depends on the method by which the pressure is applied (Agostoni and Mead, 1964; see Chapter III). In general, when a pressure is applied to the respiratory system there is a change in lung volume at which breathing takes place. Because both the compliance and the flow-resistance of the respiratory system vary with lung volume, this implies that externally applied pressures cause a change in the internal load to breathing. In addition, as the effectiveness of the respiratory muscles as pressure generators also varies with lung volume (Rahn *et al.*, 1946; Pengelly *et al.*, 1970) pressure breathing must necessarily have a profound effect on respiration.

**Immediate response to pressure breathing.**—When pressure is applied to the respiratory system there is usually an immediate change in the end-expiratory lung volume and in ventilation. In most experimental animals, the latter is caused both by changed tidal volume and respiratory frequency (Widdicombe, 1965), while in man the changes in frequency are generally less pronounced (Rahn *et al.*, 1946; Widdicombe, 1965). The human and animal responses also differ in other important ways, e.g. in man during a brief period of positive pressure breathing the respiratory muscles are nearly or completely relaxed at end-expiration (Agostoni and Mead, 1964; see Chapter III), while in anaesthetised animals such as cats and dogs this is not the case (Bjurstedt, 1953; Bishop, 1963).

The most elegant and extensive studies on the immediate effects of pressure breathing in animals have been carried out by Beverly Bishop (1963, 1964, 1967, 1968). She demonstrated that in anaesthetised cats the application of continuous positive pressure (tracheal pressure elevated throughout the respiratory cycle) results in an immediate inhibition of the diaphragm and excitation of the abdominal muscles, which persist during pressure breathing. She also reported that during expiratory pressure loading (positive pressure opposing expiration alone) there is a progressive increase in diaphragmatic activity with each succeeding loaded breath, while the response of the abdominal muscles is virtually the same as during

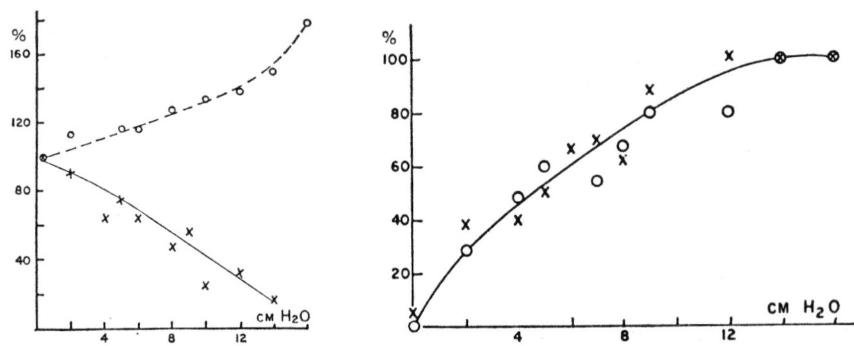


FIG. 98.—THE EFFECTS OF PRESSURE BREATHING ON DIAPHRAGMATIC AND ABDOMINAL ELECTROMYOGRAPHIC ACTIVITY IN THE CAT

*Left:* The mean inspiratory activity of the diaphragm for 10 Dial-anaesthetised cats as a function of the pressure opposing expiration (in cm $H_2O$). The ordinate compares the amplitude of the diaphragm integrated EMG with that during eupneic breathing. Diaphragm activity is augmented when expiration only is opposed by the increased pressure (open circles), whereas it is inhibited when the pressure remains elevated throughout the respiratory cycle (crosses).
*Right:* The mean expiratory activity of the abdominal muscle for 10 cats as a function of the pressure opposing expiration. All points (circles for expiratory loading and crosses for continuous pressure breathing) fall about a common curve, suggesting that abdominal muscle activity is adjusted primarily to the pressure opposing expiration. (From Bishop, 1967.)

continuous positive pressure breathing. Fig. 98 summarises the responses of the diaphragm (left) and abdominal muscles (right) during continuous pressure breathing (crosses) and expiratory pressure loading (open circles). The ordinates indicate the mean amplitude of the integrated EMGs' (for the sixth breath following the elevation of pressure), expressed as a percentage of the average discharge during normal breathing. Both during continuous and expiratory pressure breathing, tidal volume decreased and end-expiratory lung volume increased, the latter increasing somewhat more during continuous than during expiratory pressure breathing. Because bilateral vagotomy abolished abdominal activity in both conditions as well as the inhibition of the diaphragm caused by continuous pressure breathing, Bishop (1967) concluded that these reflexes are mediated by pulmonary stretch receptors. Bilateral vagotomy, however, did not always abolish the progressive diaphragmatic augmentation in response to expiratory pressure loading, confirming previous findings in rabbits by Campbell *et al.* (1964), who attributed this diaphragmatic response to reflexes originating from the intercostal muscles spindles. It is of interest to note that in the above experiments the diaphragmatic activity during the first breath following the introduction of expiratory pressure remained virtually the same as under control conditions. This supports the view that the load compensatory mechanisms mediated by the intercostal muscles spindles become effective only subsequent to the first loaded breath.

Whereas in anaesthetised animals nervous reflex mechanisms play a major role in the immediate adjustments to pressure breathing, this does not appear to be the case in conscious man. The immediate effects of positive and negative pressure applied at the mouth on end-expiratory lung volume and tidal volume in three seated men are illustrated in Fig. 99. The curves represent the average response during the first four breaths following the application of pressure. Also shown is the relaxation volume-pressure relationship of the total respiratory system which was measured by the method of "relaxation pressures" described by Rahn *et al.* (1946) (see Chapter III). It can be seen that with the application of positive or negative pressures, the end-expiratory lung volume changed in a manner predictable from the relaxation volume-pressure curve of the respiratory system, suggesting that the expiratory muscles did not participate substantially in the immediate response to pressure breathing. As shown in Fig. 99, with increasing positive pressure the tidal volume decreased progressively. Three factors almost certainly contribute to this phenomenon: (*a*) with increasing lung volume the inspiratory muscle fibres become shorter and, as is well known, the force developed by a muscle decreases as its length diminishes; (*b*) with increasing lung volume the mechanical advantage of the diaphragm (and possibly of other inspiratory muscles) in converting force into pressure diminishes (Marshall, 1962); (*c*) at high lung volumes the compliance of the respiratory system is reduced (Fig. 99), a fact which contributes to further reduction in tidal volume. During negative pressure breathing (Fig. 99), up to pressures of about $-10$ cm $H_2O$, there was a slight increase in tidal volume, probably because of the predominance of the factors *a* and *b*. At more negative pressures the tidal volume decreased, conceivably as a result of the marked reduction in total respiratory compliance and airway conductance which occurs at low lung volumes.

Although the immediate changes in tidal volume which occur in man during pressure breathing may be explained entirely by the mechanisms mentioned above, the contribution of reflex mechanisms cannot be excluded. Evidence has recently been presented indicating that inflation of of human lungs with large volumes may result in inspiratory inhibition (Hering-Breuer inflation reflex), which is abolished by bilateral vagal block (Guz *et al.*, 1964). In man the importance of vagal reflexes appears to be

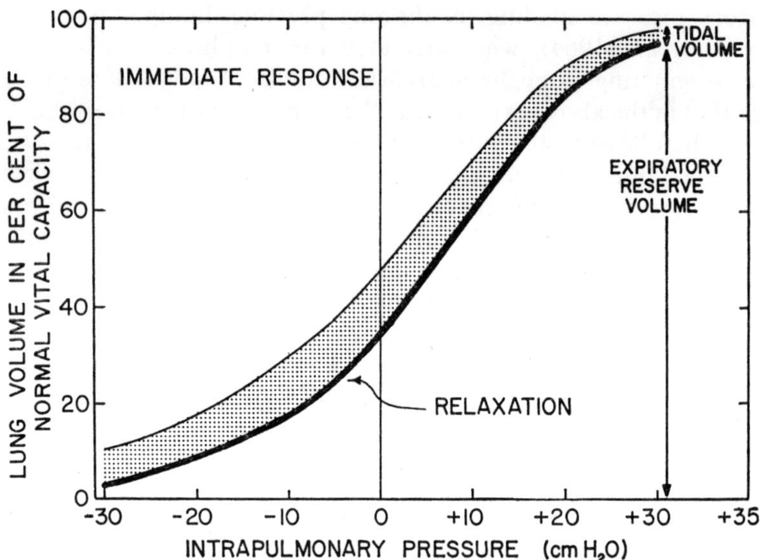


FIG. 99.—IMMEDIATE EFFECTS (FIRST FOUR BREATHS) OF PRESSURE BREATHING ON TIDAL VOLUMES AND INSPIRATORY AND EXPIRATORY RESERVE VOLUMES

Average results on three seated subjects. Relaxation curve is also shown (heavy solid curve). Note that immediately after the application of pressure the end-expiratory level follows the relaxation curve. (After Flenley, Pengelly and Milic-Emili, unpublished observations.)

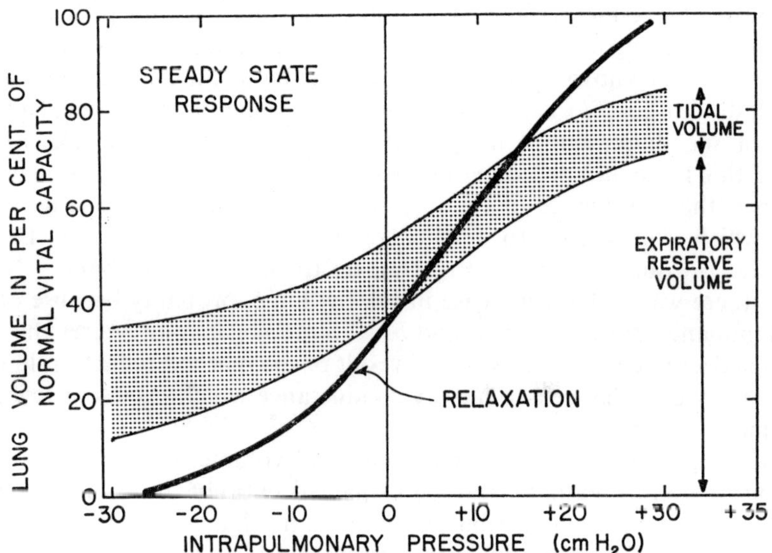


FIG. 100.—STEADY STATE RESPONSE OF PRESSURE BREATHING ON TIDAL VOLUMES AND INSPIRATORY AND EXPIRATORY RESERVE VOLUMES

Average results on ten seated subjects. Note that under steady state conditions the end-expiratory volume deviates from the relaxation curve. (After Rahn *et al.*, 1946.)

considerably less than in most animals. Indeed, in some animals lung inflation by positive pressure can cause such a prolonged apnoea that the animals die, presumably of asphyxia.

In conclusion, nervous reflex mechanisms in man appear to play a relatively minor role in the immediate adjustments to pressure breathing, which is probably almost entirely mediated by factors intrinsic to the ventilatory pump. By contrast, during prolonged pressure breathing the contribution of nervous reflexes becomes of considerable importance (see below).

**Steady state response to pressure breathing.**—If external pressures are applied to human subjects for longer periods, the changes in end-expiratory level no longer follow the relaxation curve, as demonstrated by Rahn *et al.* (1946) and more recently by Agostoni (1962) and Ernsting (1965). The results obtained by Rahn *et al.* on seated men are shown in Fig. 100. All measurements were taken under steady state conditions at least five minutes after the application of any given pressure. As the applied pressure increases, the tidal volume band tends to fall below the relaxation curve and as pressure becomes more negative the tidal volume is found to fall above the relaxation curve. This means that the expiratory muscles do not relax completely on inspiration when breathing against positive pressures, while the inspiratory muscles do not relax completely on expiration when breathing against negative pressures. Thus, during prolonged pressure breathing a reflex respiratory muscle activity comes into play which counteracts the applied pressures. The nature of the reflexes which in man stabilise lung volume during prolonged pressure breathing is not fully understood. Bjurstedt (1953) presented evidence that in anaesthetised dogs the vagi may mediate special reflex pathways, which increase the tone of the abdominal wall during positive pressure breathing. Similar results were obtained in experiments on anaesthetised cats by Bishop (1967) who also reported that continuous positive pressure breathing results in reflex inhibition of diaphragmatic activity. Bilateral vagotomy abolished the responses of both the diaphragm and abdominal muscles to continuous positive pressure breathing. Conditions relevant to conscious, seated man, however, can hardly be extrapolated from results obtained on anaesthetised, supine animals. Indeed, Agostoni (1962) reported that in seated, conscious man the activity of the diaphragm persists during positive pressure breathing. This provides an abdomino-thoracic pressure gradient, which should facilitate the venous return from the abdominal cavity. As the reduction of cardiac output appears to be the main limiting factor on positive pressure breathing (Ernsting, 1965), this gradient should increase the tolerance to higher pressures.

Besides its mechanical interest, the activity of the expiratory muscles during positive pressure breathing raises a problem concerning the regulation of breathing. As pointed out above (but see also page 223), the activity of the expiratory muscles may not be regulated by the chemoreceptor-respiratory centre control system. Accordingly, during positive pressure breathing the ventilation would be expected to become relatively independent of the chemical drive. In fact, prolonged positive pressure breathing results in hyperventilation in most human subjects as evidenced by a decrease in alveolar $CO_2$ tension (Rahn *et al.*, 1946; Ernsting, 1965).

Little is known of the mechanisms which cause the increased inspiratory muscle tone during negative pressure breathing. Whatever its nature, however, this phenomenon is useful in terms of gas exchange. In man at low lung volumes there is evidence of considerable airway closure, which causes impaired gas exchange within the lungs (Nunn *et al.*, 1965; Burger and Macklem, 1968 ;Sutherland *et al.*, 1968). Thus, the reflex inspiratory activity which comes into play during prolonged negative pressure breathing by maintaining a higher lung volume effectively decreases airway closure.

**Tolerance to externally applied pressures.**—The foremost physiological limitation in positive pressure breathing is the circulatory embarrassment caused by the elevation of intrathoracic pressure, which disturbs the normal gradient in venous pressure from the periphery to the heart. Thus, excessive positive pressures lead eventually to syncope. The majority of subjects can tolerate positive pressures of about 30 cm $H_2O$ for 30 minutes, while those accustomed to this manoeuvre can tolerate breathing up to about 60 cm $H_2O$ (Ernsting, 1965). Higher pressures lead rapidly to syncope, and may also cause lung rupture.

The foremost physiological limit to negative pressure breathing is probably pulmonary engorgement resulting from increased central blood volume and airway closure in the dependent parts of the lungs, with a consequent impairment of gas exchange. Most subjects can tolerate negative pressures of about $-30$ cm $H_2O$ for prolonged periods, while lower pressures can be endured safely only for shorter periods (Rahn *et al.*, 1946). A full account of the factors limiting the tolerance to positive and negative pressure breathing can be found elsewhere (Rahn *et al.*, 1946; Ernsting, 1965).

## Summary

The respiratory system has a remarkable capacity to maintain ventilation within relatively narrow limits despite considerable changes in mechanical loading. This mechanical stability of ventilation is provided by three main load-compensating mechanisms: (a) the chemoreceptor-respiratory centre control system, (b) nervous reflexes originating from pulmonary receptors (Hering-Breuer reflexes) and proprioceptors in the muscles of the chest wall, and (c) mechanisms intrinsic to the respiratory pump. The latter include the intrinsic properties of the respiratory muscles (force-length and force-velocity relationships), their mechanical arrangement, and the mechanical properties of the other various structures comprising the ventilatory apparatus. In conscious man, the load-compensatory adjustments which occur within the first breath following the addition of a load, particularly if elastic, are provided chiefly by the mechanisms intrinsic to the ventilatory pump, while in anaesthetised animals nervous reflexes originating from the thorax also contribute to the immediate load compensation. During prolonged application of external respiratory loads, load compensation involves all the three factors listed above both in man and in animals.

In man, the immediate response to pressure breathing depends chiefly on the mechanical properties of the ventilatory pump (including the respiratory muscles); in animals, reflexes originating from the thorax contribute substantially to the immediate response. During prolonged pressure breathing, the ventilatory response is dominated by reflexes in both man and animal.

It should be noted that the ventilatory response to mechanical loading may, at least in part, depend on consciousness, and may be modified by a learned process.

Although many of the factors which provide ventilatory stability to mechanical loading of respiration have been identified, little is known concerning their relative roles under the various conditions of daily life.

## REFERENCES

AGOSTONI, E. (1962). Diaphragm activity and thoraco-abdominal mechanics during positive-pressure breathing. *J. appl. Physiol.*, **17**, 215–220.

AGOSTONI, E., and MEAD, J. (1964). Statics of the respiratory system. *Handbook of Physiology*, Sect. 3, Vol. 1, pp. 387–409. Washington, D.C.: American Physiological Soc.

BENNETT, E. D., JAYSON, M. I. V., RUBENSTEIN, D., and CAMPBELL, E. J. M. (1962). The ability of man to detect added non-elastic loads to breathing. *Clin. Sci.*, **23**, 155–162.

BISHOP, B. (1963). Abdominal muscle and diaphragm activities and cavity pressures in pressure breathing. *J. appl. Physiol.*, **18**, 37–42.

BISHOP, B. (1964). Reflex control of abdominal muscles during positive-pressure breathing. *J. appl. Physiol.*, **19**, 224–232.

BISHOP, B. (1967). Diaphragm and abdominal muscle responses to elevated airway pressures in the cat. *J. appl. Physiol.*, **22**, 959–965.

BISHOP, B. (1968). Diaphragm and abdominal muscle activity during induced hypotension. *J. appl. Physiol.*, **25**, 73–79.

BJURSTEDT, H. (1953). Influence of the abdominal muscle tone on the circulatory response to positive pressure breathing in anaesthetized dogs. *Acta physiol. scand.*, **29**, 145–162.

BLAND, S., LAZEROU, L., DYCK, G., and CHERNIACK, R. M. (1967). The influence of the "chest wall" on respiratory rate and depth. *Resp. Physiol.*, **3**, 47–54.

BURGER, E. J., JR., and MACKLEM, P. (1968). Airway closure: demonstration by breathing 100% $O_2$ at low lung volumes and by $N_2$ washout. *J. appl. Physiol.*, **25**, 139–148.

CAMPBELL, E. J. M. (1958). *The Respiratory Muscles and the Mechanics of Breathing*. London: Lloyd-Luke.

CAMPBELL, E. J. M. (1964). Motor pathways. *Handbook of Physiology*, Sect. 3, Vol. 1, pp. 535–543. Washington, D.C.: American Physiological Soc.

CAMPBELL, E. J. M., DICKINSON, C. J., DINNICK, O. P., and HOWELL, J. B. L. (1961). The immediate effects of threshold loads on the breathing of men and dogs. *Clin. Sci.*, **21**, 309–320.

CAMPBELL, E. J. M., DICKINSON, C. J., and HOWELL, J. B. L. (1964). The immediate effects of added loads on the inspiratory musculature of the rabbit. *J. Physiol.* (*Lond.*), **172**, 321–331.

CAMPBELL, E. J. M., DINNICK, O. P., and HOWELL, J. B. L. (1961). The immediate effects of elastic loads on the breathing of man. *J. Physiol.* (*Lond.*), **156**, 260–273.

CAMPBELL, E. J. M., and HOWELL, J. B. L. (1962). Proprioceptive control of breathing. *Ciba Foundation Symposium on Pulmonary Structure and Function*, pp. 29–45. Eds. A. V. S. de Reuck and M. O'Connor. London: J. & A. Churchill.

CAMPBELL, E. J. M., and HOWELL, J. B. L. (1963). The sensation of breathlessness. *Brit. med. Bull.*, **19**, 36–40.

CHERNIACK, R. M. (1965). Work of breathing and the ventilatory response to $CO_2$. In: *Handbook of Physiology*, Sect. 3, Vol. 11, 1469–1474. Washington, D.C.: American Physiological Soc.

CHERNIACK, R. M., and SNIDAL, D. P. (1956). The effect of obstruction to breathing on the ventilatory response to $CO_2$. *J. clin. Invest.*, **35**, 1286–1290.

DAVIES, H. W., HALDANE, J. S., and PRIESTLEY, J. G. (1919). The response to respiratory resistance. *J. Physiol.* (*Lond.*), **53**, 60–69.

DELHEZ, L., and PETIT, J. M. (1966). Influence de l'application soudaine d'une résistance additionelle au débit aérien sur le régime ventilatoire de l'homme conscient. *Arch. int. Physiol. Biochem.*, **74**, 270–275.

ELDRIDGE, F., and DAVIS, J. M. (1959). Effect of mechanical factors on respiratory work and ventilatory responses to $CO_2$. *J. appl. Physiol.*, **14**, 721–726.

ERNSTING, E. (1965). The physiology of pressure breathing. In: *A Textbook of Aviation Physiology*, pp. 343–373. Ed. J. A. Gillies. Oxford: Pergamon Press.

FENN, W. O. (1963). A comparison of respiratory and skeletal muscles. *Perspectives in Biology* (Houssay Memorial Volume), pp. 293–300. Eds. C. F. Cori, V. G. FOGLIA, L. F. LELOIR and S. OCHOA. Amsterdam: Elsevier.

FREEDMAN, S. (1969). The functional capacity of the respiratory muscles in man. (Ph.D. Thesis, Univ. of London.)

FREEDMAN, S., and WEINSTEIN, S. A. (1965). Effects of external elastic and threshold loading on breathing in man. *J. appl. Physiol.*, **20**, 469–472.

GUZ, A., NOBLE, M. I. M., TRENCHARD, D., COCHRANE, H. L., and MAKEY, A. R. (1964). Studies on the vagus nerves in man: their role in respiratory and circulatory control. *Clin. Sci.*, **27**, 293–304.

GUZ, A., NOBLE, M. I. M., WIDDICOMBE, J. G., TRENCHARD, D. MUSHIN, W. W., and MAKEY, A. R. (1966). The role of vagal and glossopharyngeal afferent nerves in respiratory sensation, control of breathing and arterial pressure regulation in conscious man. *Clin. Sci.*, **30**, 161–170.

HALL, F. G., and ZECHMAN, F. (1957). Respiratory effects of increased air flow resistance in dogs. *Proc. Soc. exp. Biol.* (*N.Y.*), **96**, 329–332.

HERING, E., and BREUER, J. (1868). Die Selbsteuerung der Atmung durch den Nervus Vagus. *S.-B. Akad. Wiss. Wien, math.-nat. Kl.*, **57** (11), 672–677.

HOWELL, J. B. L. (1966). Breathlessness in Pulmonary Disease. In: *Breathlessness*, pp. 165–177. Eds. J. B. L. Howell and E. J. M. Campbell. Oxford: Blackwell.

HOWELL, J. B. L., and CAMPBELL, E. J. M. Ed. (1966). *Breathlessness*. Oxford: Blackwell.

KELLOGG, R. H., MEAD, J., LEITH, D. E., and KONNO, K. (1970). Ventilatory responses to $CO_2$ when respiratory resistance and elastance are reduced in normal subjects. (Submitted for publication.)

KILLICK, E. M. (1935). Resistance to inspiration—its effects on respiration in man. *J. Physiol.* (*Lond.*), **84**, 162–172.

KONNO, K., and MEAD, J. (1968). Static volume-pressure characteristics of the rib cage and abdomen. *J. appl. Physiol.*, **24**, 544–548.

LYNNE-DAVIES, P., COUTURE, J., WEST, D., BROMAGE, P. R., PENGELLY, L. D., and MILIC-EMILI, J. (1969). Stability of the respiratory system to added elastic loads in cats. (Manuscript in preparation).

MARSHALL, R. (1962). Relationships between stimulus and work of breathing at different lung volumes. *J. appl. Physiol.*, **17**, 917–921.

MCILROY, M. B., ELDRIDGE, F. L., THOMAS, J. P., and CHRISTIE, R. V. (1956). The effect of added elastic and non-elastic resistances on the pattern of breathing in normal subjects. *Clin. Sci.*, **15**, 337–344.

MEAD, J. (1960). Control of respiratory frequency. *J. appl. Physiol.*, **15**, 325–336.

MEAD, J. (1966). Mechanical factors in the control of breathing—three problems. In: *Breathlessness*, pp. 139–146. Eds. J. B. L. Howell and E. J. M. Campbell. Oxford: Blackwell.

MEAD, J., PETERSON, N., GRIMBY, G., and MEAD, J. (1967). Pulmonary ventilation measured from body surface movements. *Science*, **156**, 1383–1384.

MILIC-EMILI, J., and TYLER, J. M. (1963). Relation between work output of respiratory muscles and end-tidal $CO_2$ tension. *J. appl. Physiol.*, **18**, 497–504.

MONZEIN, P. (1962). Contribution a l'étude des réflexes ventilatoires d'origine proprioceptive thoraco-pulmonaire chez l'homme. (Ph.D. Thesis, Univ. of Paris.)

NUNN, J. F., COLEMAN, A. J., SACHITHANANDAN, T., BERGMAN, N. A., and LAWS, J. W. (1965). Hypoxaemia and atelectasis produced by forced expiration. *Brit. J. Anaesth.*, **37**, 3–12.

NUNN, J. F., and EZI-ASHI, T. I. (1961). The respiratory effects of resistance to breathing in anaesthetized man. *Anesthesiology*, **22**, 174–185.

OTIS, A. B., MCKERROW, C. B., BARTLETT, R. A., MEAD, J., MCILROY, M. B., SELVERSTONE, N. J., and RADFORD, E. P., JR. (1956). Mechanical factors in distribution of pulmonary ventilation. *J. appl. Physiol.*, **8**, 427–443.

PENGELLY, L. D., ALDERSON, A., and MILIC-EMILI, J. (1970). Mechanics of the diaphragm. *J. appl. Physiol.* (Submitted for publication.)

POPE, H., HOLLOWAY, R., and CAMPBELL, E. J. M. (1968). The effects of elastic and resistive loading of inspiration on the breathing of conscious man. *Resp. Physiol.*, **4**, 363–372.

RAHN, H., OTIS, A. B., CHADWICK, L. E., and FENN, W. O. (1946). The pressure-volume diagram of the thorax and lung. *Amer. J. Physiol.*, **146**, 161–178.

SEARS, T. A., and NEWSOM DAVIS, J. (1968). The control of respiratory muscles during voluntary breathing. *Ann. N.Y. Acad. Sci.*, **155**, 183–190.

SILVERMAN, L., LEE, R. C., LEE, G., DRINKER, K. M., and CARPENTER, T. M. (1943). Fundamental factors in the design of protective respiratory equipment: inspiratory airflow measurements on human subjects with and without resistance. *O.S.R.D. Report No. 1222.*

SILVERMAN, L., LEE, G., PLOTKIN, T., AMORY, L., and YANCEY, A. R. (1945*a*). Fundamental factors in the design of protective respiratory equipment: inspiratory and expiratory air flow measurements on human subjects with and without resistance at several work rates. *O.S.R.D. Report. August 1st.*

SILVERMAN, L., LEE, G., YANCY, A. R., AMORY, L., BARNEY, L. J., and LEE, R. C. (1945*b*). Fundamental factors in the design of protective respiratory equipment: a study and an evaluation of inspiratory and expiratory resistances for protective respiratory equipment. *O.S.R.D. Report No. 5339.*

SILVERMAN, L., LEE, G., PLOTKIN, T., SAWYERS, L. A., and YANCEY, A. R. (1951). Air flow measurements on human subjects with and without respiratory resistance at several work rates. *Arch. industr. Hyg.*, **3**, 461–478.

SUTHERLAND, P. W., KATSURA, T., and MILIC-EMILI, J. (1968). Previous volume history of the lung and regional distribution of gas. *J. appl. Physiol.*, **25**, 566–574.

WIDDICOMBE, J. G. (1965). Respiratory reflexes. *Handbook of Physiology*, Sect. 3, Vol. I, pp. 585–630. Washington, D.C.: American Physiological Soc.

WILEY, R. L., and ZECHMAN, F. W. (1965). Initial responses to added viscous resistance to inspiration in dogs. *J. appl. Physiol.*, **20**, 160–163.

WILEY, R. L., and ZECHMAN, F. W. (1968). Transient respiratory responses to step-changes in airflow resistance in anaesthetized cats. *Resp. Physiol.*, **6**, 105–112.

ZECHMAN, F., HALL, F. G., and HULL, W. E. (1957). Effects of graded resistance to tracheal air flow in man. *J. appl. Physiol.*, **10**, 356–362.

# Chapter XIV

# RESPIRATORY SENSATION

E. J. M. CAMPBELL AND J. NEWSOM DAVIS

## GENERAL CHARACTERISTICS

IT is common experience that, when attention is directed to breathing, the movements of the chest wall may be perceived. Breathing through an increased airway resistance can also be recognised, as a feeling of hindrance, but when carried out as a voluntary manoeuvre for a short period it is not of itself an unpleasant experience. In some circumstances, however, awareness of the movements of breathing, or of hindrance to these movements, may be associated with a sensation of breathlessness. Breathlessness or dyspnoea are commonly used synonymously to describe two different sensory phenomena: firstly, awareness that breathing is excessive, as may occur for instance at unaccustomed altitude, and which does not have an unpleasant quality; and secondly, awareness of difficulty in breathing such as occurs when breathing is mechanically hindered, and which, in the limit, approaches the sensation experienced at the breaking point of breath holding. In some circumstances, a sensation of tightness in the chest may be experienced without any measurable change in lung or chest wall mechanics.

This chapter will be concerned with the means by which movement and hindrance to movement of the chest are perceived and with the mechanisms which underlie the sensation of dyspnoea. The importance of mechanical factors relating to chest wall and lung movements in the genesis of dyspnoea will be discussed first with a theoretical analysis of the dyspnoeic stimulus, before the role of particular receptor mechanisms is considered.

## MECHANICAL FACTORS IN DYSPNOEA

Although dyspnoea may be associated with an increase in $Pa_{CO_2}$ or a decrease in $Pa_{O_2}$, the blood gases are often norm also that, while chemical factors may contribute to dyspnoea, they cannot be its only cause. A number of observations have indicated the importance of mechanical events to the genesis of dyspnoea. Most of these have been made in experiments on breath holding. During this manoeuvre, the initial period is characterised by voluntary inhibition of the diaphragm, which may be silent or tonically active. There follows involuntary rhythmic activity which increases in frequency as the break point is approached, as illustrated in Fig. 101 (Agostoni, 1963). The onset of diaphragmatic activity is dependent on the $Pa_{CO_2}$ and $Pa_{O_2}$, and appears to be unaffected by neurogenic factors related to lung volume or respiratory movements.

Breath holding time is directly related to the initial lung volume and a straight line extrapolation of this relationship to RV indicates a breath holding time at that volume of 8 seconds, approximately the normal circulation time from lung to brain (Mithoefer, 1965). During the breath hold lung volume decreases, and the smaller the final lung volume the

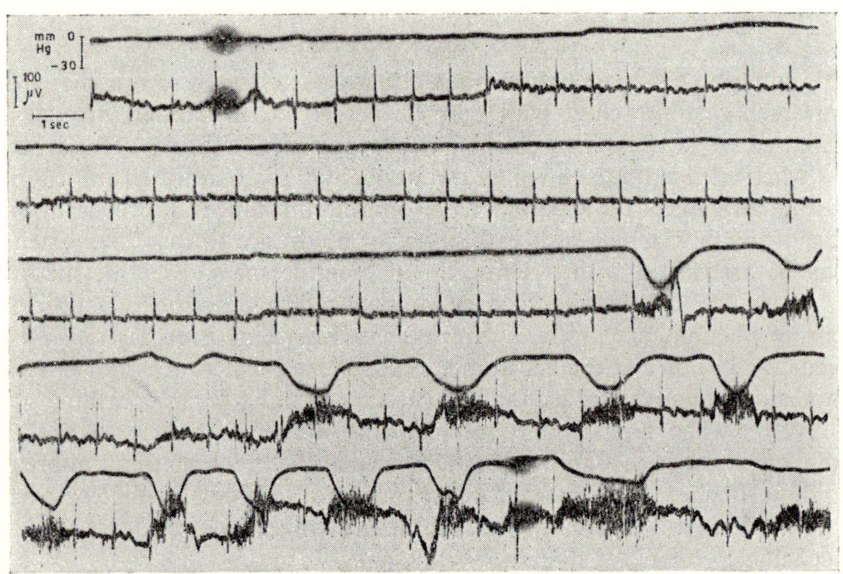


FIG. 101.—ELECTRICAL ACTIVITY OF THE DIAPHRAGM DURING BREATH HOLDING

Oesophageal pressure (upper tracing) and electrical activity of the diaphragm (lower tracing) during breath holding at resting volume after breathing air. Beginning and end of apnoea are indicated by the light signals. (From Agostoni, 1963.)

lower the $Pa_{CO_2}$ which can be tolerated (Mithoefer, 1965), although this relationship is not absolute because it is possible to perform a breath hold at a high $P_{CO_2}$ and smaller lung volume than existed at the break point of a previous breath hold (Godfrey *et al.*, 1969). A new model (Fig. 102) has recently been proposed to describe the interaction of chemical and non-chemical stimuli in breath holding (Godfrey and Campbell, 1968).

The importance of non-chemical factors in the duration of breath holding was demonstrated in a simple experiment by Fowler (1954) who showed that, after breath holding to breaking point, a further period of holding was possible if a number of breaths were allowed from a gas mixture of 8·2 per cent $O_2$ and 7·5 per cent $CO_2$ which did not decrease the chemical stimuli to breathe. Mithoefer and his colleagues (1953) studied the ability of subjects to rebreathe at a fixed frequency from a bag containing

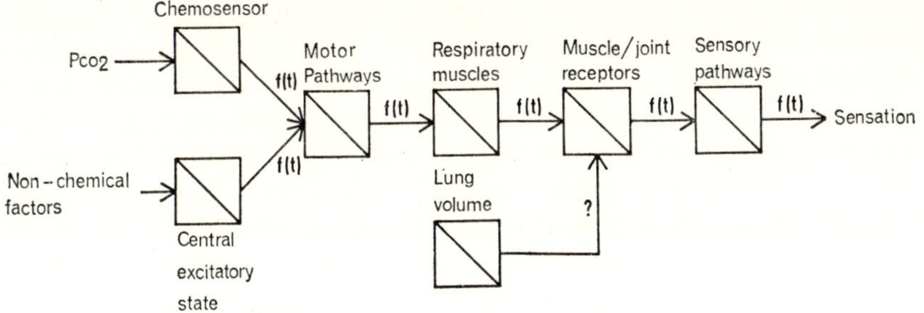


FIG. 102.—A MODEL OF THE FACTORS LIMITING BREATH HOLDING

Each square represents a function generator, that is it receives one or more input signals, modifies them in some way, and provides an output. Following the model from left to right, during breath holding the increased chemical drive ($P_{CO_2}$ in this simplification) together with the non-chemical drive (which may be a self-cycling loop) summate to produce a central excitatory state build up. This produces a demand for ventilation which is transmitted to the motor pathways of the brain and spinal cord and drives the lower motor neurones governing the respiratory muscles. These muscles in turn contract, but because the lung volume is prevented from changing by the voluntary act of breath holding, there is a disproportion between the tension developed in these muscles and the displacement produced. This discrepancy between tension and displacement is in some way transmitted to sensation by afferent impulses from the muscles and chest wall. Godfrey and Campbell suggested that static lung volume might operate reflexly by modifying the output of the motor pathways (i.e. upstream from the respiratory muscles) but as Agostoni has shown (1963) the onset of diaphragm activity is unaffected by the lung volume for a given value of $P_{CO_2}$. This suggests that the lung volume effect is downstream from the respiratory muscles. (From Godfrey and Campbell, 1968.)

oxygen, carbon dioxide and nitrogen. When inflation of the lungs was restricted by the volume of the bag, uncontrollable stimulation to breath occurred with less hypoxia and less hypercapnia than when a larger bag was employed, i.e. greater degrees of hypoxia and hypercapnia could be tolerated when there was no restriction to the size of breath. In a similar experiment where tidal volume was limited to 350 ml at a frequency of 10 breaths per minute, Wright and Branscomb (1954) found that the intolerable breathlessness which developed was entirely abolished when the subject was permitted a single inhalation from a gas mixture made up of virtually pure nitrogen, so that he was then able to return to the previous restricted breathing pattern for a further 30–120 seconds.

In their study of paralysed subjects, Opie *et al.* (1959) showed that patients who were chronically overventilated by artificial ventilation could tolerate a relatively large rise in $P_{CO_2}$ when this was produced by adding $CO_2$ to the inspired air whereas the rise tolerated when tidal volume was reduced was very much smaller. These authors concluded that a factor related to respiratory movements modified the subject's appreciation of the adequacy of ventilation.

## ANALYSIS OF THE DYSPNOEIC STIMULUS

What are the characteristics of the stimulus which provoke awareness that breathing is excessive for a particular exercise level, or that it is mechanically hindered? The recognition that breathing is excessive is probably based upon the total experience of the condition in which the increased ventilation occurs. In other words, the stimulus consists not only of the chemical drive or the increased depth and rate of respiration, but also of factors related to the context in which the increased chemical drive occurs. Thus a subject recently arrived at altitude is more aware of the moderately increased ventilation of mild exercise than he was, at sea level, of the greater ventilation of harder exercise. Such observations need not present problems requiring explanation if it is recognised that the normal individual has experience of the level of ventilation that normally accompanies—is "appropriate" to—the activity in which he is engaged. Indeed, the increase in ventilation at the beginning of exercise before there are any chemical changes in the blood suggests that this experience is used to anticipate the ventilatory requirement (Torelli and Brandi, 1961).

**Awareness of Mechanically Hindered Breathing**

The ability to detect mechanical loads added to breathing has provided some clues about the nature of the dyspnoeic stimulus. It has been found that the addition of an elastic load of about 2·5 cm $H_2O$/l or of a resistive load of about 0·75 cm $H_2O$/l/sec can be perceived (Campbell *et al.*, 1961; Bennett *et al.*, 1962; Wiley and Zechman, 1966–67; Newsom Davis, 1967). These loads are equivalent to increases of about 10–25 per cent of the normal physiological load; they are too small to cause distress but the findings do shed light on the nature of the mechanical information needed for detection. Campbell and his co-workers concluded that two forms of mechanical information were needed to perceive a load added to breathing. They reasoned that the detection of an elastic load, for example, could not be accounted for in terms of changes in pressure or volume, in force or displacement, but only if pressure *was related to* volume or force to displacement. The subject perceives that an additional load is present because the increase in volume of the chest during the breath is less than expected *when related to* the pressure developed by the respiratory muscles. Campbell *et al.* (1963) later found that the ability to detect and identify a resistive load is greater than the ability to detect and identify an elastic load. Wiley and Zechman (1966–67) have shown that, in common with other sensory modalities, perception of an added airflow resistance obeys the Weber-Fechner law in that the increment of airway resistance detected is proportional to the magnitude of the resting resistance. Thus in conditions where the resting resistance was increased, a larger added resistance was required to reach the detection threshold. This interpretation may be summarised by saying that the mechanical information required for the detection of a resistance *is* resistance (pressure ÷ flow), neither pressure nor flow alone being sufficient.

Campbell *et al.* (1961) introduced the term "length : tension appropriateness" to express this central idea. "Length" is preferable to "volume" and "tension" to "pressure" because they have simpler dimensions and facilitate interpretation in terms of neural structures; "appropriateness" is preferable to "relationship" or other terms such as "ratio" because pressure:volume and length:tension relationships change in various physiological situations, but these changes will still be recognised as "appropriate" if they are in keeping with previous experience. However, Campbell (1966) has subsequently pointed out that the use of tension was not logically necessary; information about length alone would suffice. Thus the detection of a load could be interpreted as the awareness that the change in length (volume, flow) achieved was less than the change demanded. By removing the need to postulate information specifically about force or tension this modification may contribute to the ultimate identification of the responsible nervous mechanisms.

These studies on the detection of load and their analysis do not prove that the information required for detection is the same as the dyspnoeic stimulus, but it may be (Campbell and Howell, 1963; Campbell, 1966).

## SENSORY MECHANISMS

It is hardly necessary to emphasise that dyspnoea is a complex sensation, and that there is no reason to suppose that any type or group of receptors is uniquely responsible for its genesis. Indeed, it is possible that there are several different sensations whose only common feature is their reference to the chest or their association with the act of breathing. Nevertheless, it remains pertinent to examine in as quantitative a manner as possible the receptor mechanisms subserving particular sensory modalities which may be involved in the production of dyspnoea. Such studies, necessarily carried out in man, have largely been concerned with examining the effects of interruption of different neural pathways.

### Pulmonary and Upper Respiratory Tract Receptors

Among the several different types of afferent end-organs in the lungs, there are some which are known to sense volume changes (Adrian, 1933; Knowlton and Larrabee, 1946; Widdicombe, 1954) and which also respond to the rate of change of lung volume (Davis *et al.*, 1956). Other tracheobronchial endings are excited by sudden large pressure or volume changes in the airways, and are believed to be the afferent end-organs for the mechanical cough reflex (Widdicombe, 1954, 1964). A further group of epithelial receptors, lying deep in the lungs, show the properties of lung irritant receptors, responding to gaseous irritants, histamine aerosol and anaphylaxic reaction in the lungs, large inflations and deflations and pulmonary congestion (Mills *et al.*, 1970). These receptors rarely cause coughing, but do cause reflex hyperpnoea and bronchoconstriction. Finally, a group of juxta-capillary receptors have been defined which appear to respond rather specifically to increased pulmonary congestion (Paintal, 1970).

The afferent fibres from these receptors travel to the medulla via the vagus nerves. Little is known of their projection to higher levels (see p. 249) but stimulation of the cervical vagus in the cat causes an evoked potential over the orbital surface of the frontal lobe (Bailey and Bremer, 1938; Dell and Olson, 1951; Siegfried, 1961).

These receptors are known to subserve respiratory reflexes (see Widdicombe, 1964). What is their role in respiratory sensation? Wright and

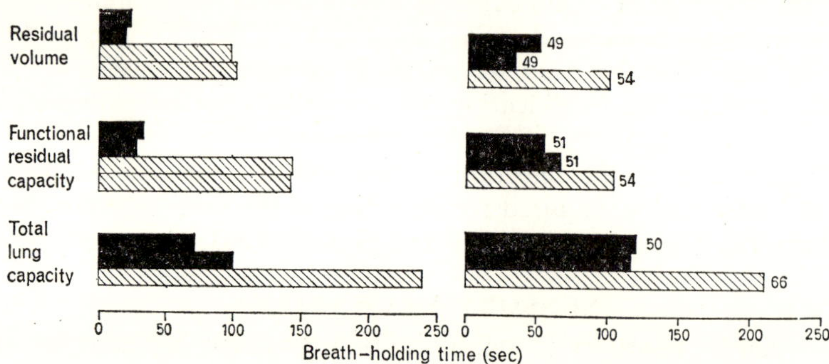


FIG. 103.—THE EFFECT OF VAGAL AND GLOSSOPHARYNGEAL NERVE BLOCK ON BREATH HOLDING TIMES IN TWO HEALTHY CONSCIOUS SUBJECTS

On air (left) and 100 per cent $O_2$ (right). Solid block rectangles are control values, and cross-hatched rectangles are the values during block of glossopharyngeal and vagus nerves bilaterally. The figures at the end of the rectangles in the study on the right represent the end-tidal $P_{CO_2}$ at breaking point. (From Guz *et al.*, 1966.)

Branscomb (1954) suggested that some of the features of dyspnoea might be accounted for by a disturbance of the pulmonary stretch reflexes, but it was not until the effects of bilateral vagal blocking in conscious man were reported (Guz *et al.*, 1966) that more direct evidence of the contribution of pulmonary receptors to respiratory sensation became available. In two conscious healthy subjects who underwent bilateral vagal (and associated glossopharyngeal) nerve block, which abolished the Hering-Breuer inflation reflex, no difference in the sensation of breathing during the block was experienced. Furthermore, breathing which was mechanically hindered by a large resistive load (31 cm $H_2O$/l/sec at 1 l/sec) was as unpleasant as it had been before the block, and the ability to detect added elastic loads was unaltered. The effect on breath holding, on the other hand, was profound (Fig. 103). Breath holding time during apnoea started at TLC was approximately doubled and the usual distress towards breaking point was alleviated. This implies that the drive to breathe during breath holding arises predominantly from lung receptors (the type of receptor involved is not clear) and is vagally mediated; but, as the authors point out, it does not necessarily mean that the unpleasant sensation itself originates from lung receptors. It is of interest that, although no information is available about respiratory muscle activity during vagal block in the two healthy subjects studied by Guz *et al.* (1966), in a third subject who had pulmonary sarcoidosis the usual respiratory efforts during a brief period of breath holding were abolished during the block, implying a reduced ventilatory drive. This raises the possibility that the distress of breath holding is not due to awareness of the pulmonary afferent discharge itself but is a consequence of the increased respiratory muscle activity which that discharge provokes.

Studies by Campbell *et al.* (1967, 1969) on the effects of muscular paralysis with *d*-tubocurarine on the duration and sensation of voluntary apnoea provide evidence in support of this. The duration of apnoea started at resting lung volume was greatly prolonged from about one and a half minutes to at least four minutes ("breaking point" was not in fact reached) and the distressing sensation in the chest was abolished. Communication during paralysis was achieved by excluding one arm from the circulation by a cuff for a period immediately following the injection; this permitted the subject to signal by finger movements. The authors conclude that frustration of the motor response to the respiratory drive rather than consciousness of the drive itself is likely to be the cause of the distress of breath holding.

The results of the experimental study by Campbell *et al.* (1967) at first seem to contradict the clinical observation that patients with respiratory muscle weakness or paralysis suffer from extreme breathlessness. Indeed, Smith and his co-workers (1947) reported that a healthy subject paralysed with *d*-tubocurarine experienced "shortness of breath" despite adequate ventilation, until a smooth rhythmical pressure was applied to the rebreathing bag at an increased frequency of 24 per minute. Campbell *et al.* (1967) emphasise, however, that their experimental situation of trying *not* to breathe or of *not trying* to breathe is different from that of the paralysed patient who is *trying* to breathe, and who is aware that his chest fails to respond to the willed movement.

Although vagal block causes no change in the sensation of normal breathing in healthy subjects, in some patients with cardiopulmonary disease it has reduced or abolished the sensation of dyspnoea present at rest (Guz *et al.*, 1970). The ventilatory pattern changed towards slow and deep breathing, and breath holding was prolonged. The effects in these patients are presumed to result from reduction of the drive to breathe.

Several lines of evidence indicate that pulmonary receptors can evoke sensation directly. First, Morton *et al.* (1950) noted that patients could localise with some accuracy the site of pain produced by electrical stimulation of the left and right main bronchus and trachea carried out through a bronchoscope. The site was usually referred to the correct side or midline, but in the longitudinal axis the error was sometimes several inches. After unilateral vagal section, stimulation of the ipsilateral main bronchus caused no pain in five subjects, and contralateral pain in three subjects, indicating that the afferent innervation is mediated by the vagus and is partially crossed (Morton *et al.*, 1951). The discomfort associated with mechanical stimulation of the trachea and main bronchi is presumably also vagally mediated.

Second, certain subjects may experience severe tightness in the chest, for example on exposure to cotton dust, under conditions in which there has been no measurable change in lung or chest wall mechanics (Howell, 1970); this might perhaps arise from lung-irritant receptors (see Mills *et al.*, 1970). Further evidence comes from studies in patients with total high cervical (C3) spinal cord transection, in whom the only afferents remaining from thoracic structures are those travelling in the vagus nerve. In one patient, pulmonary congestion following myocardial infarction gave rise to a sensation of shortness of breath (Prys-Roberts and Spalding, 1970). Another patient complained of tightness of the chest when a volume of 0·5 L was sucked rapidly out of the lungs at FRC, through a tracheostomy tube, which would be expected to cause airway distortion or collapse, whereas the expulsion of a similar volume by compression of the chest wall and abdomen was not perceived (Newsom Davis, Semple and Spencer, to be published). In this latter patient, apnoea started at resting lung volume could be tolerated for a prolonged period, similar to that of the curarised subjects, confirming that afferent information in the vagus does not directly give rise to the distress associated with breath holding. It is also of interest that this patient was quite unable to tell whether he was being ventilated or not, whereas in a control subject, with respiratory muscle paralysis due to poliomyelitis, changes in Vt of only 150 ml could be detected. Although this appears to suggest that vagal afferents do not directly subserve the perception of passive movements of the chest wall, these observations could not be confirmed by Noble *et al.* (1970) in a similar case so that further studies are necessary. The experiments of Opie *et al.* (1959) mentioned above (p. 293) were repeated in one of the patients with cervical cord transection (Newsom Davis *et al.*, to be published). The rise in $PaCO_2$ tolerated when $CO_2$ was added to the inspired air remained greater than that tolerated when the $V_T$ was reduced, as in the patients of Opie *et al.*, suggesting that vagal and not chest wall afferents are responsible for the effect.

**Muscle Receptors**

The receptors found in skeletal muscle are muscle spindles, tendon organs and Pacinian corpuscles; in addition, some afferent fibres are free ending and others terminate around blood vessels. The intercostal muscles are well supplied with muscle spindles and tendon organs; in contrast, the diaphragm has few proprioceptors, with relatively more tendon organs than muscle spindles (see Chapters VI and VII). Pacinian corpuscles and free endings are found in both muscles.

The concept of a "muscular sense" was first formulated by Charles Bell (1833) who described it as "a property internal to the frame by which we thus know the position of the members of our body". Since that time, the question of whether "muscle sense" arises in the muscles themselves has been the subject of some controversy; recent work suggests that it does not.

Sherrington (1918, 1924) thought that position sense in the eyeball was dependent upon afferent information from the extraocular muscles, and that muscle receptors subserved mental experience. This belief was subsequently widely held. Further investigation of position sense in the eyeball (Brindley and Merton, 1960) however, demonstrated that non-visual knowledge of the position of the eyes depended entirely upon judgement of the effort of will employed in attempting to move the eyes, even though the extraocular muscles are richly supplied with muscle spindles (Cooper and Daniel, 1949). Thus passive displacement of one eye causes no sensation if the cornea is occluded with an opaque cap and the conjunctiva anaesthetised.

Provins (1958) showed that local anaesthesia of the metacarpophalangeal joint of the index finger caused a gross impairment of the perception of passive movement at the joint, which was not significantly improved when the movement was resisted by active muscular contraction. Thus the receptors in the muscle, which were unaffected by the anaesthesia, did not indicate that movement had in fact occurred. Further, pulling a tendon exposed under local anaesthesia, so as to stretch the muscle, produces no sensation referable to the muscle (Gelfan and Carter, 1967).

Merton (1964), in a review concerned with muscle sense, gives an account of some experiments in which the hand was anaesthetised by a pneumatic tourniquet at the wrist. The distal joint of the thumb could still be moved because the muscles which flex and extend it lie in the forearm and were therefore above the level of the block. Active movements of the joint were carried out with an accuracy similar to that before the anaesthesia, but if the movement was resisted by holding the thumb the subject believed that the movement had occurred just the same. Thus information from the receptors in the muscles concerned in the attempted movement did not give rise to awareness that the movement had not been achieved.

We may conclude from this evidence that information from muscle spindles and tendon organs does not form the basis of postural sense in the limbs. The Pacinian corpuscles and free ending afferents may subserve pressure-pain sensation (Paintal, 1960). It is reasonable to suppose, then, that muscle spindles and tendon organs in respiratory muscles also do not give rise to the perception of movement or resistance to movement of the chest wall. Although there is no direct evidence for this, some support is given by the finding that patients with cerebellar disease, in whom the control of muscle spindles is likely to be disturbed (Granit *et al.*, 1955), showed no impairment in their ability to detect an external airway resistance (Newsom Davis, 1967).

While muscle receptors do not appear to provide specific information about movement or resistance to movement of the chest wall, they may be concerned with the distress associated with breath holding, and perhaps also when breathing is mechanically hindered. The work of Campbell *et al.* (1967, 1969) on the effects of curarisation suggests that respiratory muscle contraction is essential for this distress. Breath holding was not prolonged in patients with local anaesthetic spinal cord block at T1 (Eisele *et al.*, 1968), which blocks rib cage afferents and leaves those from the diaphragm intact, and the onset of the urge to breathe during the breath hold coincided with the onset of the diaphragmatic contraction (see also, Agostoni, 1963). Bilateral phrenic nerve block prolongs breath holding (Noble *et al.*, 1970). Thus diaphragmatic muscle receptors alone may subserve the sensation of distress accompanying breath holding. Which group of receptors might be involved is uncertain.

## Rib-cage Joint Receptors

In the limbs, the perception of movement and resistance to movement appear to depend upon an intact articular innervation (Browne *et al.*, 1954; Provins, 1958), and physiological studies of joint receptors in animals have indicated that they are able to provide the required information

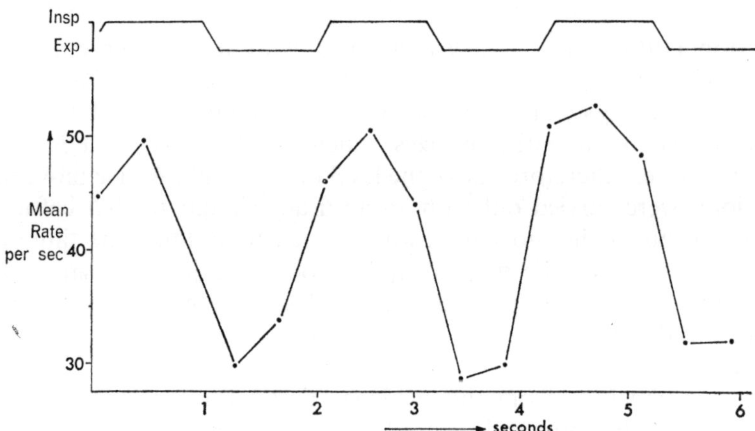


FIG. 104.—THE RESPONSE OF A SLOWLY ADAPTING RIB JOINT MECHANORECEPTOR DURING SPONTANEOUS BREATHING.

The upper trace is schematic and taken from an endotracheal pressure trace. (From Godwin-Austen, 1969.)

(Skoglund, 1956; Mountcastle and Powell, 1959). The afferent pathway for these receptors lies in the dorsal columns of the spinal cord (Skoglund, 1956). Recently Godwin-Austen (1969) has identified slowly-adapting mechanoreceptors in the costo-vertebral joints of the cat and rabbit which are excited by displacement of the ribs. The discharge frequency of these receptors is related to the position of the rib, the units accelerating to a maximum at one end of the range of movement. The response of one of these receptors during spontaneous breathing is shown in Fig. 104. Manually imposed rib movements elicited a dynamic response which was directly related to the velocity of the movement and adapted within two seconds. A few receptors of a different type were encountered in that no resting discharge was present and on rapid displacements of the rib they showed a high frequency response which adapted completely on termination of the movement. Afferent activity has been recorded in the dorsal columns in phase with respiratory movements (Boruchow and Nelson, 1959; Yamamoto *et al.*, 1960) and when the thorax is distorted (Yamamoto *et al.*, 1956).

Very little information is available about the perception of passive movements of the chest wall. It was grossly impaired in the patient mentioned above (p. 298) with total de-afferentation of the chest wall due to cord transection at C3, whereas a tracheotomised control patient with chronic poliomyelitis could detect quite small passive changes in tidal volume despite almost complete respiratory muscle paralysis.

Newsom Davis (1967) studied the ability to detect added resistive loads to breathing in a group of patients with upper spinal cord lesions in whom the modality predominantly affected was joint position sense. These patients showed a markedly impaired ability to detect added loads and it was argued that this followed loss of information from joint receptors in the rib cage. Airway resistance was not measured in these subjects, however, so the possibility remains that the detection threshold would be in the normal range when expressed as the ratio of added resistance to resting resistance (*cf.* Wiley and Zechman, 1966–7). In a further six patients with complete physiological transection of the spinal cord between C6 and C8, an impairment of detection was found in only four (Newsom Davis, 1967). Zechman *et al.* (1967) also report a normal detection threshold in a patient with cervical cord transection. Moreover, the detection of elastic loads was unimpaired in four patients tested during spinal cord block at T1 which would de-afferent the chest wall (Eisele *et al.*, 1968).

The inconsistency in these observations may be accounted for in part by sensory information from the upper respiratory tract. An added external elastic or resistive load produces a transient pressure change in the mouth and pharynx during breathing, and untrained subjects when questioned usually related the presence of a load to an intraoral or pharyngeal sensation. When oro-pharyngeal mucosal sensation was reduced by a local anaesthetic, the detection threshold of patients with partial upper spinal cord lesions was further impaired while a control group was unaffected (Newsom Davis, 1967). This indicates that patients deprived of information from rib cage receptors relied heavily upon oro-pharyngeal sensation, while for the healthy subjects this was not necessary.

It is thus probable that joint receptors in the rib cage subserve the perception of movements of the chest wall and of resistance to such movements. The suggestion that diaphragmatic muscle receptors might subserve this function, based on the effects of spinal cord block at T1 (Eisele *et al.*, 1968), would attribute to muscle receptors in the diaphragm a postural sense which they do not have elsewhere in the body, but may be substantiated by experiments in which the possibility of other sensory clues have been eliminated (see Noble *et al.*, 1970).

In order that the movements of the chest wall may be perceived, joint receptors must provide information about position, change of position, and rate of change of position. The discharge characteristics of rib-joint receptors are consistent with this (Godwin-Austen, 1969). The perception of resistance to movement is more complex. It is probably dependent not only upon joint position sense but also upon the "sense of effort" (Merton, 1964). The sensation of hindrance would then arise from the comparison of the change in joint position with the muscular effort needed to achieve it. In addition, the increased force acting at a joint when movement is carried out against a resistance causes an altered pattern of its receptors' discharge (see Skoglund, 1956), which could contribute to the sensation of resistance.

In the particular case of movements of the chest wall, a further mechanism might operate. As explained in detail in Chapter II, when the load on breathing increases, as occurs when there is a mechanical hindrance, a deformation of the rib cage results, changes in the dorso-ventral diameter lagging on those in the lateral diameter (Agostoni and Mognoni, 1966). This deformation might be expected to give rise to a phase shift between the discharge of joint receptors in different parts of the rib cage, the disparity between this discharge pattern and that occurring in the unloaded state representing the extent to which mechanical factors were hindering respiration. Perception of this mechanical hindrance would lead in turn to awareness that a greater respiratory effort was required to achieve the previous level of ventilation. During breath holding, movements of the rib cage are restricted by the closed airways, but the augmenting rhythmic respiratory muscle contraction nevertheless produces a deformation of the rib cage, which might similarly give rise to a phase shift in rib-joint receptor discharge. It seems unlikely, however, that joint receptor discharge alone accounts for the characteristic distress experienced towards breaking point.

### Other Receptors

No information is available about the contribution to respiratory sensation of cutaneous sensory endings, receptors in the connective tissue of the chest wall and intra-abdominal receptors. The characteristics of these receptors make it unlikely that they alone are able to provide the specific sensory information required for the perception of excessive ventilation or mechanically hindered breathing.

## Summary

Respiratory sensation includes the perception of movement of the chest wall and of hindrance to movement, and the sensation of respiratory distress or dyspnoea which may be associated with breath holding, excessive breathing, or mechanically hindered breathing.

Perception of movements of the chest wall probably depends upon information from rib-cage joint receptors. Mechanical hindrance to these movements may be recognised by comparison of the movement achieved with the sense of effort made, but the altered pattern of receptor discharge resulting from the deformation of the rib cage and the different forces acting on the costo-vertebral joints may also be important.

During breath holding, afferents travelling in the vagus nerves are concerned with the increased respiratory drive, but do not themselves give rise directly to the distress at breaking point. The distress depends upon respiratory muscle contraction, and appears to be mediated by muscle receptors, for it is still present when all chest wall afferents except those from the diaphragm are blocked.

Pain evoked by electrical stimulation of the trachea or main bronchi is vagally mediated. Lung receptors probably subserve the sensation of tightness in the chest which is experienced in certain clinical and experimental situations.

The dyspnoea of excessive breathing arises from an appreciation that chest wall movements are inappropriate, in terms of past experience, while that associated with respiratory muscle weakness arises from a recognition that the chest wall movement achieved is less than expected for the sense of effort made. The dyspnoea accompanying mechanical hindrance to breathing may have a basis similar to that of the distress of breath holding, but the perception of the hindrance itself may be subserved by a different mechanism.

## REFERENCES

ADRIAN, E. D. (1933). Afferent impulses in the vagus and their effect on respiration. *J. Physiol.* (*Lond.*), **7b**, 332–358.

AGOSTONI, E. (1963). Diaphragm activity during breath holding: factors related to its onset. *J. appl. Physiol.*, **18**, 30–36.

AGOSTONI, E., and MOGNONI, P. (1966). Deformation of the chest wall during breathing efforts. *J. appl. Physiol.* **21**, 1827–1832.

BAILEY, P., and BREMER, F. (1938). A sensory cortical representation of the vagus nerve. *J. Neurophysiol.*, **1**, 405–412.

BELL, SIR CHARLES (1833). *The hand: its mechanism and vital endowments as envincing design.* (Bridgewater Treatises on the Power, Wisdom and Goodness of God as manifested in the Creation) Ch. IX. Of the Muscular Sense, pp. 191–207. London: Wm. Pickering.

BENNETT, E. D., JAYSON, M. I. V., RUBENSTEIN, D., and CAMPBELL, E. J. M. (1962). The ability of man to detect added non-elastic loads to breathing. *Clin. Sci.*, **23**, 155–162.

BORUCHOW, I., and NELSON, J. (1959). Respiratory unit activity in the cervical cord of the cat. *Fed. Proc.*, **18**, 14.

BRINDLEY, G. S., and MERTON, P. A. (1960). The absence of position sense in the human eye. *J. Physiol.* (*Lond.*), **153**, 127–130.

BROWNE, K., LEE, J., and RING, P. A. (1954). The sensation of passive movement at the metatarso-phalangeal joint of the great toe in man. *J. Physiol.* (*Lond.*), **126**, 448–458.

CAMPBELL, E. J. M. (1966). The relationship of the sensation of breathlessness to the act of breathing. In: *Breathlessness*, pp. 55–64. Eds. J. B. L. Howell and E. J. M. Campbell. Oxford: Blackwell.

CAMPBELL, E. J. M., BENNETT, E. D., and RUBENSTEIN, D. (1963). The ability to distinguish between added elastic and resistive loads to breathing. *Clin. Sci.*, **24**, 201–207.

CAMPBELL, E. J. M., FREEDMAN, S., CLARK, T. J. H., ROBSON, J. G., and NORMAN, J. (1967). The effect of muscular paralysis induced by tubocurarine on the duration and sensation of breath holding. *Clin. Sci.*, **32**, 425–432.

CAMPBELL, E. J. M., FREEDMAN, S., SMITH, P. S., and TAYLOR, M. E. (1961). The ability of man to detect added elastic loads to breathing. *Clin. Sci.*, **20**, 223–231.

CAMPBELL, E. J. M., GODFREY, S., CLARK, T. J. H., FREEDMAN, S., and NORMAN, J. (1969). The effect of muscular paralysis induced by tubocurarine on the duration and sensation of breath holding during hypercapnia. *Clin. Sci.*, **36**, 323–328.

CAMPBELL, E. J. M., and HOWELL, J. B. L. (1963). The sensation of breathlessness. *Brit. med. Bull.*, **19**, 36–40.

COOPER, S., and DANIEL, P. M. (1949). Muscle spindles in human extrinsic eye muscles. *Brain*, **72**, 1–24.

DAVIS, H. L., FOWLER, W. S., and LAMBERT, E. H. (1956). Effect of volume and rate of inflation and deflation on transpulmonary pressure and response of pulmonary stretch receptors. *Amer. J. Physiol.*, **187**, 558–566.

DELL, P., and OLSON, R. (1951). Projections thalamiques, corticales et cérébelleuses des afférences viscérales vagales. *C.R. Soc. Biol.* (*Paris*), **145**, 1084–1088.

EISELE, J., TRENCHARD, D., BURKI, N., and GUZ, A. (1968). The effect of chest wall block on respiratory sensation and control in man. *Clin. Sci.*, **35**, 23–33.

FOWLER, W. S. (1954). Breaking point of breath holding. *J. appl. Physiol.*, **6**, 539–545.

GELFAN, S., and CARTER, S. (1967). Muscle sense in man. *Exp. Neurol.*, **18**, 469–473.

GODFREY, S., and CAMPBELL, E. J. M. (1968). The control of breath holding. *Resp. Physiol.*, **5**, 385–400.

GODFREY, S., EDWARDS, R. H. T., and WARRELL, D. A. (1969). The influence of lung shrinkage on breath holding time. *Quart. J. exp. Physiol.*, **54**, 129–140.

GODWIN-AUSTEN, R. B. (1969). The mechanoreceptors of the costo-vertebral joints. *J. Physiol. (Lond.)*, **202**, 737–753.

GRANIT, R., HOLMGREN, B., and MERTON, P. A. (1955). The two routes for excitation of muscle and their subservience to the cerebellum. *J. Physiol. (Lond.)*, **130**, 213–224.

GUZ, A., NOBLE, M. I. M., EISELE, L., and TRENCHARD, D. (1970). In: *Ciba Foundation Symposium on Breathing: Hering-Breuer Centenary Symposium*, pp. 315–328. London: Churchill.

GUZ, A., NOBLE, M. I. M., WIDDICOMBE, J. G., TRENCHARD, D., MUSHIN, W. W., and MAKEY, A. R. (1966). The role of vagal and glossopharyngeal afferent nerves in respiratory sensation, control of breathing and arterial pressure regulation in conscious man. *Clin. Sci.*, **30**, 161–170.

HOWELL, J. B. L. (1970). In: *Ciba Foundation Symposium on Breathing: Hering-Breuer Centenary Symposium*, pp. 287–291. London: Churchill.

KNOWLTON, G. C., and LARRABEE, M. G. (1946). A unitary analysis of pulmonary volume receptors. *Amer. J. Physiol.*, **147**, 100–114.

MERTON, P. A. (1964). Human position sense and sense of effort. *Symp. Soc. exp. Biol.*, **18**, 387–400.

MILLS, J. E., SELLICK, H., and WIDDICOMBE, J. G. (1970). In: *Ciba Foundation Symposium on Breathing: Hering-Breuer Centenary Symposium*, pp. 77–92. London: Churchill.

MITHOEFER, J. C. (1965). Breath holding. *Handbook of Physiology*, Sect. 3. Respiration Vol. II, pp. 1011–1025. Eds. W. O. Fenn and H. Rahn. Washington, D.C.: American Physiological Soc.

MITHOEFER, J. C., STEVENS, C. D., RYDER, H. W., and MCGUIRE, J. (1953). Lung volume restriction, hypoxia and hypercapnia as interrelated respiratory stimuli in normal man. *J. appl. Physiol.*, **5**, 797–802.

MORTON, D. R., KLASSEN, K. P., and CURTIS, G. M. (1950). The clinical physiology of the human bronchi. I. Pain of tracheo-bronchial origin. *Surgery*, **28**, 699–704.

MORTON, D. R., KLASSEN, K. P., and CURTIS, G. M. (1951). The clinical physiology of the human bronchi. II. The effect of vagus section upon pain of tracheo-bronchial origin. *Surgery*, **30**, 800–809.

MOUNTCASTLE, V. B., and POWELL, T. P. S. (1959). Central nervous mechanisms subserving position sense and kinesthesis. *Bull. Johns Hopk. Hosp.*, **105**, 173–200.

NEWSOM DAVIS, J. (1967). Contribution of somatic receptors in the chest wall to detection of added inspiratory airway resistance. *Clin. Sci.*, **33**, 249–260.

NOBLE, M. I. M., EISELE, J., TRENCHARD, D., and GUZ, A. (1970). In: *Ciba Foundation Symposium on Breathing: Hering-Breuer Centenary Symposium*, pp. 233–246. London: Churchill.

Opie, L. H., Smith, A. C., and Spalding, J. M. K. (1959). Conscious appreciation of the effects produced by independent changes of ventilation volume and end-tidal $P_{CO_2}$ in paralysed patients. *J. Physiol.* (*Lond.*), **149**, 494–499.

Paintal, A. S. (1960). Functional analysis of group III afferent fibres of mammalian muscles. *J. Physiol.* (*Lond.*), **152**, 250–270.

Paintal, A. S. (1970). In: *Ciba Foundation Symposium on Breathing: Hering-Breuer Centenary Symposium*, pp. 59–71. London: Churchill.

Provins, K. A. (1958). The effect of peripheral nerve block on the appreciation and execution of finger movements. *J. Physiol.* (*Lond.*), **143**, 55–67.

Prys-Roberts, C., and Spalding, J. M. K. (1970). In: *Ciba Foundation Symposium on Breathing: Hering-Breuer Centenary Symposium*, p. 249. London: Churchill.

Sherrington, C. S. (1918). Observations on the sensual role of the proprioceptive nerve supply of the extrinsic ocular muscles. *Brain*, **41**, 332–343.

Sherrington, C. S. (1924). Problems of muscular receptivity. *Nature* (*Lond.*), **113**, 892–894 and 929–932.

Siegfried, J. (1961). Topographie des projections corticales du nerf vague chez le chat. *Helv. physiol. pharmacol. Acta*, **19**, 269–278.

Skoglund, S. (1956). Anatomical and physiological studies of knee joint innervation in the cat. *Acta physiol. scand.*, **36**, Suppl. 124.

Smith, S. M., Brown, H. O., Toman, J. E. P., and Goodman, L. S. (1947). The lack of cerebral effects of *d*-tubocurarine. *Anesthesiology*, **8**, 1–14.

Torelli, G., and Brandi, G. (1961). Regulation of the ventilation at the beginning of muscular exercise. *Int. Z. angew. Physiol.*, **19**, 134–142.

Widdicombe, J. G. (1954). Receptors in the trachea and bronchi of the cat. *J. Physiol.* (*Lond.*), **123**, 71–104.

Widdicombe, J. G. (1964). Respiratory reflexes. *Handbook of Physiology*, Sect. 3. Respiration Vol. I, pp. 585–630. Eds. W. O. Fenn and H. Rahn. Washington, D.C.: American Physiological Soc.

Wiley, R. L., and Zechman, F. W. (1966–7). Perception of added airflow resistance in humans. *Resp. Physiol.*, **2**, 73–87.

Wright, G. W., and Branscomb, B. V. (1954). The origin of the sensations of dyspnoea. *Trans. Amer. clin. climat. Ass.*, **66**, 116–123.

Yamamoto, S., Miyajima, M., and Urabe, M. (1960). Respiratory neuronal activities in spinal afferents of cat. *Jap. J. Physiol.*, **10**, 509–517.

Yamamoto, S., Sugihara, S., and Kuru, M. (1956). Microelectrode studies on sensory afferents in the posterior funiculus of cat. *Jap. J. Physiol.*, **6**, 68–85.

Zechman, F. W., O'Neill, R., and Shannon, R. (1967). Effect of low cervical spinal cord lesions on detection of increased airflow resistance in man. *Physiologist*, **10**, 356.

# THE RESPIRATORY MUSCLES IN DISEASE

# Chapter XV

# DISEASES OF THE LUNG AND CHEST WALL

E. J. M. CAMPBELL

THERE have been many studies giving data on the mechanics of the lung or changes in lung volumes in a wide range of pathological conditions. This account is restricted to a description of the main types of disturbances and how they affect the kinematics, statics and dynamics of the chest wall and the respiratory muscles. It can be said at the outset that these subjects have as yet been little studied. For more detailed accounts of the other aspects of respiratory disorders in these conditions see Bates and Christie (1964) or the *Handbook of Physiology* (Fenn and Rahn, 1965).

## OBSTRUCTIVE CONDITIONS

In patients with asthma, bronchitis and emphysema the intrathoracic airways narrow, particularly during expiration, either because of disease processes in their walls or because of the decrease in the lung recoil. The effects on pulmonary mechanics have been intensively studied, but the effects on the chest wall have been little studied. The compliance of the chest wall has been reported to be reduced (Cherniack and Hodson, 1963), but their results may have been affected by a variable degree of muscular relaxation.

Airways obstruction leads to an increase in the residual volume and the functional residual capacity (RV and FRC) and a reduction in the inspiratory capacity (IC) and vital capacity (VC). In patients in whom the airways obstruction is chronic, certain changes are recognisable in the shape and movements of the thoracic cage, but the mode of their production and the importance of the type, severity and duration of the disease in their production are uncertain (Campbell, 1969; Godfrey *et al.*, 1969). In brief, the chest is said to become barrel shaped, to lose its normal expansion and to move up and down en bloc. In more detail, the mechanics can be described as follows: the ribs are more horizontal and in the upper chest their movement in the coronal plane ("bucket-handle movement") is diminished. The diaphragm is lower in position and its costal fibres are oblique rather than vertical; they therefore pull the costal margin inwards rather than outwards. The low position of the diaphragm may, by its action on the central tendon and the base of the pericardium, also be responsible for the inspiratory descent of the trachea which is commonly present. The scalene muscles contract forcibly during quiet inspiration and the sternomastoids are readily used when breathing is increased. The abdominal expiratory muscles are not used (Campbell and Friend, 1955).

As implied above, these changes are not exactly explicable. They are certainly more closely related to the secondary increase in lung volume rather than to the airway obstruction, but the relationship is not simple. They are not present in a young healthy person breathing artificially at a larger lung volume than is usually present in a florid case. Two probable additional factors are slow plastic changes in the configuration of the rib cage when it is sustained at an increased volume and secondly the changed patterns of muscular force as the ribs become horizontal, the diaphragm depressed and the upper chest accessory muscles increasingly active.

It should be noted that the average alveolar pressure (integrated with respect to time) in patients with airway obstruction is above atmospheric. This slight pressure operating for a long time may increase the antero-posterior diameter. Systematic studies on the changes of shape and on the movements of the chest wall in these conditions must be performed before the factors causing these abnormalities can be interpreted. The recent studies of the kinematics of the chest wall (Chapter II) indicate the approaches that could be employed. On inspection of a severe case the movements of various parts of the chest are obviously out of phase particularly when breathing is increased. This increases the work of breathing in a manner that is not registered by the conventional analysis of pressure, volume and flow changes (Chapter V), and may contribute to the sensation of dyspnoea (Chapter XIV).

The effects of chronic airway obstruction on the ordinary muscles of breathing are uncertain. The diaphragm appears to contract more forcibly than normal, but because of its poor mechanical advantage it probably contributes less to the external mechanical work of breathing. The effects on the intercostals are even less known. Campbell (1954) noticed that EMG activity in the upper spaces recorded with surface electrodes was surprisingly slight compared with the palpable tautening of the space during inspiration and suggested much of this tautening may be mechanically transmitted from the scalene muscles. The maximum breathing capacity (MBC) is greatly reduced, the work of breathing increased and the efficiency of breathing reduced (Cherniack, 1959), but these patients can sustain a ventilation equal to their maximum breathing capacity (MBC) for many minutes (Clark *et al.*, 1969). Burns and Howell (1969) studied a group of patients with chronic airways obstruction who were more breathless than could be accounted for by the severity of their pulmonary disease. In these patients breathlessness was less clearly related to exertion and there were a number of features which emphasised psychological and psychiatric factors in the genesis of dyspnoea.

## NON-OBSTRUCTIVE (RESTRICTIVE) CONDITIONS

### Intrathoracic Conditions

Lesions such as consolidation, fibrosis or pleural effusion diminish the movement of the overlying chest wall, but no detailed studies of kinematics have been reported. The overall effect of these conditions on the ventilatory function is to cause reduction in VC (usually affecting principally the IC), little change in FRC, some reduction in MBC, but rarely and only terminally ventilatory failure with hypercapnia. The work of breathing is increased.

The effect of such lesions, whether localised or generalised, on the respiratory muscles is unknown. Their action might be expected to be increased by the added mechanical load, but it is also possible that persistent failure of a part of the chest to expand may lead to reduction in the activity of the local respiratory muscles. It is also possible that localised disease may inhibit the action of the respiratory muscles reflexly. If these factors were all operative, the resultant action of the respiratory muscles and the movements of the chest in any given region would represent a complex resultant of local and general, mechanical and reflex, acute and chronic factors. This subject has not been studied.

### Conditions Affecting the Chest Wall

Lesions of the rib cage affect the mechanics of breathing if they reduce the firmness or the mechanical uniformity of the chest wall. Thus the distorted cage of kyphoscoliosis causes severe reduction in ventilatory capacity often leading to ventilatory failure with hypercapnia (Hanley *et al.*, 1958; Bergofsky *et al.*, 1959) and so does the crushed or stove-in chest. On the other hand, although ankylosing spondylitis may cause some reduction in VC it causes little reduction in MBC and never, if uncomplicated, causes ventilatory failure (Travis *et al.*, 1960; Zorab, 1962; Miller and Sproule, 1964). The reduction in VC is due both to a decrease in TLC and to an increase in RV; the FRC is also increased. Kinematic studies of all these conditions are needed.

Increase in abdominal mass as by pregnancy (Prowse and Gaensler, 1965; Gee *et al.*, 1967) or obesity (Dempsey *et al.*, 1966; Holley *et al.*, 1967) characteristically reduces the expiratory reserve volume (ERV) and FRC. At the eighth month of pregnancy there is about a 25 per cent reduction in FRC and a 40 per cent reduction in ERV. Adiposity sufficient to double the body weight causes similar changes. Pregnancy has little effect on the ventilatory capacity. Obesity reduces the compliance of the chest wall (Naimark and Cherniack, 1960), but the effects on dynamics are variable. Although ventilatory failure with hypercapnia may be associated with obesity in the Pickwickian syndrome (Burwell *et al.*, 1956) it is uncertain that the cause of the underbreathing is mechanical (Campbell, 1965). In many patients with this syndrome the obesity is mild (Richter *et al.*, 1957; Rodman *et al.*, 1962; McNicol and Pride, 1965). Conversely, gross obesity (over 30 stone; 190 kg) is compatible with a normal $P_{CO_2}$ (Said, 1960; Cullen and Formel, 1962; Billiet *et al.*, 1964). Lourenço (1969) has recently demonstrated that the increase in integrated EMG activity in the diaphragm per millimetre rise in $Pa_{CO_2}$ was much less in obese subjects who were hypoventilating than in obese subjects with normal ventilation; total chest compliance showed no significant differences between the two groups. This suggests that an inability to increase respiratory muscle activity sufficiently to overcome the load caused by obesity is a major factor in the genesis of hypoventilation in these subjects. The effects of ascites are uncertain because few patients have been studied in whom the lungs themselves were unaffected by the causative disease (Abelmann *et al.*, 1954).

## REFERENCES


ABELMANN, W. H., FRANK, N. R., GAENSLER, E. A. and CUGELL, D. W. (1954). Effects of abdominal distention by ascites on lung volumes and ventilation. *Arch. intern. Med.*, **93**, 528–540.

BATES, D. V., and CHRISTIE, R. V. (1964). *Respiratory Function in Disease: Introduction to the Integrated Study of the Lung.* Philadelphia: W. B. Saunders Co.

BERGOFSKY, E. H., TURINO, G. M., and FISHMAN, A. P. (1959). Cardiorespiratory failure in kyphoscoliosis. *Medicine*, **38**, 263–317.

BILLIET, L., VAN DE WOESTIJNE, K. P., and GYSELEN, A. (1964). In: *L'Exploration Fonctionelle Pulmonaire*, pp. 999. Eds. H. Denolin, P. Sadoul and N. G. M. Orie. Paris: Flammarion.

BURNS, B. H., and HOWELL, J. B. L. (1969). Disproportionately severe breathlessness in chronic bronchitis. *Quart. J. Med.*, **38**, 277–294.

BURWELL, C. S., ROBIN, E. D., WHALEY, R. D., and BICKELMANN, A. G. (1956). Extreme obesity associated with alveolar hypoventilation—a Pickwickian syndrome. *Amer. J. Med.*, **21**, 811–818.

CAMPBELL, E. J. M. (1954). The muscular control of breathing in man. (Ph.D. Thesis, Univ. of London.)

CAMPBELL, E. J. M. (1965). Respiratory failure. *Brit. med. J.*, **1**, 1451–1460.

CAMPBELL, E. J. M. (1969). Physical signs of diffuse airways obstruction and lung distension. *Thorax*, **24**, 1–3.

CAMPBELL, E. J. M., and FRIEND, J. (1955). Action of breathing exercises in pulmonary emphysema. *Lancet*, **1**, 325–329.

CHERNIACK, R. M. (1959). Respiratory effects of obesity. *Canad. med. Ass. J.*, **80**, 613–616.

CHERNIACK, R. M., and HODSON, A. (1963). Compliance of the chest wall in chronic bronchitis and emphysema. *J. appl. Physiol.*, **18**, 707–711.

Clark, T. J. H., Freedman, S., Campbell, E. J. M., and Winn, R. R. (1969). The ventilatory capacity of patients with chronic airways obstruction. *Clin. Sci.*, **36**, 307–316.

Cullen, J. H., and Formel, P. F. (1962). The respiratory defects in extreme obesity. *Amer. J. Med.*, **32**, 525–531.

Dempsey, J. A., Reddan, W., Rankin, J., and Balke, B. (1966). Alveolar-arterial gas exchange during muscular work in obesity. *J. appl. Physiol.*, **21**, 1807–1814.

Fenn, W. O., and Rahn, H., Eds (1965). *Handbook of Physiology*, Sect. 3. Respiration, Vol. II. Washington, D.C.: American Physiological Soc.

Gee, J. B. L., Packer, B. S., Millen, J. E., and Robin, E. D. (1967). Pulmonary mechanics during pregnancy. *J. clin. Invest.*, **46**, 945–952.

Godfrey, S., Edwards, R. H. T., Campbell, E. J. M., Armitage, P., and Oppenheimer, E. A. (1969). Repeatability of physical signs in airways obstruction. *Thorax*, **24**, 4–9.

Hanley, T., Platts, M. M., Clifton, M., and Morris, T. L. (1958). Heart failure of the hunchback. *Quart. J. Med.*, **27**, 155–171.

Holley, H. S., Milic-Emili, J., Becklake, M. R., and Bates, D. V. (1967). Regional distribution of pulmonary ventilation and perfusion in obesity. *J. clin. Invest.*, **46**, 475–481.

Lourenço, R. V. (1969). Diaphragm activity in obesity. *J. clin. Invest.*, **48**, 1609–1614.

McNicol, M. W., and Pride, N. B. (1965). Unexplained underventilation of the lungs. *Thorax*, **20**, 53–65.

Miller, J. M., and Sproule, B. J. (1964). Pulmonary function in ankylosing spondylitis. *Amer. Rev. resp. Dis.*, **90**, 376–382.

Naimark, A., and Cherniack, R. M. (1960). Compliance of the respiratory system and its components in health and obesity. *J. appl. Physiol.*, **15**, 377–382.

Prowse, C. M., and Gaensler, E. A. (1965). Maternal physiology. Respiratory and acid base changes during pregnancy. *Anesthesiology*, **26**, 381–392.

Richter, T., West, J. R., and Fishman, A. P. (1957). The syndrome of alveolar hypoventilation and diminished sensitivity of the respiratory center. *New Engl. J. Med.*, **256**, 1165–1170.

Rodman, T., Resnick, M. E., Berkowitz, R. D., Fennelly, J. F., and Olivia, J. (1962). Alveolar hypoventilation due to involvement of the respiratory center by obscure disease of the central nervous system. *Amer. J. Med.*, **32**, 208–217.

Said, S. I. (1960). Abnormalities of pulmonary gas exchange in obesity. *Ann. intern. Med.*, **53**, 1121–1129.

Travis, D. M., Cook, C. D., Julian, D. G., Crump, C. H., Helliesen, P., Robin, E. D., Bayles, T. B., Burwell, C. S. (1960). The lungs in rheumatoid spondylitis. *Amer. J. Med.*, **29**, 623–632.

Zorab, P. A. (1962). The lungs in ankylosing spondylitis. *Quart. J. Med.*, **31**, 267–280.

# Chapter XVI

# DISEASES OF THE NERVOUS SYSTEM

J. NEWSOM DAVIS

## SUPRASPINAL DISORDERS

### APRAXIA

THE respiratory muscles are subject both to a voluntary (willed) and an involuntary descending influence. Rarely, disease of the cerebral hemispheres may interfere with voluntary control. In the limbs, a disturbance in the ability to perform familiar purposive movements when the motor and sensory mechanisms required for the act remain intact has been termed an "apraxia". Hebertson and his colleagues (1959) have described a similar disorder for certain respiratory movements in a proportion of patients who had Cheyne-Stokes respiration. While these patients responded promptly and consistently to non-respiratory commands, they showed a striking lack of response to respiratory commands such as "hold your breath", and "blow". The writer has also seen a similar case who was unable, on command, to take a deep breath, cough, sniff or hold his breath, although normal reflex coughing occurred and deep breathing could be induced by increasing dead space. Other commands, for instance that he should brace his shoulders back, were well followed. The patients of Hebertson *et al.* (1959) also showed a lack of awareness of their respiratory symptoms, despite the presence of Cheyne-Stokes respiration, a disorder which the authors term "respiratory anosognosia".

Difficulty with volitional control of breathing, which may not be a true apraxia, has also been reported in patients with post-encephalitic Parkinsonism (Kim, 1968).

### RESPIRATORY DYSRHYTHMIA

**Periodic Respiration (Cheyne-Stokes Respiration)**

In this disorder, tidal volume is subject to a smooth, incremental-decremental change, usually separated by periods of apnoea. It should be distinguished from the respiratory dysrhythmia in which there is irregular variation of tidal volume and frequency from breath to breath (see below).

The association of periodic breathing with disease of the cerebral hemispheres was described in the original case report by Cheyne (1818), and has been confirmed by many writers since then (Jackson, 1895;

Grimmer *et al.*, 1939; Talbert *et al.*, 1954). The mechanism of Cheyne-Stokes respiration (CSR) is controversial. Douglas and Haldane (1909) believed it to be due to the effects of hypoxia on the respiratory centre. Brown and Plum (1961) have argued for a primary neurogenic origin for the disorder, while acknowledging that circulatory delay and hypoxaemia, which were not present in all their cases, may accentuate the periodicity. The increased ventilatory response to $CO_2$ found in patients with CSR is attributed to a hypersensitivity of the respiratory centre deprived of the inhibitory influence of the cerebral cortex. However, hypersensitivity to $CO_2$ accounts only in part for the disorder, since it does not explain why some patients with such hypersensitivity do not develop periodic breathing (Heyman *et al.*, 1958; Brown and Plum, 1961). A raised respiratory threshold for $CO_2$ may be a contributory cause (Brown and Plum, 1961). Lange and Hecht (1962), on the other hand, favour a delay in the lung to brain circulation time as the fundamental cause of CSR, the majority of their group of patients having no evidence of organic neurological disease. It has been shown that a circulatory delay introduced between heart and brain in dogs gives rise to CSR (Guyton *et al.*, 1956). The abnormal $CO_2$ response in patients with CSR was thought to be due to altered respiratory mechanics secondary to congestive heart failure (Lange *et al.*, 1968).

**Post-hyperventilation Apnoea**

In patients with bilateral disease of the cerebral hemispheres, post-hyperventilation apnoea is commonly prolonged (in healthy subjects, apnoea does not usually exceed 12 seconds), and has proved to be a useful bedside indication of diffuse impairment of cerebral function (Plum *et al.*, 1962).

**Hyperpnoea**

Many clinical reports have described the association of hyperpnoea with brain stem infarction or haemorrhage (see Steegman, 1951; Plum and Swanson, 1959). Pupillary abnormalities are common, often with oculomotor palsies. Respiratory frequency may range from 28 to 38 per minute with alveolar minute ventilation four times the normal and an arterial pH of up to 7·52, as in the series of Plum and Swanson (1959). A psychogenic cause for hyperventilation is, of course, also well recognised (Guze *et al.*, 1952). In epidemic encephalitis lethargica, tachypnoea was the commonest respiratory disorder, according to Turner and Critchley (1925, 1928), being either continuous or paroxysmal, and in some patients occurring only in sleep. Swallowing could temporarily inhibit it. Occasionally, there was a synchronous contraction of limb muscles. Respiratory rates of up to 70/minute could occur. In one patient, Turner and Critchley (1925) recorded a plasma pH of 7·57, and a similar value was recorded in another case by Harrop and Loeb (1923). Hyperventilation was the presenting feature of a patient who subsequently died from infiltration of mid- and upper pons and of midbrain by a malignant cerebral reticulosis (Lange and Laszlo, 1965). Medulla and spinal cord were spared. Neurogenic hyperventilation in a patient with acute intermittent porphyria has also been described (Baker and Messert, 1967), developing during sleep.

The post-mortem changes in the brain stem in six patients with neurogenic hyperventilation were described by Plum and Swanson (1959). The dominant finding was necrosis of the central portion of the pons, in four cases due to infarction and in two to haemorrhage, with involvement of the corticospinal tracts and basal pontine nuclei. However, pulmonary and circulatory changes secondary to the brain stem damage could not be excluded as the stimulus to respiration, and may indeed account for all cases of apparently "neurogenic" hyperventilation in comatose patients (Plum, 1966).

### Irregular Respiration

Cluster breathing, apneusis or gasping may be present. In cluster breathing, several breaths are separated by a period of apnoea but the smooth incremental-decremental pattern of Cheyne-Stokes respiration is lacking. An inspiratory position (apneusis) or expiratory position (expiratory apneusis) may be held for prolonged periods. When apneusis occurs at every breath, a gasping respiration results.

Irregular breathing was seen in acute cases of encephalitis lethargica (Turner and Critchley, 1925) and sometimes returned to normal during sleep. Pontine haemorrhage may give rise to a marked irregularity of breathing (Steegman, 1951; Plum and Alvord, 1964), which is often a terminal event. Both the patients described by Plum and Alvord (1964) showed severe inspiratory and expiratory apneusis, and cluster breathing was present in one of them. There was no gross abnormality of the blood gases, but in one patient the ventilatory response to $CO_2$ was reduced, and the expiratory "cramps" were abolished by $CO_2$ breathing. Apneusis has also been described in an infant in whom small pontine haemorrhages were found at post-mortem (Kirkwood and Myers, 1923). Irregularity of rate and depth of respiration occurs in acute bulbar poliomyelitis (Sarnoff *et al.*, 1951), characteristically apparent only during drowsiness in the early stages (Plum and Swanson, 1958) and later leading to profound hypoventilation.

Baker *et al.* (1950) correlated the respiratory dysfunction in patients with bulbar poliomyelitis with the medullary changes at autopsy. Damage to the small reticular neurones in the ventro-lateral region of the reticular formation in the upper medulla was found in patients who had severe respiratory symptoms characterised by varying rate and depth, and apnoeic periods. However, a loss of these small neurones of up to 30 per cent might be associated with no disorder of respiration. Similar changes in the reticular formation were reported in two cases by Finley (1931) and by Plum and Swanson (1958).

In two patients with apneustic breathing, basilar artery thrombosis had caused almost complete brain stem transection at the level of the trigeminal motor nucleus, infarction in particular extending laterally to involve the lateral reticular formation at the level of the locus caeruleus (Plum and Alvord, 1964). The medial pontine tegmentum was spared.

## Diaphragmatic Flutter

In diaphragmatic flutter, contractions occur at a frequency which is usually between 1 and 8 per second. It is distinguished from hiccup (singultus) by the absence of the characteristic inspiratory sound. It may affect one or both halves of the diaphragm, and is occasionally seen in association with palatal myoclonus (Guillain, 1938; Langworthy and Grimmer, 1939). In these latter cases, degeneration of the olivary complex in the brain stem has been present (Guillain, 1938).

Previous cases reported in the literature have been reviewed by Rigatto and Medeiros (1962). In most cases, no underlying disorder was apparent. In a few patients, there was a history of epidemic encephalitis lethargica within four years of the onset of the flutter. In a number of other cases phrenic nerve irritation was implicated, and attributed variously to cervical rib, enlarged mediastinal glands, pleurisy and peritoneal adhesions. Flutter in association with apneustic respiration has been described in pregnancy (Ting *et al.*, 1963). Flutter could be precipitated in these subjects by certain manoeuvres such as coughing, deep breathing, eating, sneezing and exercise. Swallowing and supraclavicular pressure occasionally suppressed an attack. The mechanism of this disorder is unknown.

## Hiccup

Earlier accounts of hiccup have been predominantly concerned with the classification of aetiological factors and with its treatment (Mayo, 1932; Noble, 1934; Bailey, 1943; Samuels, 1952; Butt *et al.*, 1961), sometimes with emphasis on its historical aspects (see Riddell, 1930). Even in recent reports, the extent to which muscles other than the diaphragm are involved has not been established. No attempt will be made here to tabulate the extensive list of precipitating factors but rather to indicate the fields from which the phenomenon may be evoked.

Stimulation in the receptive field of the vagus, such as by visceral manipulation at operation (Wylie and Churchill-Davidson, 1966) and such as occurs in many gastro-intestinal disorders, may cause hiccup. Irritation of the phrenic nerve, for instance by neurofibroma or pericarditis, may also give rise to the phenomenon. It has been reported in association with herpes zoster affecting lumbar (Brooks, 1931) and thoracic segments (Efrati, 1956). Lesions of the central nervous system which may give rise to hiccup involve the brain stem, and include tumours (Stotka *et al.*, 1962), encephalitis, vascular disease and syphilis. Toxic causes (e.g. uraemia) have usually been presumed to act at a central level.

Electromyographic recording of hiccup provides some information about its nature (Newsom Davis, 1970). The brief discharge which characterises the condition can be recorded not only in the diaphragm but also in the inspiratory intercostal muscles. Its intensity increases during hypocapnia following hyperventilation and diminishes as rhythmic breathing resumes. During voluntary activation of an expiratory muscle, there is a brief expiratory inhibition coincident with the hiccup, which becomes progressively less effective as the drive to the expiratory muscle increases. The characteristics of hiccup are consistent with an origin in a supraspinal mechanism distinct from that serving rhythmic breathing.

Hiccup may be abolished by certain stimuli involving the pharynx, such as swallowing iced water and traction of the tongue. Salem and his colleagues (1967) have reported an encouragingly high success rate in curing hiccup in both conscious and anaesthetised subjects by stimulation of the pharynx with a nasal catheter. The authors argue that the afferent discharge in the vagus nerves evoked by the stimulus may block specifically those afferent impulses responsible for hiccup. Alternatively, pharyngeal stimulation may terminate hiccup by its effect upon the central respiratory control mechanism, for it is known from animal studies that such stimulation may change the respiratory pattern (Takagi *et al.*, 1966). Other effective stimuli may be presumed to act centrally in a similar manner; breath holding, rebreathing, fear, anger and sexual intercourse have all been known to abolish the disorder.

## Central Alveolar Hypoventilation

This condition is due to a disorder of the central neural mechanism concerned with controlling ventilation. In central alveolar hypoventilation (CAH), $Pa\text{CO}_2$ is raised and $Pa\text{O}_2$ is low. Polycythaemia and a low blood pH are usually present. The vital capacity is often normal or only slightly reduced and the ventilatory rate is low in some patients. The disorder is characterised further by the return of the blood gases to normal values on voluntary hyperventilation, and by an impaired or absent ventilatory response to inhaled $CO_2$ and sometimes to exercise.

Patients with a past history of encephalitis lethargica have later been found to have CAH (Smith, 1956; Efron and Kent, 1957; Garlind and Linderholm, 1958; Strieder *et al.*, 1967). Parkinsonism is not necessarily present. CAH may also follow other forms of encephalitis (Cohn and Kuida, 1962; Seriff, 1965; Campbell, 1965), including poliomyelitis. In the latter disease it may occur both in the acute stage (Sarnoff *et al.*, 1951; Plum and Swanson, 1958) and convalescent periods (Lukas and Plum,

1952; Linderholm and Werneman, 1956; Cherniack *et al.*, 1957; Fishman *et al.*, 1966). CAH has also been reported in two patients with sarcoidosis involving the central nervous system (Daum *et al.*, 1965), following surgery to the brain stem (Severinghaus and Mitchell, 1962), in cases of treated syphilis, cervical syringomyelia (Rodman *et al.*, 1962), Arnold-Chiari malformation (Campbell, 1965), and in a man with mental retardation attributed to neonatal cyanosis (Seriff, 1965). Neurological signs were present in a number of further cases but no diagnosis could be made (Ratto *et al.*, 1955; Bedell and Rosenberg, 1961; Naeye, 1961; Fraser *et al.*, 1963). Hughes (1967) has described a patient with hypoventilation secondary to brain stem tumour. Bulbar palsy due to motor neurone disease may also cause CAH.

Benaim and Worster-Drought (1954) reported a patient with dystrophia myotonica who had alveolar hypoventilation ($Paco_2$ 67·4 mm Hg, $Pao_2$ 57 mm Hg and pH 7·29), and secondary polycythaemia. Alveolar minute ventilation was between 7 and 8 litres, but could be increased voluntarily to 45 litres. Myotonic discharges were recorded from a number of muscles including the intercostals, and the diaphragm was seen on fluoroscopy to have a considerably reduced inspiratory excursion with jerky relaxation. A number of further patients with this disease who developed alveolar hypoventilation have since been reported (Bashour *et al.*, 1955; Kilburn, Eagan and Sieker, 1959; Kilburn, Eagan and Heyman, 1959; Cannon, 1962; Gillam *et al.*, 1964). In the series of nine patients studied by Kilburn, Eagan, Sieker and Heyman (1959) there was an impaired ventilatory response to $CO_2$. Furthermore, in three patients who inhaled 100 per cent oxygen for five minutes, the $Pco_2$ increased significantly with a reduction in minute ventilation. On the basis of these observations, it was suggested that a disorder of the central mechanism was contributing to the ventilatory disturbance of these patients. It is of interest in this connection that hypersomnolence may occur in dystrophia myotonica (Phemister and Small, 1961), and that patients may often suffer an abnormal depression of respiration with anaesthetic agents (Kaufman, 1960; Gillam *et al.*, 1964). It is also possible that the afferent discharge of the disordered muscle spindles found in this disease is contributing to the ventilatory disorder.

Autopsy has been abnormal in four out of the five patients with CAH examined. Increased capillaries were found in the medullary region in two patients (Seriff, 1965) and neuronal degenerative changes maximal in the medulla in the others (Naeye, 1961; Fraser *et al.*, 1963).

There remain a few patients in whom no neurological disease was apparent (Pare and Lowenstein, 1956; Richter *et al.*, 1957; Lawrence, 1959; Rodman *et al.*, 1962; Tsitouris and Fertakis, 1965). A post-mortem examination carried out in one of these patients showed no abnormalities in the medulla (Lawrence, 1959).

CAH shares some characteristics with the physiological changes in healthy subjects occurring during drowsiness and sleep, when a significant rise in $P_{CO_2}$ and fall in $P_{O_2}$ may occur (Mills, 1953; Birchfield *et al.*, 1958). It is not unexpected therefore that patients with CAH often may hypoventilate most severely during sleep (Fraser *et al.*, 1963; Fishman *et al.*, 1966).

Microscopic examination of the medulla has indicated rather specific sites of damage in polio patients with ventilatory failure (Baker *et al.*, 1950; see page 238). Such changes have been absent in the few autopsies carried out in other patients with CAH. The impaired response of the central mechanism to a $CO_2$ stimulus certainly accounts for the observed changes in ventilation, but the precise nature of the central disorder is not known.

## Respiratory Mechanics

In hemiplegia, movements of the chest may be affected on the ipsilateral side; this has been both observed clinically and confirmed by recordings (Jackson, 1895; Kolb and Kleyntjens, 1937; Robinson *et al.*, 1950; Fluck, 1966). In quiet breathing, upper chest movement was decreased by a mean value of about 10 per cent on the side of the hemiplegia, but the lower chest movement was symmetrical (Fluck, 1966). During voluntary deep breathing, the reduction of upper chest movement on the hemiplegic side rose to over 15 per cent, and movement of the lower chest was also now found to be 10 per cent less than the unaffected side. In contrast to this, during involuntary deep breathing there was no asymmetry of chest movement.

Parkinsonism can give rise to dyspnoea and respiratory dysfunction (Nugent *et al.*, 1958; Keltz, 1965; Neu *et al.*, 1967). Vital capacity may be reduced with an increase in FRC and RV. No evidence of alveolar hypoventilation attributable to the disease was present in these patients. In post-encephalitic Parkinsonism, the amplitude of tidal volume may vary less than in healthy subjects and the respiratory frequency is higher (Kim, 1968). Unilateral Parkinsonism is associated with impaired volitional chest movements on the ipsilateral side which, in contrast to hemiplegic patients, does not improve with involuntary deep breathing (Kolb and Kleyntjens, 1937). Delhez and Petit (1961) demonstrated electromyographically a continuous activity of an antagonistic type in many accessory respiratory muscles of patients with Parkinsonism, with the discharge pattern in the diaphragm relatively unaffected. These observations also in part explain the monotonous speech of these patients. The studies of Draper and his co-workers (1960) have shown that respiratory muscle activity is finely controlled during speech. The loudness of the voice is dependent upon the sublaryngeal pressure (see p. 106), and small fluctuations in pressure are associated with stressed syllables. Simultaneous recording of expiratory intercostal activity reveals that these muscles are directly concerned with the control of the pressure fluctuations, stressed syllables being preceded by a burst of electromyographic discharge. The impairment of the fine control of muscle present in Parkinsonism would thus, in the respiratory muscles, be expected to lead to a voice of limited and inflexible volume.

The involuntary movements of chorea, choreoathetosis and the dystonias may be superimposed upon respiratory movements (see Robinson *et al.*, 1950) and are particularly apparent during speech.

In cerebellar disease, vital capacity, inspiratory reserve volume and maximum breathing capacity may all be reduced (Hormia, 1957). The explosive speech from which these patients may suffer implies a disordered control of the expiratory muscles.

## SPINAL DISORDERS

### Tract Lesions

Localised cord lesions above the level of the phrenic outflow (C4) may result in total respiratory paralysis. From post-mortem studies of patients who had undergone high cervical cordotomies, Nathan (1963) concluded that many or all of the fibres of the descending respiratory tract lie in the anterior part of the lateral columns (see Chapter XII). Respiratory failure is now a recognised complication of bilateral high cervical cordotomy (Belmusto *et al.*, 1963). Occasionally ventilatory failure may occur in multiple sclerosis when there is severe spinal cord involvement (Guthrie *et al.*, 1952).

Lesions of the cervical cord below C5 or of the upper thoracic cord may cause a decreased vital capacity, lung compliance, expiratory reserve volume, and maximum breathing capacity (Hemingway *et al.*, 1958; Stone and Keltz, 1963) and the cough reflex will be weakened. Bergofsky (1964) reported alveolar hypoventilation in some patients with cervical cord lesions, but in other series (McKinley *et al.*, 1969) blood gases and dynamic lung compliance have been normal. This discrepancy might be due to a decreased incidence of respiratory complications in the latter series. McKinley *et al.* (1969) found that the thoracic component of ventilation ranged from 22 per cent to 90 per cent of total ventilation in their chronic quadriplegic patients; this was attributed to the activity of the accessory muscles of respiration. The oxygen cost of breathing in tetraplegics is also slightly raised (Silver, 1963). Respiratory rate and airway resistance are unaffected. Guttmann and Silver (1965) have recorded intercostal electromyographic activity in phase with late inspiration and with early expiration in patients with total cord lesions above T1. This activity probably represents a stretch reflex in the paralysed muscles.

### Anterior Horn Cell Disease

The respiratory muscle paralysis of poliomyelitis may lead to ventilatory failure, but frequently this is complicated by the bulbar effects of the disease (Sarnoff *et al.*, 1951; Brody *et al.*, 1964). Motor neurone disease (amyotrophic lateral sclerosis), which involves both lower and upper motor neurones, may be associated with dyspnoea and ventilatory failure (Feltman *et al.*, 1952; Miller *et al.*, 1957; Paul and Appenzeller, 1962), and is the commonest cause of death in this illness. The alveolar hypoventilation which may be present can result either from the neurogenic wasting of the respiratory muscles or from the involvement of the respiratory neurones in the brain stem (see above). Tetanus, in which the normal inhibitory influences on the anterior horn cell are blocked, may lead to severe respiratory embarrassment during the muscle spasms.

## DISORDERS OF NERVE, NEUROMUSCULAR TRANSMISSION AND MUSCLE

Ventilatory insufficiency is a common accompaniment of these disorders. In peripheral neuropathy, it may be acute in onset, requiring assisted ventilation (Hewer *et al.*, 1968), or relapsing as in some cases of hypertrophic polyneuritis (Harris and Newcomb, 1929). Ventilatory failure is often a terminal event in patients with chronic neuropathies. Myasthenia gravis can also produce acute respiratory failure. In muscular dystrophy and polymyositis, ventilatory insufficiency is common. Alveolar hypoventilation, however, is not a usual feature of non-myotonic dystrophies, even when ventilatory insufficiency is profound (Kilburn, Eagan, Sieker and Heyman, 1959), although occasional cases occur (Neustadt *et al.*, 1964; Buchsbaum *et al.*, 1968). In dystrophia myotonica, central alveolar hypoventilation may occur (see p. 319). The muscular abnormality in this latter condition usually does not lead to significant reduction in the vital capacity (Kilburn, Eagan, Sieker and Heyman, 1959; Gillam *et al.*, 1964; Lee and Hughes, 1964). The maximum voluntary ventilation may be impaired, however, and the maximum expiratory pressure that can be achieved is severely reduced (Gillam *et al.*, 1964).

Unilateral lesions of the phrenic nerve may produce transient dyspnoea and slight impairment of respiratory function (Fackler *et al.*, 1967). The effects of sudden bilateral lesions may be profound (see McCredie *et al.*, 1962). Among the less well known causes of phrenic nerve palsy reported in the literature are included neuralgic amyotrophy (Cape and Fincham, 1965), a sequel of anti-tetanus serum (Smith and Smith, 1955; McCredie *et al.*, 1962), herpes zoster (Spiers, 1929; Borstoff, 1966) and non-penetrating trauma to the cervical nerve roots (Larson and Evans, 1963). In many patients, phrenic nerve palsy may be an isolated finding of obscure origin.

## REFERENCES

Bailey, H. (1943). Persistent hiccough. *Practitioner*, **150**, 173–177.

Baker, A. B., Matzke, H. A., and Brown, J. R. (1950). Poliomyelitis. III. Bulbar poliomyelitis; a study of medullary function. *Arch. Neurol. Psychiat. (Chic.)*, **63**, 257–281.

BAKER, N. H., and MESSERT, B. (1967). Acute intermittent porphyria with central neurogenic hyperventilation. *Neurology* (*Minneap.*), **17**, 559–566.

BASHOUR, F., WINCHELL, P., and REDDINGTON, J. (1955). Myotonica atrophica and cyanosis. *New Engl. J. Med.*, **252**, 768–770.

BEDELL, G. N., and ROSENBERG, R. S. (1961). Direct method for respiratory centre evaluation in man. *J. appl. Physiol.*, **16**, 928–933.

BELMUSTO, L., BROWN, E., and OWENS, G. (1963). Clinical observations on respiratory and vasomotor disturbance as related to cervical cordotomies. *J. Neurosurg.*, **20**, 225–232.

BENAIM, S., and WORSTER-DROUGHT, C. (1954). Dystrophia myotonica with myotonia of the diaphragm causing pulmonary hypoventilation with anoxaemia and secondary polycythaemia. *Med. ill.* (*Lond.*), **8**, 221–226.

BERGOFSKY, E. H. (1964). Mechanism for respiratory insufficiency after cervical cord injury. *Ann. intern. Med.*, **61**, 435–447.

BIRCHFIELD, R. I., SIEKER, H. O., and HEYMAN, A. (1958). Alterations in blood gases during natural sleep and narcolepsy. *Neurology* (*Minneap.*), **8**, 107–112.

BRODY, A. W., O'HALLORAN, P. S., CONNOLLY, J. J., JR., WANDER, H. J., and SCHWERTLEY, F. W. (1964). Ventilatory mechanics and strength in subjects paralyzed after poliomyelitis. *Dis. Chest*, **46**, 263–275.

BROOKS, W. D. W. (1931). Zoster, hiccup and varicella. *Brit. med. J.*, **2**, 298–299.

BROSTOFF, J. (1966). Diaphragmatic paralysis after herpes zoster. *Brit. med. J.*, **2**, 1571–1572.

BROWN, H. W., and PLUM, F. (1961). The neurologic basis of Cheyne-Stokes respiration. *Amer. J. Med.*, **30**, 849–860.

BUCHSBAUM, H. W., MARTIN, W. A., TURINO, G. M., and ROWLAND, L. P. (1968). Chronic alveolar hypoventilation due to muscular dystrophy. *Neurology* (*Minneap.*), **18**, 319–327.

BUTT, H. R., JR., HAMELBERG, W., and JACOBY, J. (1961). Hiccup: its possible cause and treatment in anaesthesia. *Anesth. Analg. Curr., Res.*, **40**, 181–185.

CAMPBELL, E. J. M. (1965). Respiratory failure. *Brit. med. J.*, **1**, 1451–1460.

CANNON, P. J. (1962). The heart and lungs in myotonic muscular dystrophy. *Amer. J. Med.*, **32**, 765–775.

CAPE, C. A., and FINCHAM, R. W. (1965). Paralytic brachial neuritis with diaphragmatic paralysis. *Neurology* (*Minneap.*), **15**, 191–193.

CHERNIACK, R. M., EWART, W. B., and HILDES, J. A. (1957). Polycythaemia secondary to respiratory disturbances in poliomyelitis. *Ann. intern. Med.*, **46**, 720–727.

CHEYNE, J. (1818). A case of apoplexy, in which the fleshy part of the heart was converted into fat. *Dublin Hosp. Reps.*, **2**, 216–223.

COHN, J. E., and KUIDA, H. (1962). Primary alveolar hypoventilation associated with Western equine encephalitis. *Ann. intern. Med.*, **56**, 633–644.

DAUM, J. J., CANTER, H. G., and KATZ, S. (1965). Central nervous system sarcoidosis with alveolar hypoventilation. *Amer. J. Med.*, **38**, 893–898.

DELHEZ, L., and PETIT, J. M. (1961). Quelques modalités de l'activité des muscles respiratoires chez le parkinsonien. (Controle électromyographique.) *Rev. franç. Étud. clin. biol.*, **6**, 580–584.

DOUGLAS, C. G., and HALDANE, J. S. (1909). The causes of periodic or Cheyne-Stokes breathing. *J. Physiol.* (*Lond.*), **38**, 401–419.

DRAPER, M. H., LADEFOGED, P., and WHITTERIDGE, D. (1960). Expiratory pressures and air flow during speech. *Brit. med. J.*, **1**, 1837–1843.

EFRATI, P. (1956). Obstinate hiccup as a prodromal symptom in thoracic herpes zoster: prompt cure after injection of Largactil. *Neurology* (*Minneap.*), **6**, 601–602.

EFRON, R., and KENT, D. C. (1957). Chronic respiratory acidosis due to brain disease. *Arch. Neurol. Psychiat.* (*Chic.*), **77**, 575–587.

FACKLER, C. D., PERRET, G. E., and BEDELL, G. N. (1967). Effect of unilateral phrenic nerve section on lung function. *J. appl. Physiol.*, **23**, 923–926.

FELTMAN, J. A., NEWMAN, W., SCHWARTZ, A., STONE, D. J., and LOVELOCK, F. J. (1952). Cardiac failure secondary to ineffective bellows action of the chest cage. *J. clin. Invest.*, **31**, 762–769.

FINLEY, K. H. (1931). The neuroanatomy in respiratory failure: report of two cases. *Arch. Neurol. Psychiat.* (*Chic.*), **26**, 754–783.

FISHMAN, A. P., GOLDRING, R. M., and TURINO, G. M. (1966). General alveolar hypoventilation: a syndrome of respiratory and cardiac failure in patients with normal lungs. *Quart. J. Med.*, **35**, 261–275.

FLUCK, D. C. (1966). Chest movements in hemiplegia. *Clin. Sci.* (1966), **31**, 383–388.

FRASER, R. S., SPROULE, B. J., and DVORKIN, J. (1963). Hypoventilation, cyanosis and polycythaemia in a thin man. *Canad. med. Ass. J.*, **89**, 1178–1182.

GARLIND, T., and LINDERHOLM, H. (1958). Hypoventilation syndrome in a case of chronic epidemic encephalitis. *Acta med. scand.*, **162**, 333–349.

GILLAM, P. M. S., HEAF, P. J. D., KAUFMAN, L., and LUCAS, B. G. B. (1964). Respiration in dystrophia myotonica. *Thorax*, **19**, 112–120.

GRIMMER, R. V., HESSER, F. H., and LANGWORTHY, O. R. (1939). Rhythmic variation of respiratory excursion with bilateral injury of cortical efferent fibres. *Arch. Neurol. Psychiat.* (*Chic.*), **42**, 862–871.

GUILLAIN, G. (1938). The syndrome of synchronous and rhythmic palato-pharyngo-laryngo-oculo-diaphragmatic myoclonus. *Proc. roy. Soc. Med.*, **31**, 1031–1038.

GUTHRIE, T. C., KURTZKE, J. F., and BERLIN, L. (1952). Acute respiratory failure in multiple sclerosis and its management. *Ann. intern. Med.*, **37**, 1197–1203.

GUTTMANN, L., and SILVER, J. R. (1965). Electromyographic studies on reflex activity of the intercostal and abdominal muscles in cervical cord lesions. *Paraplegia*, **3**, 1–22.

GUYTON, A. C., CROWELL, J. W., and MOORE, J. W. (1956). Basic oscillating mechanism of Cheyne-Stokes breathing. *Amer. J. Physiol.*, **187**, 395–398.

GUZE, S. B., GABBARD, J., ROOS, A., and SASLOW, G. (1952). Chronic psychogenic hyperventilation. *Arch. Neurol. Psychiat.* (*Chic.*), **67**, 434–440.

HARRIS, W., and NEWCOMB, W. D. (1929). A case of relapsing interstitial hypertrophic polyneuritis. *Brain*, **52**, 108–116.

HARROP, G. A., and LOEB, R. F. (1923). Uncompensated alkalosis in encephalitis. *J. Amer. med. Ass.*, **81**, 452–454.

HEBERTSON, W. M., TALBERT, O. R., and COHEN, M. E. (1959). Respiratory apraxia and anosognosia. *Trans. Amer. neurol. Ass.*, 176–179.

HEMINGWAY, A., BORS, E., and HOBBY, R. P. (1958). An investigation of the pulmonary function of paraplegics. *J. clin. Invest.*, **37**, 773–782.

HEWER, R. L., HILTON, P. J., CRAMPTON SMITH, A., and SPALDING, J. M. K. (1968). Acute polyneuritis requiring artificial ventilation. *Quart. J. Med.*, **37**, 479–491.

HEYMAN, A., BIRCHFIELD, R. I., and SIEKER, H. O. (1958). Effects of bilateral cerebral infarction on respiratory centre sensitivity. *Neurology* (*Minneap.*), **8**, 694–700.

HORMIA, A. L. (1957). Respiratory insufficiency as a symptom of cerebellar ataxia. *Amer. J. med. Sci.*, **233**, 635–640.

HUGHES, J. M. B. (1967). Central respiratory failure reversed by treatment. *Brain*, **90**, 675–680.

JACKSON, J. H. (1895). Neurological fragments XV. Superior and subordinate centres of the lowest level. *Lancet*, **1**, 476–478.

KAUFMAN, L. (1960). Anaesthesia in dystrophia myotonica. *Proc. roy. Soc. Med.*, **53**, 183–188.

KELTZ, H. (1965). The effect of respiratory muscle dysfunction on pulmonary function. *Amer. Rev. resp. Dis.*, **91**, 934–938.

KILBURN, K. H., EAGAN, J. T., and HEYMAN, A. (1959). Cardiopulmonary insufficiency associated with myotonic dystrophy. *Amer. J. Med.*, **26**, 929–935.

KILBURN, K. H., EAGAN, J. T., SIEKER, H. O., and HEYMAN, A. (1959). Cardiopulmonary insufficiency in myotonic and progressive muscular dystrophy. *New Engl. J. Med.*, **261**, 1089–1096.

KIM, R. (1968). The chronic residual respiratory disorder in post-encephalitic Parkinsonism. *J. Neurol. Neurosurg. Psychiat.*, **31**, 393–398.

KIRKWOOD, W. D., and MYERS, B. (1923). A case of inspiratory apnoea in a new-born infant. *Lancet*, **2**, 65–68.

KOLB, L. C., and KLEYNTJENS, F. (1937). A clinical study of the respiratory movements in hemiplegia. *Brain*, **60**, 259–274.

LANGE, L. S., and LASZLO, G. (1965). Cerebral tumour presenting with hyperventilation. *J. Neurol. Neurosurg. Psychiat.*, **28**, 317–319.

LANGE, R. L., and HECHT, H. H. (1962). The mechanism of Cheyne-Stokes respiration. *J. clin. Invest.*, **41**, 42–52.

LANGE, R. L., BOTTICELLI, J. T., CARLISLE, R. P., TSAGARIS, T. J., and HORGAN, J. D. (1968). Observation and simulation of the circulation, acid-base balance, and response to $CO_2$ in Cheyne-Stokes respiration. *Circulation*, **37**, 331–344.

LANGWORTHY, O. R., and GRIMMER, R. V. (1939). A physiological study of the movements in palatal myoclonus. *Bull. Johns Hopk. Hosp.*, **65**, 101–111.

LARSON, R. K., and EVANS, B. H. (1963). Eventration of the diaphragm. *Amer. Rev. resp. Dis.*, **87**, 753–756.

LAWRENCE, L. T. (1959). Idiopathic hypoventilation, polycythaemia and cor pulmonale. *Amer. Rev. resp. Dis.*, **80**, 575–581.

LEE, F. I., and HUGHES, D. T. D. (1964). Systemic effects in dystrophia myotonica. *Brain*, **87**, 521–536.

LINDERHOLM, H., and WERNEMAN, H. (1956). On respiratory regulation in poliomyelitis convalescents. *Acta med. scand.*, Suppl. **316**, 135–157.

LUKAS, D. S., and PLUM, F. (1952). Pulmonary function in patients convalescing from acute poliomyelitis with respiratory paralysis. *Amer. J. Med.*, **12**, 388–396.

MAYO, C. W. (1932). Hiccup. *Surg. Gynec. Obstet.*, **55**, 700–708.

McCREDIE, M., LOVEJOY, F. W., and KALTREIDER, N. L. (1962). Pulmonary function in diaphragmatic paralysis. *Thorax*, **17**, 213–217.

McKINLEY, A. C., AUCHINCLOSS, J. H., GILBERT, R., and NICHOLAS, J. J. (1969). Pulmonary function, ventilatory control, and respiratory complications in quadriplegic subjects. *Amer. Rev. resp. Dis.*, **100**, 526–532.

MILLER, R. D., MULDER, D. W., FOWLER, W. S., and OLSEN, A. M. (1957). Exertional dyspnoea: a primary complaint in unusual cases of progressive muscular atrophy and amyotrophic lateral sclerosis. *Ann. intern. Med.*, **46**, 119–125.

MILLS, J. N. (1953). Changes in alveolar carbon dioxide tension by night and during sleep. *J. Physiol.* (*Lond.*), **122**, 66–80.

NAEYE, R. L. (1961). Alveolar hypoventilation and cor pulmonale secondary to damage to the respiratory centre. *Amer. J. Cardiol.*, **8**, 416–419.

NATHAN, P. W. (1963). The descending respiratory pathway in man. *J. Neurol. Neurosurg. Psychiat.*, **26**, 487–499.

NEU, H. C., CONNOLLY, J. J., JR., SCHWERTLEY, F. W., LADWIG, H. A., and BRODY, A. W. (1967). Obstructive respiratory dysfunction in Parkinsonian patients. *Amer. Rev. resp. Dis.*, **95**, 33–47.

NEUSTADT, J. E., LEVY, R. C., and SPIEGEL, I. J. (1964). Carbon dioxide narcosis in association with muscular dystrophy. *J. Amer. med. Ass.*, **187**, 616–617.

NEWSOM DAVIS, J. (1970). In: *Ciba Foundation Symposium on Breathing: Hering-Breuer Centenary Symposium*, pp. 201–203. London: Churchill.

NOBLE, E. C. (1934). Hiccup. *Canad. med. Ass. J.*, **31**, 38–41.

NUGENT, C. A., HARRIS, H. W., COHN, J., SMITH, C. C., and TYLER, F. H. (1958). Dyspnoea as a symptom in Parkinson's syndrome. *Amer. Rev. Tuberc.*, **78**, 682–691.

PARE, P., and LOWENSTEIN, L. (1956). Polycythaemia associated with disturbed function of the respiratory center. *Blood*, **11**, 1077–1084.

PAUL, G. R., and APPENZELLER, O. (1962). Dyspnoea as the presenting symptom in amyotrophic lateral sclerosis. *Dis. Chest*, **42**, 558–562.

PHEMISTER, J. C., and SMALL, J. M. (1961). Hypersomnia in dystrophia myotonica. *J. Neurol. Neurosurg. Psychiat.*, **24**, 173–175.

PLUM, F. (1966). Breathlessness in neurological disease: the effects of neurological disease on the act of breathing. In: *Breathlessness*, pp. 203–222. Eds. J. B. L. Howell and E. J. M. Campbell. Oxford: Blackwell Scientific Publications.

PLUM, F., and ALVORD, E. C. (1964). Apneustic breathing in man. *Arch. Neurol.* (*Chic.*), **10**, 101–112.

PLUM, F., BROWN, H. W., and SNOEP, E. (1962). Neurologic significance of post-hyperventilation apnoea. *J. Amer. med. Ass.*, **181**, 1050–1055.

PLUM, F., and SWANSON, A. G. (1958). Abnormalities in central regulation of respiration in acute and convalescent poliomyelitis. *Arch. Neurol. Psychiat.* (*Chic.*), **80**, 267–285.

PLUM, F., and SWANSON, A. G. (1959). Central neurogenic hyperventilation in man. *Arch. Neurol. Psychiat.* (*Chic.*), **81**, 535–549.

RATTO, O., BRISCOE, W. A., MORTON, J. W., and COMROE, J. H., JR. (1955). Anoxaemia secondary to polycythaemia and polycythaemia secondary to anoxaemia. *Amer. J. Med.*, **19**, 958–965.

RICHTER, T., WEST, J. R., and FISHMAN, A. P. (1957). The syndrome of alveolar hypoventilation and diminished sensitivity of the respiratory centre. *New Engl. J. Med.*, **256**, 1165–1170.

RIDDELL, W. R. (1930). Hippocrates and hiccup. *Med. J. Rec.*, **132**, 40–41.

RIGATTO, M., and MEDEIROS, N. P. De. (1962). Diaphragmatic flutter. Report of a case and review of literature. *Amer. J. Med.*, **32**, 103–109.

ROBINSON, P. K., MOSBERG, W. H., JR., and LOWE, R. C. W. (1950). Observations on diaphragmatic movement in some neurological disorders. *J. Neurol. Neurosurg. Psychiat.*, **13**, 296–306.

RODMAN, T., RESNICK, M. E., BERKOWITZ, R. D., FENNELLY, J. F., and OLIVIA, J. (1962). Alveolar hypoventilation due to involvement of the respiratory centre by obscure disease of the central nervous system. *Amer. J. Med.*, **32**, 208–217.

SALEM, M. R., BARAKA, A., RATTENBORG, C. C., and HOLADAY, D. A. (1967). Treatment of hiccups by pharyngeal stimulation in anaesthetized and conscious subjects. *J. Amer. med. Ass.*, **202**, 32–36.

SAMUELS, L. (1952). Hiccup. A ten-year review of anatomy, aetiology and treatment. *Canad. med. Ass. J.*, **67**, 315–322.

SARNOFF, S. J., WHITTENBERGER, J. L., and AFFELDT, J. E. (1951). Hypoventilation syndrome in bulbar poliomyelitis. *J. Amer. med. Ass.*, **147**, 30–34.

SERIFF, N. S. (1965). Alveolar hypoventilation with normal lungs: the syndrome of primary or central alveolar hypoventilation. *Ann. N.Y. Acad. Sci.*, **121**, 691–705.

SEVERINGHAUS, J. W., and MITCHELL, R. A. (1962). Ondine's curse—failure of respiratory centre automaticity while awake. *Clin. Res.*, **10**, 122.

SILVER, J. R. (1963). The oxygen cost of breathing in tetraplegic patients. *Paraplegia*, **1**, 204–214.

SMITH, H. P., and SMITH, H. P., JR. (1955). Phrenic paralysis due to serum neuritis. *Ann. intern. Med.*, **19**, 808–813.

SMITH, L. H., Ed. (1956). Medical Grand Rounds, Massachusetts General Hospital. *Amer. Practit.*, **7**, 1165–1172.

SPEIRS, G. O. (1929). Encephalitic diaphragmatic spasm treated by phrenicotomy. *J. nerv. ment. Dis.*, **69**, 407–413.

STEEGMAN, A. T. (1951). Primary pontile haemorrhage; with particular reference to respiratory failure. *J. nerv. ment. Dis.*, **114**, 35–65.

STONE, D. J., and KELTZ, H. (1963). The effect of respiratory muscle dysfunction on pulmonary function. Studies in patients with spinal cord injuries. *Amer. Rev. resp. Dis.*, **88**, 621–629.

STOTKA, V. L., BARCAY, S. J., BELL, H. S., and CLARK, F. B. (1962). Intractable hiccup as the primary manifestation of brain stem tumour. *Amer. J. Med.*, **32**, 313–315.

STRIEDER, D. J., BAKER, W. G., BARINGER, J. R., and KAZEMI, H. (1967). Chronic hypoventilation of central origin. A case with encephalitis lethargica and Parkinson's syndrome. *Amer. Rev. resp. Dis.*, **96**, 501–507.

TAKAGI, Y., IRWIN, J. V., and BOSMA, J. F. (1966). Effect of electrical stimulation of the pharyngeal wall on respiratory action. *J. appl. Physiol.*, **21**, 454–462.

TALBERT, O. R., CURRENS, J. H., and COHEN, M. E. (1954). Cheyne-Stokes respiration. Observations (clinical, experimental, pathological) in 17 patients on the role of the nervous system. *Trans. Amer. neurol. Ass.*, pp. 226–228.

TING, E. Y., KARLINER, J. S., and WILLIAMS, M. H., JR. (1963). Diaphragmatic flutter associated with apneustic respiration. *Amer. Rev. resp. Dis.*, **88**, 833–838.

TSITOURIS, G., and FERTAKIS, A. (1965). Alveolar hypoventilation due to respiratory centre dysfunction of unknown cause. *Amer. J. Med.*, **39**, 173–178.

TURNER, W. A., and CRITCHLEY, M. (1925). Respiratory disorders in epidemic encephalitis. *Brain*, **48**, 72–104.

TURNER, W. A., and CRITCHLEY, M. (1928). The prognosis and the late results of postencephalitic respiratory disorders. *J. Neurol. Psychopath.*, **8**, 191–208.

WYLIE, W. D., and CHURCHILL-DAVIDSON, H. C. (1966). *A Practice of Anaesthesia*, 2nd edit., pp. 60–61. London: Lloyd-Luke.

# Chapter XVII

# CLINICAL ASSESSMENT

E. J. M. CAMPBELL AND J. NEWSOM DAVIS

A COMPREHENSIVE description of techniques for the examination of the respiratory muscles and the mechanics of breathing in patients would include a largely derivative account both of physical diagnosis and of respiratory physiology. This chapter is therefore a summary with only sufficient expansion to give perspective and to emphasise methods, some of which, in our experience, are unfamiliar to most clinicians.

## MECHANICS OF THE CHEST WALL

The shape and movements of the chest wall may be assessed by the conventional methods of inspection and fluoroscopy; in some cases measurements with a caliper may be useful. Inspection and palpation are often unreliable as means of assessing even qualitatively the movements of the chest wall. Simple methods for recording movements have been developed (Agostoni *et al.*, 1965; Konno and Mead, 1967; Mead *et al.*, 1967). They may be used alone, but generally are more useful if simultaneous records of lung volume changes are taken (see Chapter II).

The major emphasis in bedside examination as traditionally taught and practised is directed to the signs of localised abnormality affecting one side or part of the chest. The changes in shape and movement in generalised airways obstruction with distension (pp. 309–310; Campbell, 1969) deserve more attention.

The statics and dynamics of the chest wall are best assessed in routine practice by spirometry using the vital capacity to indicate a maximum static volume displacement and the forced expired volume in one second ($FEV_1$) or the maximum breathing capacity to indicate dynamic capacity. Both of these are, of course, affected by pulmonary mechanics and allowance must be made for this. A more detailed analysis can be provided by plethysmography and the registration of oesophageal pressure (pp. 50–52), but these methods are not as applicable to the chest wall as they are to the lungs because of the complexity of the mechanical relationship of the respiratory muscles and the other structures of the chest wall (Chapter III, p. 48). The delineation of the static and dynamic pressure volume characteristics of the chest wall as a whole can at present be measured only during total muscular paralysis; all the methods which depend upon voluntary relaxation or the assumption of relaxation under certain conditions—for example at the end of expiration or at functional residual capacity—are suspect. Measurement of the abdominal (gastric) pressure is the only means to separate the contribution of inspiratory and expiratory muscles when they act simultaneously (see Chapters III and IV).

## INDIVIDUAL MUSCLE GROUPS

### The Diaphragm

*Examination.*—Although the most accurate assessment of diaphragm function is obtained by fluoroscopy and electrical or mechanical techniques, some estimate of its behaviour can be made by simple clinical inspection. The interpretation of the findings depends upon an appreciation of the two actions of the diaphragm: the downward traction on the central tendon which causes the dome of the diaphragm to descend; and the much smaller effect of traction on the ribs (see Chapter II) which, if the diaphragm is in its normal position, raises and everts the costal margin.

In the supine subject breathing naturally, the protrusion of the anterior abdominal wall with inspiration and its recession with expiration are caused respectively by the active descent of the diaphragm and its passive ascent due to the elastic recoil of the lungs and the weight of the abdominal contents. During quiet breathing in the supine posture the abdominal muscles are usually inactive and their behaviour need not be taken into account in the interpretation of the movements of the anterior abdominal wall. Some subjects, however, consciously breathe in an abnormal manner when they are being examined and may produce unusual patterns of movement of the anterior abdominal wall by using the abdominal muscles. They can usually be made to breathe naturally and the behaviour of the abdominal muscles can be checked by palpation as described later.

The full excursion of the diaphragm between maximum inspiration and maximum expiration can be assessed by "tidal percussion" of the upper border of liver dullness. It must be realised, however, that about half this excursion is dependent on the passive ascent of the diaphragm produced by the contraction of the abdominal muscles during maximum expiration.

The costal margin of many healthy people does not move perceptibly during quiet breathing, but it moves upwards and outwards during a deep inspiration, mainly owing to the action of other muscles, but partly to the action of the diaphragm. Its interpretation is not therefore straightforward. The presence of a normal movement of the costal margin is compatible with the presence of either a normal or a paralysed diaphragm. Normal in this context refers both to position and activity. However, if the diaphragm is contracting but is abnormally low in position, it opposes the outward movement of the costal margin, which may even move paradoxically during natural breathing—that is to say, it may move inwards during inspiration. Paradoxical movement of the costal margin commonly occurs in patients with chronic airway obstruction even when they are at rest and not dyspnoeic, as judged by the behaviour of their accessory muscles. This paradoxical movement is probably due to traction of the costal fibres of the diaphragm which, owing to the increased lung volume in such patients, are flattened, and thus pull inwards rather than upwards.

The downward and inward movement of the costal margin during forced expiration is produced by the contraction of the abdominal muscles and gives little information about the diaphragm.

To summarise: the best clinical assessment of the action of the diaphragm is provided by observation of the movement of the anterior abdominal wall during spontaneous breathing in the supine position which indicates the caudal movement due to its own contraction. Such manoeuvres as tidal percussion during deep breathing or measurement of the maximum chest expansion at the costal margin help in the assessment of the full excursion of the diaphragm but do not give information about its contractile activity. A paradoxical (inward) movement of the costal margin with inspiration probably indicates flattening of the costal part of the diaphragm.

The clinical diagnosis of unilateral paralysis of the diaphragm is unreliable. Immobility of the anterior abdominal wall on one side during quiet breathing and a lag in the upward and outward movement of the costal margin of the same side during a deep inspiration are suggestive of paralysis on that side. However, the descent of the diaphragm on the normal side may cause symmetrical protrusion of the anterior abdominal wall, and the action of the intercostal muscles on the paralysed side may cause the costal margin to move normally. Therefore, while careful clinical examination may enable paralysis to be suspected, absence of paralysis cannot be affirmed by normal findings.

*Fluoroscopy.*—The most useful radiological method of examining the diaphragm is fluoroscopy. Observation of the pattern of its movements is more informative than measurement of its excursion. During quiet breathing rhythmic vertical movements of the diaphragm are smooth and the dome maintains an evenly rounded contour. Many factors, such as increased resistance to the movement of air, partial or complete paralysis of the diaphragm, an abnormally low situation or localised disease affecting either its pleural or peritoneal surface, may affect this behaviour, and interpretation of any abnormality depends on a full evaluation of the clinical and radiological findings.

Measurements of diaphragmatic excursion during quiet or deep breathing may be misinformative because they depend so much on the action of other muscles and on the movements of the vertebral column. Thus it is not unusual to find the diaphragm higher at the end of a deep inspiration than it was at the resting position because of extension of the spine during deep inspiration (see Chapter II).

The best simple test for paralysis of the diaphragm is to observe its movement during a sudden sharp sniff. The lowering of the intrapleural pressure produced by the contraction of the other muscles of inspiration draws the paralysed diaphragm upwards. This observation is commonly referred to as a "paradoxical rise on sniffing". Unfortunately, the interpretation of the sniffing test is sometimes equivocal. A slight unilateral paradoxical movement on sniffing occurs in 6 per cent of normal people and the paradoxical movement should be greater than 2 cm before paralysis is confidently diagnosed (Alexander, 1966). During quiet breathing a paralysed diaphragm may move paradoxically. Frequently, however, it is immobile or may move in a normal direction although usually to a diminished extent, because of extension of the vertebral column and the consequent fall in abdominal pressure (see Chapter III).

*Electromyography.*—Electrical techniques may provide a more precise assessment of phrenic nerve and diaphragm function than fluoroscopy. Conduction in the phrenic nerve can be readily assessed by stimulating the nerve in the neck, at the posterior border of the sterno-mastoid muscle at the level of the thyroid cartilage, and recording the diaphragm muscle action potential either with an oesophageal electrode (Delhez, 1965) or, more conveniently, with surface electrodes over the lateral chest wall in the 8th or 9th interspaces (Newsom Davis, 1967). The twitch evoked in the muscle by an adequate stimulus is clearly visible and palpable. With this technique the range of phrenic nerve conduction times in a group of control subjects was 6·1–9·2 msec. Total paralysis of the phrenic nerve due to local disease can be shown by the failure to evoke a diaphragmatic muscle action potential. Marked slowing of conduction in the nerve, as indicated by a prolonged conduction time, can be demonstrated in patients with peripheral neuropathy, and in many cases is present before there is any significant reduction in vital capacity; this method thus provides the earliest indication of involvement of the phrenic nerve by the disease process (Newsom Davis, 1967). Sampling of diaphragmatic electrical activity with a needle electrode will only be necessary in special circumstances to establish the nature of impaired function (e.g. neuropathic or myopathic), but can be achieved safely through the 10th intercostal space in the mid-axillary line (Taylor, 1960; Fink, 1961).

*Transdiaphragmatic pressure.*—Simultaneous recording of oesophageal and gastric pressure may provide useful information on the transdiaphragmatic pressure (see Chapters III, IV and VI).

### The Intercostals

*Examination.*—In Chapter VII the conclusion was reached that the external and interchondral intercostals raise and evert the ribs during inspiration and increase the tension of the intercostal spaces. However, the diaphragm raises and everts the lower ribs, and the accessory muscles of inspiration raise the upper ribs and stretch the intercostal tissues. Sometimes the action of the accessory muscles causes a marked increase in tension in the intercostal spaces with almost no activity in the intercostal muscles as judged electromyographically (Campbell, personal observations). However, provided the accessory muscles are not contracting vigorously, expansion of the middle ribs (the fourth to seventh or eighth) and a palpable increase in tension of these spaces probably indicate that the intercostal muscles are functioning normally.

*Fluoroscopy.*—This enables the movements of the ribs, particularly in the coronal plane ("bucket handle movement") to be recorded. Interpretation of such movements in terms of intercostal activity must of course be made only after all other factors affecting the movements of the ribs have been taken into consideration.

*Electromyography.*—EMG assessment of the intercostal muscles may occasionally be helpful clinically. In order to minimise the risk of pneumothorax, bipolar needles should be used (see Taylor, 1960), so that the precise location of the needle tip may be defined by monitoring the electrical activity on a loudspeaker during insertion. In this way, activity in the individual intercostal layers may be distinguished. It should be remembered that, when the intercostal muscles are paralysed by a high spinal cord lesion but the diaphragm is still intact, phasic activity can be recorded in the intercostal muscles. This is a stretch reflex caused by diaphragmatic contraction (Guttmann and Silver, 1965).

### The Accessory Muscles of Inspiration

The scaleni, trapezii and particularly the sternomastoids can be examined readily by palpation. Sometimes in dyspnoeic patients the prominent folds produced by these muscles give the impression of a persistent contraction. This appearance is often deceptive because electromyographic examination will usually show that they do in fact relax completely during expiration (Campbell and Friend, 1955).

### The Abdominal Muscles

*Examination* of the movements of the anterior abdominal wall is of little value in studying the behaviour of the abdominal muscles because in the great majority of circumstances these movements are produced by the changes in intra-abdominal pressure consequent to the action of the diaphragm. Palpation may also be deceptive because the stretching of the abdominal wall produced by descent of the diaphragm is usually entirely passive. However, hardening of the lateral muscles (the obliques and transversi) during expiration is good evidence of activity in these muscles. Such hardening is best felt by pressing gently with the fingertips between the mid- and anterior-axillary lines. Palpation of the rectus abdominis is liable to be misleading because an apparent contraction of this muscle is often merely due to an increase in tension of the rectus sheath produced by the contraction of the lateral muscles.

Of all the respiratory muscles those of the abdominal wall are the easiest to examine by *electromyography*. Moreover, the interpretation of any recorded activity is easier than that of most other respiratory muscles. Analysis of the effects of abdominal muscle contraction may be made by recording *gastric pressure*; this should be done simultaneously with that of changes in lung volume and of the dorso-ventral diameter of the abdomen if movements of extension or flexion of the vertebral column are occurring (see Chapter IV).

## OVERALL FUNCTION

It is important to distinguish between two types of disorder affecting the respiratory muscles: first, muscle paralysis due to disorders of the descending pathways, anterior horn cell, peripheral nerve, neuromuscular transmission or muscle, and second, disease of the central control mechanism (see Chapter XVI). While the second type of disorder is relatively uncommon, it is also more easily overlooked.

### Respiratory Muscle Paralysis

A common problem here is to determine the degree of involvement of the respiratory muscles. It is convenient to recognise two grades of affection.

1. *Reduced functional reserve but resting ventilation adequate.*—The clinical signs which may be observed initially during this stage are a diminished ability to take a voluntary deep breath or cough and evidence of paralysis of one or more of thc respiratory muscle groups. At the transition between this stage and ventilatory failure, there are respiratory distress, restlessness and difficulty in sleeping, an increased respiratory rate, a rising pulse rate and loss of the ability to hold the breath, count or speak for more than a few seconds.

The most useful test during this stage is the vital capacity. A reduction to about a quarter of the predicted value usually indicates that the need for assisted ventilation is imminent and this should not, of course, be withheld on the grounds that ventilatory failure has not yet occurred. Additional tests are the measurement of maximum inspiratory and expiratory pressures. The maximum breathing capacity is not of much value and may tire the patient.

In the interpretation of these tests the following points have to be borne in mind. First, test performance may be reduced by difficulty in voluntary participation, and by disorders of the lungs or pulmonary circulation. Secondly, tests of the depth and force of inspiration are more useful in assessing ventilatory reserve because breathing is essentially an inspiratory act. Tests of the depth or force of expiration are less valuable for this purpose but are useful in evaluating the efficacy of the cough mechanism, which is of course often vitally important. Thirdly, the ordinary muscles of breathing may be weak, but the tests may be well performed by the use of accessory muscles not used during natural breathing. Finally, local weakness producing regional inadequacy of ventilation, or weakness of coughing causing retention of secretions may allow underventilation or atelectasis of parts of the lung. This may cause the usual clinical and radiological signs, but it should also be suspected if the alveolar-arterial $O_2$ pressure difference increases, implying that some parts of the lung are being inadequately ventilated in relation to their blood flow. Sometimes these changes are produced by aspiration of secretions from the upper respiratory tract due to pharyngo-laryngeal weakness or insensitivity of the cough mechanism rather than by weakness of the respiratory muscles themselves.

2. *Ventilatory failure.*—By common usage, this term means a ventilation which is insufficient to maintain a normal arterial $Pco_2$. The definitive and only reliable method of diagnosis is by estimation of the arterial $Pco_2$ either directly or indirectly. It is impossible to diagnose or exclude ventilatory failure or $CO_2$ retention by clinical examination. The signs associated with $CO_2$ retention, such as tachycardia, hypertension, sweating, twitching, and papilloedema, are non-specific and untrustworthy; on the whole, cyanosis is more likely to be due to regional reduction of ventilation than to overall inadequacy of ventilation.

## Disordered Central Control

The clinical findings when the central control mechanism is involved by disease are usually more subtle than in muscle paralysis. In chronic cases, there may be a history suggestive of alveolar hypoventilation, such as morning headaches due to $CO_2$ retention, or failure to breathe after an anaesthetic. Clinical signs of bulbar palsy, in particular a wasted and fasciculating tongue, should always suggest the possibility of this disorder. The subject may breathe at a low frequency or irregularly but this is often only apparent when he is drowsy or asleep. Other indications of ventilatory failure may be present but, as emphasised above, these signs do not invariably accompany $CO_2$ retention.

Measurements of the vital capacity and other tests appropriate for muscle paralysis are dangerously misleading in disordered central control, for the common finding is that the voluntary movements of the chest are little, if at all, impaired. Some indication of the disorder is provided by observing or recording respiratory movements during sleep, but the diagnosis can only be established conclusively by demonstrating a raised $Paco_2$ in the absence of a pulmonary cause. Because the stimulus of an arterial puncture may improve ventilation, samples obtained through an indwelling arterial catheter are more likely to be reliable, and should be taken when the subject is at rest and, when possible, asleep. The ventilatory response to $CO_2$ is a useful confirmatory test because it is usually, but not invariably, reduced.

In acute disorders, however, where involvement of the respiratory control mechanism in the brain stem is expected (see Chapter XVI), a disturbance of respiratory rhythm is likely to be the earliest sign. In such cases, it may be hazardous to wait until hypoventilation can be demonstrated before taking appropriate measures, for total cessation of breathing may occur suddenly, usually in sleep.

## REFERENCES

AGOSTONI, E., MOGNONI, P., TORRI, G., and SARACINO, F. (1965). Relation between changes or rib cage circumference and lung volume. *J. appl. Physiol.*, **20**, 1179–1186.

ALEXANDER, C. (1966). Diaphragm movements and the diagnosis of diaphragmatic paralysis. *Clin. Radiol.*, **17**, 79–83.

CAMPBELL, E. J. M. (1969). Physical signs of diffuse airways obstruction and lung distension. *Thorax*, **24**, 1–3.

CAMPBELL, E. J. M., and FRIEND, J. (1955). Action of breathing exercises in pulmonary emphysema. *Lancet*, **1**, 325–329.

DELHEZ, L. (1965). Modalités, chez l'homme normal, de la réponse électrique des piliers du diaphragme à la stimulation électrique des nerfs phréniques par des chocs uniques. *Arch. int. Physiol.*, **73**, 832–839.

FINK, B. R. (1961). Electromyography in general anaesthesia. *Brit. J. Anaesth.*, **33**, 555–559.

GUTTMANN, L., and SILVER, J. R. (1965). Electromyographic studies on reflex activity of the intercostal and abdominal muscles in cervical cord lesions. *Paraplegia*, **3**, 1–22.

KONNO, K., and MEAD, J. (1967). Measurement of the separate volume changes of rib cage and abdomen during breathing. *J. appl. Physiol.*, **22**, 407–422.

MEAD, J., PETERSON, N., GRIMBY, G., and MEAD, J. (1967). Pulmonary ventilation measured from body surface movements. *Science*, **156**, 1383–1384.

NEWSOM DAVIS, J. (1967). Phrenic nerve conduction in man. *J. Neurol. Neurosurg. Psychiat.*, **30**, 420–426.

TAYLOR, A. (1960). The contribution of the intercostal muscles to the effort of respiration in man. *J. Physiol. (Lond.)*, **151**, 390–402.

# INDEX

# INDEX